The Influence of Petrus Ramus
Edited by Feingold · Freedman · Rother

SCHWABE PHILOSOPHICA

I

HERAUSGEGEBEN VON
HELMUT HOLZHEY UND WOLFGANG ROTHER

JEREMIA 23, 29

SCHWABE & CO AG · VERLAG
BASEL

THE INFLUENCE OF PETRUS RAMUS

STUDIES IN SIXTEENTH AND SEVENTEENTH CENTURY
PHILOSOPHY AND SCIENCES

EDITED BY
MORDECHAI FEINGOLD, JOSEPH S. FREEDMAN
AND WOLFGANG ROTHER

SCHWABE & CO AG · VERLAG
BASEL · 2001

Publication assisted by a subsidy from the Swiss National Science Foundation

Printing-number: 550 copies

Press: Schwabe & Co. AG · Basel/Muttenz
ISBN 3-7965-1560-6

Table of Contents

Mordechai Feingold, Joseph S. Freedman, Wolfgang Rother
Preface 7

Wolfgang Rother
Ramus and Ramism in Switzerland 9

Theo Verbeek
Notes on Ramism in the Netherlands... 38

Thomas Elsmann
The Influence of Ramism on the Academies of Bremen and Danzig:
A Comparison... 54

Joseph S. Freedman
Melanchthon's Opinion of Ramus
and the Utilization of Their Writings in Central Europe 68

Riccardo Pozzo
Ramus' Metaphysics and its Criticism by the Helmstedt Aristotelians 92

Erland Sellberg
The Usefulness of Ramism 107

Mordechai Feingold
English Ramism: A Reinterpretation... 127

Elizabethanne Boran
Ramism in Trinity College, Dublin in the Early Seventeenth Century 177

Kees Meerhoff
"Beauty and the Beast": Nature, Logic and Literature in Ramus 200

Guido Oldrini
The Influence of Ramus' Method on Historiography and Jurisprudence 215

Ian Maclean
Logical Division and Visual Dichotomies:
Ramus in the Context of Legal and Medical Writing 228

Nicholas Jardine and Alain Segonds
A Challenge to the Reader: Ramus on *Astrologia* without Hypotheses 248

Index of Names 267

Preface

The present volume originated in a workshop, held in the *Bibelsaal* of the Herzog August Library of Wolfenbüttel on 1 and 2 July 1997 and devoted to an examination of the influence of Petrus Ramus. The papers deal with the diverse use made of Ramus both by followers and adversaries in the religious and scholarly controversies that raged throughout (mainly) Protestant Europe during the late sixteenth and early seventeenth centuries. In terms of organization the volume is divided along regional and thematic issues. The impact of Ramist ideas on Calvinism informs a first group of papers concentrating on various Reformed territories: Switzerland (Wolfgang Rother), the Netherlands (Theo Verbeek), Bremen and Gdansk (Thomas Elsmann). A second group examines Lutheran responses to Ramus: Joseph Freedman analyzes Melanchthon's view, Riccardo Pozzo evaluates the criticism by the Helmstedt Aristotelians, and Erland Sellberg scrutinizes Ramus' reception at the Swedish universities. Two further contributions probe the Ramist debate in the British Isles: in England (Mordechai Feingold) and in Ireland (Elizabethanne Boran). The section devoted to thematic issues places Ramus within Renaissance disciplines where his impact was most keenly felt: logic and literature (Kees Meerhoff), historiography and jurisprudence (Guido Oldrini), law and medicine (Ian Maclean), and astronomy (Nicholas Jardine and Alain Segonds).

Appropriately, the book is published in the city in which Ramus appears to have spent the happiest years of his life, and to which he dedicated his *Basilea ad Senatum Populumque Basiliensem* as a token of his gratitude. To add to the poignancy, it is produced by the same publishing house which printed not only some of his own works, but also several by his friend and most faithful follower, Johannes Thomas Freigius who, in turn, served occasionally as reader for the press. We are indebted to the publisher, Dr. Urs Breitenstein, who has committed himself to carry on the tradition established by the founder of the firm, Johannes Petri.

Special thanks are due to Professor Dr. Friedrich Niewöhner of the Herzog August Library for his invitation and our very friendly welcome at Wolfenbüttel, as well as to the Swiss National Science Foundation for a generous subvention. We should like to thank Dr. Helga Robinson-Hammerstein for her translation from the German of the contributions by Rother and Elsmann and to Mark James Lilley for his translation from the French of Meerhoff's text. We are also grateful to Peter Sharratt for his encouragement and for some very helpful suggestions.

Cambridge, Mass., Montgomery, Basle, August 2000

Mordechai Feingold, Joseph S. Freedman, Wolfgang Rother

Ramus and Ramism in Switzerland

WOLFGANG ROTHER

1. Ramus in Switzerland: His visit to Reformed towns (1568-1570)

The influence of Petrus Ramus on the intellectual life of Switzerland dates back to his journey there[1] which he started almost exactly four years before his violent death. Having left Paris in August 1568 he reached Basle[2] in October, via Strasbourg and Freiburg in Breisgau; in Strasbourg the rector of the Academy Johannes Sturm (1507-1589) had received him. Ramus entertained friendly relations with the medical and universal scholar Theodor Zwinger (1533-1588), who had been his student in Paris in the early 1550s and who, since 1565, was professor of Greek in Basle. Zwinger introduced Ramus to local scholars and Ramus praised them in his *Basilea,*[3] which he wrote half a year after his departure from Basle to record his thanks for the friendly reception with which he had been honoured.[4]

1 On Ramus' stay in Switzerland see Charles Waddington: *Ramus. Sa vie, ses écrits et ses opinions,* Paris 1855, pp. 192-199.

2 In addition to the pages in Waddington see also Auguste Bernus: *Pierre Ramus à Bâle. 1568-1569. Étude historique suivie d'une lettre inédite de Ramus,* Paris 1890 [Extrait du *Bulletin historique et littéraire* publié par la Société de l'Histoire du Protestantisme français, 15 octobre 1890] 23 pp.

3 *Petri Rami Basilea ad senatum populumque Basiliensem* [Lausanne: Johannes Probus] 1571. Quotations are taken from: *Petrus Ramus: Basilea. Eine Rede an die Stadt Basel aus dem Jahre 1570.* Lateinisch und Deutsch. Übersetzt und eingeleitet von Hans Fleig, Basle 1944.

4 Memini (viri Basilienses) semperque dum vivam meminero, quam jucunda, quam liberalis, quam humana Basiliensis habitatio nobis acciderit. Ibid. p. 24.

In addition to Zwinger[5] – to whose son Jakob (born 15 August 1569) Ramus became godfather – Ramus entertained close relations with several other Basle scholars as well. Foremost among these was the jurist and polyhistor Johannes Thomas Freigius (1543-1583),[6] who later became one of the most ardent Ramists. Mention should also be made of the Aristotelian Johannes Hospinian (1513-1575), professor of Greek, logic and theology. Ramus' praise of Hospinian includes a critique of Aristotelianism and an admonition to adopt a practice-related dialectic.[7] Then there was the humanist scholar Hieronymus Wolf (1516-1580),[8] who had edited the works of Isocrates and Demosthenes; Thomas Platter (1499-1582), who presided as rector over the *Paedagogium* of Basle from 1540 to 1578, and

5 Zwingerus [...], nostrae quondam scholae alumnus, et ideo verecundius voce nostra memorandus, ne magister e discipuli studiis laudem repetere videatur. Sed tamen accurata Graecae literaturae professio studiosis fructuosa, academiae universae honorifica, civibus perjucunda: sed multo magis humanae vitae theatrum, laboris infiniti, ex infinitis authoribus conflatum; sed quoniam medicinae se praecipue dedidit, Galenus praecipuis partibus, Hippocrates totus, Zwingeri industriam doctrinamque etiam tacente me, amplissime demonstrabunt. Ibid. p. 40.

6 Joannes Thomas Freigius rhetoricae secundus a Caelio filio [*i.e. Caelius Augustinus Curio, son of Caelius Secundus Curio*] professor est. Minime tamen professionis humanitate contentus, altiorem jurisprudentiae facultatem consequutus, pro aetate spem facit ampliorum vigiliarum. Ibid. p. 41.

7 Tu vero (mi Joannes Hospiniane) non solum literis Latinis et Graecis, sed assiduis in Aristotelis organo vigiliis et editis et meditatis; sed professionis fide; sed officii quadam religione, egregiam laudem consequutus es. Sed tu, tuique professionis hujus collegae caeteri, partibus infinitis ampliorem laudem consequemini, si dialecticum ut artificium, sic artificii usum in Virgilio et Homero, in Cicerone et Demosthene, juventuti proponatis, et quotidianae imitationis exemplo et opere informetis. Partiamini tempora professionis logicae: minimam temporis partem logicae praeceptis, maximam usui, meo consilio, tribuetis. Sin vobis Aristotelis judicium atque exemplum luculentius est, professionis partem philosophiae et praeceptis, partem eloquentiae et exemplis assignabitis: ut tamen memineritis, usum artis, non artem usus finem esse. Nec interea leges academiae vereamini. Academia enim pro ista culpa (qua, pro logica mortua in libris, vivam in ipso opere logico, proque plumbea auream feceritis) praemium coronamque decernet: vobisque authoribus Rodolphos Agricolas Basileam adductos, liberali gratoque animo accipiet. Ibid. pp. 41-42.

8 [...] Hieronymus Wolfius [...] nostrae aetatis fama gloriaque, academiae quondam nostrae fama gloriaque fuerunt. Ibid. p. 53.

his son, the medical doctor Felix Platter (1536-1614);[9] and finally among the jurists Basilius Amerbach (1534-1591).[10] Ramus presented his theology,[11] which was anchored on a thoroughly Scripture-based Christianity, in his *Commentariorum de Religione christiana libri quatuor,* a work largely written in Basle[12] and published posthumously (Frankfort 1576). The *Basilea* contains a significant observation on the theological instruction which he received at the time from Simon Sulzer (1508-1585) and Ulrich Koch (1525-1585). He describes this instruction as "the pure theology of the Divine Word" taken "from its sources" which he had always "sought" and which he now "believed to have found".[13] Such a testimony is interesting because both Sulzer – who had visited Luther in Wittenberg in 1538 and who had since 1548 been working in Basle for the implementation of the *Confessio Augustana* – and his brother-in-law Koch

9 His Thomas Platerus successit: uberrimi seminarii proventu, cum innumerabiles annorum plurimorum tempore grammaticos genuit, tum filium Felicem Platerum sic informavit, ut filii exemplo, singularis patris doctrina possit intellegi. Perspexi enim, et collaudavi singulare Felicis ingenium, etque ad omnes artes mirifico naturae impetu incitatum: tornandi artifex, sed animi caussa, singularis est, est et fundendi artifex singularis, sed medicinae, quam facit gratia. Anatomica pleraque studiosius digessit, herbarum ac stirpium infinitam copiam ac sylvam quotidie novam exquirit [...]. Ibid. pp. 36-37.

10 [...] Basilius Amerbachius non solum jurisprudentia, sed vera virtutis fama, civis primarius [...]. Ibid. p. 33.

11 On the study of Ramus' theology see Paul Lobstein: *Petrus Ramus als Theologe. Ein Beitrag zur Geschichte der protestantischen Theologie,* Strasbourg 1878, 86, (2) pp.

12 Lobstein: *Ramus als Theologe* (as note 11) p. 3. Bernus: *Ramus à Bâle* (as note 2) p. 5.

13 Quid theologiae professio, cujusmodo aut qualisnam fuit? Theologia nempe de professionibus ordinariis ut gradu summa est, sic divinis laudibus est amplissima. Professores ordinarii duo sunt, professionis genere mihi imprimis laudato atque optato. Hic Simon Sulcerus et Huldricus Coccius nobis auditi assidue atque auscultati sunt. Nec enim consolatio mihi major in his civilium tempestatum miseriis ulla fuit, quam ut assiduas utriusque praelectiones auscultarem, cum alter testamentum vetus hebraice, alter novum graece interpretaretur. Tum enim voti compos fieri mihi visus sum, pro quo prooemium Parisiensis academiae pene capitale nobis exstitisset: cum nostris theologis optarem, ut tali professione, puriorem verbi divini theologiam e suis fontibus ita haurirent, et ita hauriendam discipulis ostenderent. Ramus: *Basilea,* ed. Fleig (as note 3) p. 47.

stood, on account of their Lutheranism, in opposition to the Basle Reformed Church. For his part, Ramus entertained good relations with the Reformed clergy of Basle as well, notably with pastor Johannes Brandmüller (1533-1596).[14]

From Basle Ramus made excursions to Zurich and Bern in 1569. In Zurich Heinrich Bullinger (1504-1575) received him and carried on a theological correspondence with him.[15] Ramus also met the theologians Rudolph Gwalter (1519-1586) who was married to Zwingli's daughter Regula, and Ludwig Lavater (1527-1586), one of Bullinger's sons-in-law, to whom, on 22 or 31 July 1569, Ramus addressed a letter from Basle concerning his reading of the *Confessio Basiliensis,* especially the article on the Holy Supper.[16] Ramus submitted a draft of the *Commentariorum de Religione christiana libri quatuor* to the Zurich theologians, including Josias Simler (1530-1576), another son-in-law of Bullinger. As a sign of its admiration of, and sympathy for, Ramus the city of Zurich resolved to surprise him with a feast and a banquet. When the bewildered Ramus asked whether one was celebrating a wedding, Bullinger answered: yes this is the wedding of our city and you.[17] This marriage, however, remained without issue and consequence. The teaching of philosophy, especially logic, in Zurich seems to have remained strictly Aristotelian until the middle of the seventeenth century and no Ramist influences can be discerned there.[18] In Bern Ramus was received, among others, by Bene-

14 Sed a divino isto Oecolampadii animo divelli non possum. Deque eo vehementius cogitans, Joannem Brandmüllerum pene praeterii: quo tamen, si Oecolampadianae sinceritatis germanus successor suo tempore requiratur, germaniorem equidem vix ullum Basiliensibus augurer. Itaque, Brandmüllere, augurium nostrum doctrinae et vitae, ut coepisti, egregia virtute confirmato. Ibid. p. 52.

15 Ramus' letters to Bullinger (1 September 1571, 3 March 1572, 19 March 1572), kept in the *Staatsarchiv des Kantons Zürich* (E II 381: 1300-1301. 1328. 1334), are published by Waddington: *Ramus* (as note 1) pp. 433-435. 437-438. 438-440.

16 The letter is printed in Bernus: *Ramus à Bâle* (as note 2) pp. 17-23 (with French translation). Further letters are preserved in the *Zentralbibliothek Zürich.* To Lavater: Basileae, pridie cal. dec. 1568. To Gwalter: Lausannae, non. sept. 1570; Lutetiae, 10 cal. aug. 1571; Lutetiae, cal. iul. 1572; Lutetiae, 17 iul. 1572.

17 Waddington: *Ramus* (as note 1) pp. 197-198.

18 See Wolfgang Rother: The Teaching of Philosophy at Seventeenth-Century Zurich, in: *History of Universities* 11 (1992) pp. 59-74.

dictus Aretius (the Greek form of Marti, ca. 1522-1574), a theologian who had imbibed some science education, and with whom Ramus later entered into a correspondence.[19]

In October 1569 Ramus left Basle for Heidelberg proceeding in the following spring, via Frankfort, Nuremberg, Augsburg and Bern, to Geneva. From Bern on 28 August 1569, he had written a letter to Theodor Beza (1519-1605) in Geneva, criticising the latter's Aristotelianism. In his response of 20 September 1569, Beza, an opponent of the dialectic of Ramus, expressed his disapproval of Ramus' *Animadversiones* against Aristotle.[20] Despite this Ramus was cordially received in Geneva in June 1570 and delivered public lectures there. Several students became adherents of his new logic and on 10 June 1570 they asked him to continue his lectures. But Ramus had already left the town on account of an epidemic and arrived at Lausanne on 20 July 1570. He gave some lectures on logic there, which attracted large audiences, but by the end of August Ramus left Lausanne and returned to Paris via Lyon.[21]

Since Ramus was soon banned from teaching of any kind in Paris, he sought to secure a professorship in Geneva, but Beza refused. On 1 December 1570 he wrote that he greatly admired Ramus' erudition and oratorical skills – though he differed from him totally on certain issues. He pointed out, however, that there were two reasons why he could not grant his request. First, there was no vacancy in the Academy and the pecuniary resources were scarce, so that he found it impossible to increase the number of professorships. The second obstacle, Beza continued, consisted in the resolution to follow Aristotle's doctrines, without departing even an inch from them, as well in the teaching of logic as in all the other subjects. Of course, he concluded, Ramus would be welcome as a guest in Geneva.[22] This letter deprived Ramus of all hope. The unflinching Aristotelianism of Genevan orthodoxy, represented by Beza, exercised a sustained influence on the teaching of philosophy there, and the Academy

19 Waddington: *Ramus* (as note 1) p. 198. A letter of 1571, preserved in the *Stadtbibliothek Zofingen* (Kanton Aargau), is printed in ibid. p. 432.

20 Ibid. pp. 212-213.

21 For Ramus' stay in Geneva and Lausanne see ibid. pp. 213-215.

22 French translation of the letter in ibid. pp. 229-230.

remained Aristotelian until the advance of Cartesianism around 1670. As in the case of Zurich, there was practically no Ramist influence in Geneva.[23]

2. The first and most committed Swiss Ramists: Theodor Zwinger and Johannes Thomas Freigius

Two scholars stand out among those whom Ramus befriended during his stay in Switzerland, Theodor Zwinger (1533-1588) and Johannes Thomas Freigius (1543-1583). Ramist thinking exercised a lasting influence on both and they, in turn, contributed in various ways to the dissemination of Ramist philosophy and method.

Zwinger[24] commenced his studies at the Basle Arts Faculty in 1548 but shortly thereafter left for France and became an employee in a printing office in Lyon. In 1551 he travelled to Paris where, for some time, he studied with Ramus with whom he entertained a lively correspondence after Ramus' stay at Basle.[25] In 1553 he was briefly back in Basle. He subsequently proceeded to study medicine at Padua and Venice, was awarded the medical doctorate in Padua in 1559, and in the same year was received into the Basle College of Physicians and the Medical Faculty. In 1565[26] he became

23 The only evidence for a possible influence is a Genevan edition of Talaeus' *Rhetorica: Audomari Talaei Rhetorica e P. Rami* [...] *Praelectionibus observata. Recens in usum Scholarum hac forma distinctius & emendatius excusa,* Geneva: Petrus Roverius 1602.

24 For Zwinger's life see Hans Jacob Leu: *Allgemeines Helvetisches, Eydgenössisches, Oder Schweitzerisches Lexicon,* Zurich 1747-1765, vol. 20, pp. 553-555. Christian Gottlieb Jöcher: *Allgemeines Gelehrten=Lexicon,* Leipzig 1750-1751, vol. 4, p. 2246. *Athenae Rauricae sive Catalogus Professorum Academiae Basiliensis ab a. MCCCCLX. ad a. MDCCLXXVIII. cum brevi singulorum Biographia,* Basle 1778, pp. 208-211. *Allgemeine Deutsche Biographie,* Berlin [2]1967-1971, vol. 45, pp. 543-544. *Die Matrikel der Universität Basel,* Basle 1951-1975, vol. 2, p. 55.

25 The *Universitätsbibliothek Basel* preserves about 30 letters of Ramus written between 1550 and 1572, most of them to Zwinger between 1570 and 1572. Some of these letters to Zwinger are published in Waddington: *Ramus* (as note 1) pp. 423-431. 436-437.

26 Wrong date (1561) in Leu: *Schweitzerisches Lexicon* (as note 24) vol. 20, p. 554.

professor of Greek, in 1571 of practical philosophy and in 1580 of theoretical medicine.

His manifold studies and his publications show Zwinger to be a universal scholar. Most important in this respect is his *Theatrum vitae humanae,*[27] a very successful and highly regarded work, that was fulsomely praised by Ramus as a "labour without end".[28] It was an encyclopaedia composed following Ramist method, arranged by definitions and divisions and conceived didactically with comprehensive overviews in the form of dichotomous tables. The *Theatrum vitae humanae* and many other works by Zwinger distinguished him as a scholar who applied Ramist method to various branches of learning. He did, however, retain his independence in his critical attitude towards his Parisian teacher and friend as far as Aristotle was concerned. As professor of practical philosophy he lectured on the *Nicomachean Ethics* and the *Politics* of Aristotle, both of which he prepared and presented didactically in the interest of clearer instruction.[29] The key concepts of these presentations are, on the one hand, the *tabulae perpetuae,* the continuous dichotomous tables, and, on the other, the *methodice illustratus,* arranged by definitions and divisions, with a methodical-didactic approach. *Methodus* is also the key concept of his book on Cato's

27 *Theatrum Vitae Humanae, Omnium fere eorum, quae in hominem cadere possunt, Bonorum atque Malorum Exempla historica, Ethicae philosophiae praeceptis accomodata, et in XIX. libros digesta, comprehendens* [...], Basle: Johannes Oporin 1565. Further editions: Basle: Officina Frobeniana 1571. Basle: Eusebius Episcopius 1586. 1596. *Theatrum Humanae Vitae* [...] *Novem Voluminibus locupletarum, interpolatum, renovatum Jacobi Zwingeri Fil. recognitione* [...], 4 vols., Basle: Sebastian Henricpetri 1604.

28 [...] sed multo magis humanae vitae theatrum, laboris infiniti, ex infinitis authoribus conflatum [...]. Ramus: *Basilea,* ed. Fleig (as note 3) p. 40.

29 *Aristotelis* [...] *de moribus ad Nicomachum libri decem: tabulis perpetuis, quae Commentariorum loco esse queant, explicati & illustrati,* Basle: Johannes Oporin [1566]. *Aristotelis Ethicorum Nicomachiorum libri decem ex Dion. Lambini interpretatione Graecolatini* [...] *Argumentis atque Scholiis, Tabulis quinetiam Novis, methodice illustrati,* Basle: Eusebius Episcopius 1582. *Aristotelis Politicorum libri octo ex Dion. Lambini & P. Victorii interpretationib. puriss. Graecolatini* [...] *Argumentis atque Scholiis, Tabulis quinetiam in tres priores Libros illustrati,* Basle: Eusebius Episcopius 1582. See also: *Morum Philosophia poetica ex Veterum utriusque linguae Poetarum thesauris*, Basle: Eusebius Episcopius 1575.

De agricultura and Varro's *Rerum rusticarum libri tres*[30] and of his introduction to travelling.[31] The tables and commentaries on Galen[32] and Hippocrates[33] are applications of the Ramist method to medicine.

Johannes Thomas Freigius[34] from Freiburg in Breisgau is the other Basle scholar whose thinking was influenced by his personal encounter with Ramus. He studied law and languages in Freiburg. From 1562 he taught dialectic and became professor of Latin grammar in 1564. In 1567 he refused to take the oath on the Tridentine Decrees and left for Basle where he obtained the doctorate in law in 1568. The encounter with Ramus in the same year led to a close personal relationship – in his *Vita Rami* he speaks of frequent walks with him[35] – and was crucial for his further academic development. In 1570 Freigius was appointed professor of rhetoric in Basle, but he returned to Freiburg in the same year to teach dialectic and practical philosophy there. Since he presented not the philosophy of Aristotle but that of Ramus, he was forced to resign his professorship in July 1575, whereupon he proceeded to give lectures in his house. However, in December 1575 he was crossed off the Freiburg *matricula* and the students were forbidden to

30 *Methodus Rustica Catonis Atque Varronis Praeceptis aphoristicis per Locos communes digestis,* Basle: Petrus Perna [1576].

31 *Methodus apodemica in eorum gratiam. Qui cum fructu et quocunque tandem uitae genere peregrinari cupiunt,* Basle: Eusebius Episcopius 1577. Another edition: Strasbourg 1594.

32 *Tabulas & commentarium in artem medicinalem Galeni,* Basle 1561.

33 *Hippocratis Coi Asclepiadeae gentis sacrae coryphaei viginti duo commentarii Tabulis illustrati,* Basle: Eusebius Episcopius 1579.

34 On Freigius' life see Leu: *Schweitzerisches Lexicon* (as note 24) vol. 7, pp. 317-318. Jöcher: *Allgemeines Gelehrten=Lexicon* (as note 24) vol. 2, pp. 737-738. *Athenae Rauricae* (as note 24) pp. 295-296. *Allgemeine Deutsche Biographie* (as note 24) vol. 7, pp. 341-343. *Matrikel Basel* (as note 24) vol. 2, pp. 172-173.

35 Humanitate igitur illa perpetuo erga me deinceps usus est: saepeque in media via complexus, praecipue vero in illa Sampetraei viridarij de ambulatiuncula: ubi etiam adhortationes eius ad me factas adhuc memori animo recordor. Petri Rami Vita, in: *Petri Rami Praelectiones in Ciceronis Orationes octo consulares. Una cum Ipsius Vita per Ioann. Thomam Freigium collecta,* Basle: Petrus Perna 1575, p. 9. Another edition: Basle: Petrus Perna 1580, p. 9. *Petri Rami et Audomari Talaei Collectaneae Praefationes, Epistolae, Orationes; quibus adjunctae sunt Petri Rami Vita (per Ioann. Thomam Freigium) cum testamento, ejusdem Basilea* [...], Marburg 1599; another edition: Marburg 1609.

attend his private classes. In 1576 Freigius returned to Basle where he made his living as copy editor and proof-reader for Henricpetri. On 30 November 1576 he was appointed rector of the *Gymnasium* at Altdorf where his Ramism soon became odious – as it had previously been in Freiburg – and he was forbidden to teach Ramist philosophy. Freigius relinquished his post in 1582, returned to Basle and worked again for Henricpetri. On 16 January 1583 he died of the plague.

Freigius is responsible for the dissemination of Ramist philosophy and method in two significant ways: as editor of the works of Ramus (and Talaeus),[36] and as an extremely productive author,[37] who pursued the aim of

36 Freigius published the following works of Ramus: *Ciceronianus,* Basle: Petrus Perna 1573. With two prefaces by Freigius, one preface by Ramus and two letters by him, Paris, 16 August 1572. *Ciceronianus. Editio postrema* [Edited by Johannes Thomas Freigius], Frankfort: A. Wechel 1580. – *Liber de militia C. Iulii Caesaris. Cum Praefatione Ioannis Thomae Freigii,* Basle: Sebastian Henricpetri [1574]. Further editions: 1575. Frankfort: haeredes A. Wecheli 1584. – *Liber de moribus veterum Gallorum. Cum Praefatione Ioannis Thomae Freigij*, Basle: Sebastian Henricpetri [1574]. – *Dialecticae Institutiones* [...] *item Aristotelicae animadversiones a prima editione nuspiam hac methodo visae* [...] *opera Ioann. Thomae Freigii,* Basle: Sebastian Henricpetri [1575]. – *Petri Rami Praelectiones in Ciceronis Orationes octo consulares. Una cum Ipsius Vita per Ioann. Thomam Freigium collecta,* Basle: Petrus Perna 1575. Another edition: Basle: Petrus Perna 1580. In both editions the *Life of Ramus* is on pp. 4-46 and an edition of the *Basilea* on pp. 45[sic]-60. – *Professio Regia. Hoc est, Septem artes liberales, in Regia cathedra, per ipsum apodictico Parisijs docendi genere propositae, et per Ioan. Thomam Freigium In tabulas perpetuas, ceu στρώματα quaedam, relatae: ac ad publicam omnium Rameae Philosophiae studiosorum usum editae,* Basle: Sebastian Henricpetri 1576. – Freigius published also: *Audomari Talaei quem Petri Rami Theseum dicere iure possis Opera elegantioris methodicae philosophiae studiosis pernecessaria,* Basle: Petrus Perna 1575. Another edition: 1576.

37 All works that could be traced will be mentioned in their proper context. In addition he also published: Paralipomena, ad Aemylium & Ferronum adiecta, usque ad annum Christi M.D.LXIX, in: *Historia* [...] *Pauli Aemylij* [...], Basle: Sixtus Henricpetri 1569. – *Methodica actionum iuris repetitio: Ad ordinem iurisconsulti Triboniani instituta: ut iuris tyronibus loco introductionis ad alios prolixiores autores intelligendos esse possit* [...] *Cum Praefatione Ioannis Tho. Freigij,* Basle: Sixtus Henricpetri 1569. – *Ciceronianus* [...], *in quo: Ex Ciceronis monumentis, Ratio instituendi Locos communes demonstrata: & Eloquentia cum Philosophia coniuncta, descripta est Libris decem,* Basle: Sebastian Henricpetri [1575]. Further

establishing Ramism in the various branches of learning and scholarship. His works are didactically motivated and distinguished by the claim to present all knowledge methodically in the Ramist sense.

The earliest of Freigius' works, the *Trium artium logicarum schematismi,*[38] which had been published during Ramus' stay in Basle and which consists exclusively of dichotomous tables, purports to penetrate all *artes* in a logical-methodical manner – in this case grammar, rhetoric and dialectic. The *Paedagogus*[39] exemplifies the pedagogical-didactic motivation of Ramism. It presents an overview of the entire field of knowledge. Each representation of an individual branch of knowledge which is called *ars* in accordance with Ramist understanding, always starts with a definition which is followed by divisions according to the Ramist schema. First comes grammar which is defined as *ars bene loquendi* (p. 1); it is subdivided into Latin, Greek and Hebrew grammar, followed by dialogues to facilitate the acquisition of the French language (pp. 81-124). Then follow rhetoric as the *ars bene dicendi* (p. 126), which consists of *elocutio* and *pronuntiatio,* logic as *ars bene disserendi* (p. 134), arithmetic as *ars bene numerandi* (p. 145), music as *ars bene canendi* (p. 159), geometry as *ars bene metiendi* (p. 219), architecture as *ars domos & aedificia varia extruendi* (p. 247), then mechanics (pp. 264-268) and physics as *doctrina de natura rerum* (p. 270). Finally ethics is defined as *ars bene beateque vivendi* (p. 287), economy as *ars bene dispensandi* (p. 290), politics as *ars bene gubernandi* (p. 293), followed by the apodemic as *ars bene peregrinandi ad consequendam Politicam rerum cognitionem* (p. 295). He

editions: 1579. 1596. – *Εὐφημία Ioan. Thomae Freigii, Basilea ad Scholae Noricae in Altdorfio Rectoratum discedentis. Ad Civitatem & Academiam Basiliensem,* Basle: [Sebastian Henricpetri] 1576. – *Rhetorica, Poetica, Logica, ad usum Rudiorum in Epitomen redactae,* Nuremberg 1580. Further editions: 1582. 1585. – *Ἐπιλοίμια seu Liber Tristium. Tempore pestis, summa moestitia & dolore scriptus,* Basle: Sebastian Henricpetri 1583. – *In XII P. Virgilii Aeneid. Libros, Tabulae: Omnibus Poeticae Artis studiosis utiles & necessariae,* Basle: Sebastian Henricpetri [1587].

38 *Trium artium logicarum, grammaticae, dialecticae & rhetoricae brevis succinctique schematismi, ex eius generis autoribus melioribus in usum studiosorum & tyronum* [...], Basle: Sixtus Henricpetri 1568.

39 *Paedagogus. Hoc est, Libellus ostendens qua ratione prima artium initia pueris quam facillime tradi possint,* Basle: Sebastian Henricpetri 1582.

speaks of the study of the Ancients (*De antiquitatis studio,* pp. 297-303), of military discipline (*De polemica seu disciplina rei militaris,* pp. 304-310), of history as *memoria rerum in hoc mundo gestarum* (p. 310) and, at the very end, of jurisprudence as *ars aequi et boni* (p. 313) and of medicine as *ars bene medendi* (p. 342). Freigius had written individual works on many of these *artes;* they were all distinguished by their uniform application of Ramist method, and their presentation in the form of dichotomous tables is conspicuous. Some of these writings appear under the title *Quaestiones:* on logic and ethics,[40] on geometry and stereometry,[41] on physics[42] and on economy and politics.[43] What is striking in all these works is the insertion of genuine Aristotelian material into the basic pattern of the Ramist method. This is a procedure on which Freigius reflects – for example with respect to the first part of the *Quaestiones Logicae et Ethicae* – where he claims to have adapted the logic of Aristotle to the Ramist practice of *bene disserendi.*[44] One of the chief contributions made by Freigius is the application of the Ramist method to the legal sciences in his logic for jurists,[45] in the *Partitiones,*[46] which consist of dichotomous tables, and in the *Synopsis* of the Pandects[47] – *synopsis,* of course,

40 *Quaestiones Εώθιναὶ καὶ Δειλιναὶ: seu Logicae & Ethicae. Cum Analysi Logica, Ethica & Politica in 15. Ciceronis orationes,* Basle: Sebastian Henricpetri 1576. First edition: 1574.

41 *Questiones geometricae et stereometricae in Euclidis & Rami Στοιχείωσιν in quibus Logica veterum Mathematicorum illustratur & demonstratur,* Basle: Sebastian Henricpetri 1583.

42 *Quaestiones physicae, in quibus, Methodus doctrinam Physicam legitime docendi, describendique, rudi Minerva descripta est, Libris XXXVI,* Basle: Sebastian Henricpetri [1579], 1295 pp. Another edition: 1585.

43 *Quaestiones Oeconomicae et Politicae,* Basle: Sebastian Henricpetri [1578]. Another edition: 1591 [in reality: 1581].

44 Primum volumen continens [...] Organum: Aristotelis legibus conformatum, & ad usum bene disserendi accomodatum, denique exemplis Poetarum illustratum (separate title page in front of part 1).

45 *De Logica Iureconsultorum libri duo,* Basle: Sebastian Henricpetri [1582].

46 *Partitiones Iuris utriusque,* Basle: Sixtus Henricpetri 1571. Second edition: Basle: Sebastian Henricpetri [1581]. – *Partitiones feudales,* Basle: Sebastian Henricpetri 1581.

47 *Paratitla seu Synopsis Pandectarum iuris civilis,* Basle: Sebastian Henricpetri [1583].

was a key concept of Ramist methodology[48] – as well as in the revised edition of the lectures on the Pandects by Ulrich Zasius (1461-1535).[49] There is also pedagogical-didactic motivation behind the German language *Neüwe Practica Iuris.*[50]

Finally mention must also be made of Freigius' apology in the preface to his *Mosaicus*[51] that appeared in print shortly after his death. There he defended himself against the allegation that he had rejected the philosophy of Aristotle. On the contrary, Freigius proclaimed, he had given expositions in Freiburg and Altdorf of the logic, ethics and politics of Aristotle from their Greek sources and had analysed them methodically.[52] He attacked neither

48 See e.g.: *Historiae Synopsis. Seu Praelectionum historicarum in Altorfiano Norimbergensium Gymnasio delineatio,* Basle: Sebastian Henricpetri [1580]. Another edition: 1600.

49 *Zasius Ioan. Thomae Freigii. I. C. Hoc est: In Pandectas Iuris civilis Commentarii a Cl. I. C. Udalrico Zasio olim diffuse tractati: Nunc vero: Secundum leges Methodicas in compendium redacti: multis apodicticis aucti, elencticisque exonerati: ad theorian & praxin accomodatiores redditi. Accesserunt: Textus Florentinus: infinitae distinctiones seu Partitiones: Axiomata, Syllogismi, Paratitla, Germanica verborum civilium interpretatio. Et Titulorum qui in Pandectis habentur, continuatio nova,* Basle: Sebastian Henricpetri 1576.

50 *Neüwe Practica Iuris Und Formulen oder Concepten allerley / in zwen theil abgetheilet. Auß welchen der erst in sich begreifft die Practick / so im Rechten gebreüchlich unnd üblich / in ein besondere lustige und richtige ordnung / so bißher nie gebraucht worden / verfertiget / und nach den Lateinischen Partitionibus iuris, vormals in Tabelweiß offentlich aufgegangen / oder auch nach ordnung Keyserlichen Institutionen / gestellet unnd angeordnet. Der Ander Theil Hält inn sich etliche fürtreffliche Teutsche Consilia unnd Rhatschläg des Ehrwürdigen / Hochgelehrten / Weitberhümpten / wol erfahrenen Herrn Doctoris Ulrici Zasii seliger gedechtnuß / vormals in Teutscher sprach nie an tag gegeben. Jetzunder erst gemeinem nutz zu gutem in Truck verfertiget,* Basle: Sebastian Henricpetri [1574].

51 *Mosaicus, Continens Historiam Ecclesiasticam, 2494 annorum, ab orbe condito usque ad Mosis mortem,* Basle: Sebastian Henricpetri 1583. At the end of the unpaginated preface (sig. a2-b7 = 27 pp.) the date is indicated as: mense Decembri [...] 1582.

52 Quid igitur minus verum est aut consentaneum quam dicere, Freigius Aristotelem explodit: & idem Logica, Ethica, Politica Aristotelis integra publice non modo in Friburgensi & illa ipsa Altorfiana Academia & quidem in fontibus ipsis Graece explicavit: verumetiam editis libris commendavit & studiosam iuventutem methodica Analysi, pro sua tenuitate ad ea legenda adiuvit? Ibid. sig. a6v.

the old nor the new *Peripatetics* – among whom he, remarkably, counted Ramus himself.[53]

3. Swiss Ramus Editions

In his *Basilea* Ramus praises the Basle printers,[54] who, already before his visit as well as notably during and also after, had published a large number of editions of his works.[55] The following appeared in Basle before his journey to Switzerland: his commentaries on Cicero (1553),[56] the *Institutiones dialecticae* (1554),[57] the *Ciceronianus* (1557)[58] and the *Arithmetica* (1567).[59] During his stay in Basle his didactic writings were likewise printed: in 1569 the *Scholae in liberales artes,*[60] the *Grammatica,*[61] the *Rhetorica* by

53 At ego me cum Peripateticis neque antiquis illis doctis & eloquentibus hominibus, neque cum recentioribus recte & sane philosophantibus (quales numero etiam Vivetem, Ramum, Patricium, in primisque Zuingerum, cui Aristoteles optime notus et honoratus est) bellum habere affirmo. Ibid. sig. a7v-a8r.

54 Verum sit aliquis in laudibus academiae nostrae modus: et tamen ex hac tot tamque praestantium ingeniorum parente, vel hospite academia, Typographiam natam et educatam, id est, alimoniam plurimorum civium, opulentiamque certissimam praeterire non possumus. Hinc Amerbachii, Frobenii, Episcopii, Petrei, Curiones, Herwagii, Isengrini, Cratandri, Oporini, Bechelii, Guarini, Bryllingeri, Pernae, caeterique nobiles typographi Basileam non tantum opibus, sed egregia liberalis industriae apud omnes nationes fama decorarunt. Ramus: *Basilea,* ed. Fleig (as note 3) p. 55.

55 For this section I have checked the holdings of the *Universitätsbibliothek Basel,* the *Catalogue* in Waddington: *Ramus* (as note 1) pp. 441-477, which is, however, incomplete for the Basle imprints, Walter J. Ong: *Ramus and Talon Inventory,* Cambridge, Mass. 1958, Frank Hieronymus: *1488 Petri – Schwabe 1988. Eine traditionsreiche Basler Offizin im Spiegel ihrer frühen Drucke,* Basle 1997.

56 Enarrationes in 2. et 3. orationem Ciceronis de lege agraria, in orationem pro Rabirio perduellionis reo, in quattuor Catilinarias, in: *In omnes M. T. Ciceronis orationes* [...] *Enarrationes,* Basle: Johannes Oporin 1553.

57 *Institutionum dialecticarum libri tres* [...] *A. Talaei praelectionibus illustrati,* Basle: Nicolaus Episcopius junior 1554.

58 *Ciceronianus,* Basle 1557.

59 *Arithmeticae libri tres,* Basle 1567.

60 *Scholae in liberales artes,* Basle: Eusebius Episcopius et Nicolai fratris haeredes 1569.

61 *Grammaticae libri quattuor,* Basle: Eusebius Episcopius et Nicolai fratris haeredes 1569. *Rudimenta Grammaticae Latinae,* Basle: Eusebius Episcopius et Nicolai fratris haeredes 1569.

Talaeus with the lectures by Ramus,[62] the *Dialectica,*[63] the *Scholae mathematicae,*[64] the *Geometria,*[65] the *Arithmetica*[66] and its German translation by Christian Wurstisen (Urstitius, 1544-1588).[67] In the same year his correspondence with the Tübingen Aristotelian Jakob Schegk (1511-1587) was published in Basle.[68] The *Basilea,* written in Geneva, was published in Lausanne in 1571[69] as was the *Defensio pro Aristotele* against Schegk.[70] In the year of his death there followed one further edition of the *Dialectica.*[71] As posthumous Basle imprints one finds in 1573 further editions of the *Ciceronianus*[72] and the *Rhetorica,*[73] in 1574 one edition of *De moribus veterum Gallorum,*[74] in 1574 and 1575 one edition each of *De militia Caesaris,*[75] while in 1574, 1576 and 1577 one

62 *Audomari Talaei Rhetoricae libri duo, P. Rami Praelectionibus illustrati,* Basle: Eusebius Episcopius et Nicolai fratris haeredes 1569.

63 *Dialectica, Audomari Talaei praelectionibus illustrata,* Basle: Eusebius Episcopius et Nicolai fratris haeredes 1569.

64 *Scholarum mathematicarum libri unus et triginta,* Basle: Eusebius Episcopius et Nicolai fratris haeredes 1569, XVI, 320 pp.

65 *Geometriae libri septem et viginti,* Basle: Eusebius Episcopius et Nicolai fratris haeredes 1569.

66 *Arithmeticae libri duo,* Basle: Eusebius Episcopius et Nicolai fratris haeredes 1569.

67 *Ein sehr Nutzliche unnd Kunstreiche Arithmetick / oder Rechenkunst / auß warem Grund / durch Petrum Ramum /* [...] *verteutscht / durch M. Christianum Wurstisium* [...], Basle: Samuel Apiarius 1569.

68 *P. Rami et Iacobi Schecii epistolae, in quibus de logicae artis institutione agitur,* [Basle] 1569.

69 *Basilea ad senatum populumque Basiliensem,* Anno 1571 [Lausanne: Johannes Probus].

70 *Defensio pro Aristotele adversus Iacobum Schecium,* Lausanne: Iohannes Probus 1571.

71 *Dialectica, Audomari Talaei praelectionibus illustrata,* Basle: Eusebius Episcopius et Nicolai fratris haeredes 1572.

72 *Ciceronianus,* Basle: Petrus Perna 1573. With two prefaces by Freigius, a preface by Ramus and two letters by him, Paris, 16 August 1572.

73 *Audomari Talaei Rhetoricae libri duo, P. Rami Praelectionibus illustrati,* Basle: Eusebius Episcopius et Nicolai fratris haeredes 1573.

74 *Liber de moribus veterum Gallorum. Cum Praefatione Ioannis Thomae Freigij,* Basle: Sebastian Henricpetri [1574].

75 *Liber de militia C. Iulii Caesaris. Cum Praefatione Ioannis Thomae Freigii,* Basle: Sebastian Henricpetri [1574]. Another edition: 1575.

edition each of the *Ciceronianus et Brutinae questiones*[76] were published. In 1575 appeared one edition each of the *Aristotelicae Animadversiones*[77] and the *Praelectiones in Ciceronis Orationes octo consulares,*[78] and in 1576 the *Professio Regia.*[79] There followed in 1577 a further edition of the *Dialectica*[80] and an edition of the *Ciceronianus* together with *Grammatica, Rhetorica* and *Dialectica,*[81] in 1578 further editions of the *Scholae in liberales artes,*[82] of the *Scholae mathematicae*[83] and the *Rhetorica.*[84] In 1580 there appeared once more the *Arithmetica,*[85] the *Geometria*[86] and the *Praelectiones in Ciceronis Orationes octo consulares;*[87] in 1582 once more the *Scholae in liberales artes,*[88] in

76 *Ciceronianus et Brutinae quaestiones,* Basle 1574. Further editions: Basle: Petrus Perna 1576. 1577.

77 *Aristotelicae Animadversiones* [...] *opera Ioan. Thomae Freigii,* Basle: Sebastian Henricpetri [1575].

78 *Petri Rami Praelectiones in Ciceronis Orationes octo consulares. Una cum Ipsius Vita per Ioann. Thomam Freigium collecta,* Basle: Petrus Perna 1575, pp. 4-46: Life of Ramus, pp. 45[sic]-60: Basilea.

79 *Professio Regia. Hoc est, Septem artes liberales, in Regia cathedra, per ipsum apodictico Parisijs docendi genere propositae, et per Ioan. Thomam Freigium In tabulas perpetuas, ceu στρώματα quaedam, relatae: ac ad publicam omnium Rameae Philosophiae studiosorum usum editae,* Basle: Sebastian Henricpetri 1576. Contains: Ciceronianus, Grammatica latina, graeca, gallica. Rhetorica (Talaei), Dialectica, Physica, Ethica (= Liber de moribus veterum Gallorum, etc.), Geometria.

80 *Dialectica, Audomari Talaei praelectionibus illustrata,* Basle 1577.

81 *Ciceronianus, cum grammatica, rhetorica, dialectica,* Frankfort and Basle 1577.

82 *Scholae in liberales artes,* Basle: Eusebius Episcopius et Nicolai fratris haeredes 1578.

83 *Scholarum mathematicarum libri unus et triginta,* Basle: Eusebius Episcopius et Nicolai fratris haeredes 1578.

84 *Audomari Talaei Rhetoricae libri duo, P. Rami Praelectionibus illustrati,* Basle: Eusebius Episcopius et Nicolai fratris haeredes 1578.

85 *Arithmeticae libri duo,* Basle: Eusebius Episcopius et Nicolai fratris haeredes 1580.

86 *Geometriae libri septem et viginti,* Basle: Eusebius Episcopius et Nicolai fratris haeredes 1580.

87 *Petri Rami Praelectiones in Ciceronis Orationes octo consulares. Una cum Ipsius Vita per Ioann. Thomam Freigium collecta,* Basle: Petrus Perna 1580, pp. 4-46: Life of Ramus, pp. 45[sic]-60: Basilea.

88 *Rameae Scholae, et Defensio Petri Rami,* Basle 1582.

1585 again the *Dialectica*[89] and in 1589 the *Rhetorica.*[90] In Bern editions of the *Arithmetica*[91] were printed in 1596 and 1599. These were followed in 1601 by an edition of the *Rhetorica,*[92] which was also published in Geneva a year later.[93] Ludwig Lutz, the main proponent of Ramism in Basle in the seventeenth century, edited the *Basilea.*[94] Finally, in 1616 there appeared in Bern editions, *in usum Scholarum,* of the *Rhetorica* and the *Dialectica,* one year later of the *Arithmetica,* in 1619 once more of the *Rhetorica,* while in 1670 the *Rhetorica* and *Dialectica* were again published.[95]

89 *Dialectica, Audomari Talaei praelectionibus illustrata,* Basle 1585.

90 *Audomari Talaei Rhetorica e P. Rami Praelectionibus observata* [...], Basle: Conrad Waldkirch 1589.

91 *Arithmetica* [...] *cum explicationibus lectiss. Rud. Snellii,* Bern 1596. Another edition: 1599.

92 *Audomari Talaei Rhetorica e P. Rami* [...] *Praelectionibus observata. Recens in usum Scholarum Bernensium cum Scholiis Ioann. Piscatoris distinctius excusa,* Bern: Ioannes le Preux 1601.

93 *Audomari Talaei Rhetorica e P. Rami* [...] *Praelectionibus observata. Recens in usum Scholarum hac forma distinctius & emendatius excusa,* Geneva: Petrus Roverius 1602.

94 *Basilea ad senatum populumque Basiliensem, scripta sub discessum P. Rami: cum is Lutetiae Parisiorum repeteret; paulo post tempore in laniena Bartholomaeana cum aliis Evangelii Christi martyribus, misere ac nefarie trucidatus, anno 1572,* [Amberg] 1606, Preface by Ludwig Lutz.

95 *Audomari Talaei Rhetorica e P. Rami* [...] *Praelectionibus observata. Recens in usum Scholarum hac forma distinctius & emendatius excusa,* Bern: Abraham Weerlin 1616, 94 pp. in 12°. – *Dialecticae Libri Duo. Recens in usum Scholarum hac forma distinctius & emendatius excusi,* Bern: Abraham Weerlin 1616, 84 pp. in 12°. – *Arithmeticae libri duo; nunc primum hac manuali forma* [...] *editi,* Bern: Abraham Weerlin 1617, 81 pp. in 12°. – *Audomari Talaei Rhetorica e P. Rami* [...] *Praelectionibus observata. Recens in usum Scholarum hac forma distinctius & emendatius excusa,* Bern: Thomas Villerianus, impensis M. Iohannis Willii et Iohannis Benthemii 1619. – *Audomari Talaei Rhetorica e P. Rami* [...] *Praelectionibus observata. Recens in usum Scholarum hac forma distinctius & emendatius excusa,* Bern: Georg Sonnleitner 1670. – *Dialectica* [...] *distinctius et emendatius excusa,* Bern: Georg Sonnleitner 1670, 77, (18) pp. in 8°.

4. The Reception of Ramus in the Seventeenth Century

a) Basle

The teaching of logic at the University of Basle during the first half of the seventeenth century[96] was strongly influenced by Ludwig Lutz (1577-1642) who, after his father's appointment to a parish in Mühlhausen in Alsace, lived for a year (1587-1588) in the house of Theodor Zwinger (1533-1588), Ramus' pupil and friend. In 1596 he stood in for the Basle professor of mathematics Peter Ryff (1552-1629), who had written an influential handbook of geometry according to the Ramist method.[97] Lutz taught later at *Gymnasia* in Southern Germany, and in 1611 he became *Professor Organi Aristotelis* in Basle. Lutz was a teacher first and foremost: he continued the tradition of the didactically motivated Ramism of Zwinger and Freigius in Basle, and by his handbooks he ensured the lasting effect of the Ramist method at the universities and *Gymnasia* which persisted, in some places, well into the eighteenth century.

The *Artis Logicae Praecepta,*[98] as Lutz emphasized in his dedication, serve as an introduction to the study of the *Organum Aristotelis.*[99] In the *Praeludium de philosophia et artibus liberalibus* Lutz defines philosophy as "knowledge of divine and human things, in so far as the human intellect

96 See in general: Wolfgang Rother: *Die Philosophie an der Universität Basel im 17. Jahrhundert. Quellen und Analyse,* Ph. D. thesis, Zurich 1980, 957 pp. (Typescript); for the teaching of logic and rhetoric cf. pp. 269-310. 459-495. 691-706. See also: Wolfgang Rother: Zur Geschichte der Basler Universitätsphilosophie im 17. Jahrhundert, in: *History of Universities* 2 (1982) pp. 153-191.

97 *Quaestiones Geometricae in Euclidis et P. Rami Στοιχείωσιν In usum Scholae Mathematicae collectae* [...] *quibus geodaesian adiecimus per usum radii geometrici,* Frankfort 1600. Second edition: 1621. Third edition: 1649.

98 *Artis Logicae Praecepta: Pro usu Tyronum Philosophiae in Academiâ Basileensi,* Basle: Sebastian Henricpetri [1620], 162, (3), (2) pp. in 8°. For the use in the Basle *Paedagogium: Artis logicae rudimenta pro usu Gymnasii Literarii Balileensis,* Basle: Sebastian Henricpetri [before 1625], 38 pp. in 8°.

99 [...] ad meliorem fructuosioremque Organi Aristotelici audientiam [...]. *Artis Logicae Praecepta* (as note 98) p. 4.

can acquire it by itself"[100] and advocates an encyclopaedic notion of philosophy.[101] He then divides philosophy into *instrumentalis* and *realis* with further subdivisions. The structure and content of the *Praecepta* are Ramist. Logic is defined as *ars bene disserendi;*[102] its parts are *inventio* and *dispositio,* again defined and subdivided in the Ramist fashion.[103] The *Praecepta* were used for forty years in the instruction of logic in the first class and were replaced in 1660 by Franco Burgersdijk's *Institutionum logicarum libri duo* – an eclectic Aristotelian-Ramist compendium, which means that even after 1660 Ramism was still present in Basle.

The penetration of Aristotelian material by Ramist method is exemplified by Lutz's bilingual edition with commentary of the *Organum, Tabulis synopticis perpetuis illustratum,*[104] which, between 1619 and 1659, formed the basis of logic teaching in the second class. (In 1659 the *Professio Organi Aristotelis* was made a chair of history.) Lutz related the *Organum* to the *dispositio* or doctrine of judgement, the parts thereof to the individual books of the *Organum,* namely the noetic or axiomatic judgement to the books *De categoriis* and *De interpretatione,* and the dianoetic or syllogistic judgement to the *Topica,* the *Sophistici elenchi* and the *Analytica.*[105] The

[100] Philosophia est notitia rerum divinarum ac humanarum, quantam quidem ratio humana per se-ipsam consequi potest. Ibid. p. 5.

[101] Sic que philosophia omnium artium liberalium cognitionum simulatque usum includit, quam Graeci Ἐγκυκλοπαιδίαν, quasi circulum seu orbem liberalis eruditionis cognominant. Ibid. p. 5.

[102] Ibid. p. 9.

[103] Inventio est prior logicae pars de inquirendis argumentis ad thema quodcunque explicandum idoneis. Ibid. p. 9. – Dispositio est altera logicae pars de argumentis cum quovis proposito themate rite disponendis; ad recte de illo iudicandum. Ibid. pp. 51-52.

[104] *Aristotelis Organum, Graece & Latine; Commentario analytico & paraphrastico, Tabulis quinetiam synopticis perpetuis, illustratum, ac Discentium usibus accomodatum,* Basle: Sebastian Henricpetri 1619, (29), 906, (32) pp. in 4°. In the left column is the Greek text, in the right column the Latin translation by Julius Pacius, at the bottom the paraphrase and commentary by Lutz. Then with a separate title page: *Organi Aristotelis nova per Tabulas analyticas perpetuas Expositio,* Basle: Sebastian Henricpetri [1619], 232, (8) pp. in 4°.

[105] The book is divided into two parts with separate title pages: (1) Organi Aristotelis Pars prior: Libros continens, ad Judicium rationis noeticum pertinentes; De Categoriis, & De Interpretatione. Quibus de more praemissa est Porphyrii Isagoge.

aphoristic presentation in 1636 of the *Organum* in the form of theses is also intended to serve didactic purposes.[106]

Lutz's *Artis Rhetoricae Praecepta,*[107] used as a textbook until well into the eighteenth century, departs in an essential respect from the Ramist view of the relationship between logic and rhetoric. Although the definition of rhetoric as *ars bene dicendi* is taken over verbatim from Talaeus' *Rhetorica,*[108] it is not divided – as in the case of Talaeus – into two books (*De elocutione, De pronuntiatione*), but into four parts, that is *inventio, dispositio, elocutio, et pronuntiatio sive actio,* which means that *inventio* and *dispositio* are duplicated. The definition of rhetorical *inventio* as "excogitation of true or likely things which make the cause plausible"[109] is taken verbatim from the *Rhetorica ad C. Herennium* (I 2),[110] as is the definition of the *dispositio sive collocatio* as an "ordering and distribution of the invented things, which demonstrates what is to be collocated by what topoi"[111] – in this context Lutz also refers to Bartholomäus Keckermann's *Systema rhetoricae* (Hanau 1608).[112] A further departure from Ramus is taken by Lutz in his declaration of the purpose of rhetoric as persuasion (*ad persuadendum*)[113] – this is in agreement with Aristotle in the *Ars rhetorica* (I 2) and

(2) Organi Aristotelis Pars posterior: Libros ad Judicium rationis dianoeticum pertinentes; Topicos quidem VIII. cum appendice De Sophisticis Elenchis; Analyticos vero IV. complectens.

106 *Organi Aristotelis Thetica & Aphoristica Expositio; materia pro logicis exercitationibus futura,* Basle: Georg Decker 1636, 220 pp. in 4°.

107 *Artis Rhetoricae Praecepta: Pro usu Tyronum Philosophiae in Academiâ Basileensi,* Basle: Sebastian Henricpetri [1620?], 120, (5) pp. in 8°. Another edition: Basle: Bertschi 1661, 114, (4) pp. in 8°. For the use in the Basle *Paedagogium: Artis Rhetoricae Rudimenta,* Basle: Sebastian Henricpetri [before 1625], 38 pp. in 8°.

108 *Audomari Talaei Rhetorica* I 1 (for the Basle editions see above pp. 21-24).

109 Inventio, prima pars Rhetoricae, est rerum verarum aut verisimilium excogitatio, quae causam probabilem reddant. *Artis Rhetoricae Praecepta* (as note 107) p. 4.

110 There are 10 references between 1668 and 1689, see Rother: *Philosophie in Basel* (as note 96) p. 705, n. 114.

111 Dispositio (sive Collocatio), secunda Rhetoricae pars, est ordo et distributio rerum inventarum; quae demonstrat, quid quibusque in locis sit collocandum. *Artis Rhetoricae Praecepta* (as note 107) p. 38.

112 In Book I 7 (p. 97): [...] dispositio, quae est apta ordinatio & connexio inventorum.

113 Bene dicere est scienter, perite et ornate dicere: hoc est, apposite ad persuadendum. *Artis Rhetoricae Praecepta* (as note 107) p. 3.

with Cicero in *De oratore* (I [138]). Like Ramus and against the *Rhetorica ad C. Herennium* (I 2) and Quintilian's *Institutiones oratoriae* (III 3), Lutz excludes the faculty of memory (*memoria*) from logic, on the grounds that it is not so much a skill as based on natural gifts and training.[114]

The printed disputations are an important indicator of Ramus' effect on Basle. They appeared in great numbers[115] from the late 1590s and they testify to a broad reception of Ramist philosophy and method in Basle. It is conceivable that students acquired their knowledge from reading Ramus' *Dialectica* which had been printed in Basle; it is more likely, however, that they read other Ramist authors, e.g. Amandus Polanus von Polansdorf[116] or Bartholomäus Keckermann,[117] since it is these who are cited. After 1620 it is most probable that the *Artis Logicae Praecepta* and the *Artis Rhetoricae Praecepta* by Lutz were the sources of the reception of Ramism.

In the theses up to the middle of the seventeenth century the Ramist definition of logic or dialectic as *ars bene disserendi* is often repeated.[118] Occasionally it is compared with other definitions, e.g. by Lutz:[119] he launches an attack on certain *Peripatetics* who, like Jacopo Zabarella,[120] define logic as *habitus instrumentalis, quo vero a falso discernitur* or as *ars ratiocinandi vel docendi*. Furthermore the Ramist division of logic into *inventio* and *dispositio* is as uncontroverted as it is frequently repeated.[121] As source of this division Lutz cites the Stoics in one instance and after them

114 Vulgo superadditur quinta, Memoria: quam, firmam animi perceptionem, rerum & verborum, definiunt. Verum haec, non tam artis, ac praesertim dispositionis; quam naturae & exercitationis opus est & effectum. Ibid. p. 4.

115 See the catalogue in Rother: *Philosophie in Basel* (as note 96) pp. 342-449.

116 *Logicae libri duo iuxta naturalis methodi leges conformati,* Herborn 1590. Basle 1598. 1599. Cited by Lutz: *Themata logica controversa,* Basle 1612, thes. 1. *Positiones philosophicae controversae,* Basle 1613, thes. log. 1. *Problemata philosophica miscellanea,* Basle 1617, thes. log. 1.

117 *Systema logicae, tribus libris adornatum* [...] I 1, Hanau 1600. Cited by Lutz: *Themata logica controversa,* Basle 1612, thes. 1.

118 There are 16 references between 1620 and 1649, see Rother: *Philosophie in Basel* (as note 96) p. 692, n. 16.

119 *Themata logica controversa,* Basle 1612, thes. 1.

120 *De natura logica libri duo* I 13, in: *Opera logica,* Venice 1578.

121 There are 16 references between 1613 and 1634, see Rother: *Philosophie in Basel* (as note 96) p. 694, n. 29.

Ramus,[122] and in another passage he even adds – with a reference to the Ramus commentary by George Downame[123] – the name of Aristotle.[124] Lutz's attempt to synchronise the structure of the *Organum Aristotelis* with that of the Ramist dialectic is exemplified by his claim that Zarabella's division of logic into general and particular is also right, since the *logica generalis* is concerned with the tools of *bene disserendi* in general and the *logica specialis* with the *modus bene disserendi* in particular. He continues by saying that Aristotle had treated the *logica generalis* in *De categoriis* and *De interpretatione,* and the *logica specialis* in the *Analytica,* the *Topica* and the *Sophistici elenchi.*[125] A further example of the Aristoteleo-Ramism in Basle and also an example of the relativity of historical classification, is the disputation by Johannes Rudolf Wettstein on Aristotle's *De interpretatione.*[126] He relies on a number of authors whom he calls *Peripatetici,* whose works, however, represent an Aristoteleo-Ramist eclecticism, e.g. Otto Casmann's *Logicae Rameae et Melanchthonianae collatio et exegesis* (Hanau [2]1599), Haloinus Gothofredus' *Thetica logicae Aristotelicae et Rameae synopsis* (Herborn 1602), Amandus Polanus von Polansdorf's *Syntagma logicum Aristotelico-Ramaeum ad usum inprimis theologicum accomodatum* (Basle 1605 and 1611), Konrad Dieterich's *Institutiones dialecticae ex probatissimis Aristotelis et Rami interpretibus* (Giessen 1609) and Lutz's *Organi Aristotelis Thetica & Aphoristica Expositio* (Basle 1636). Following these authors Johannes Jakob Werenfels was able to cite the Aristotelian definition of syllogism and describe it as agreeing with that of Ramus.[127]

The theses on rhetoric, understood as *ars bene dicendi,*[128] repeat as far as the content is concerned Lutz's *Ars Rhetoricae Praecepta,* but they are interesting on account of several arguments which are not found in Lutz. The

122 *Themata logica controversa,* Basle 1612, thes. 8.

123 *Commentarii in P. Rami* [...] *dialecticam* I 2, Frankfort 1601.

124 Lutz: *Problemata miscellanea ex logicis deprompta,* Basle 1615, thes. 8.

125 Ibid. thes. 9.

126 *Disputatio logica exhibens positiones quasdam ex libro Aristotelis περὶ ἑρμηνείας praecipue desumptas,* Basle 1643.

127 *Disputatio logica de natura et partibus syllogismi,* Basle 1616, thes. 7 and 8.

128 Rhetorica est ars bene dicendi. Bene dicere est scienter, perite et ornate dicere: hoc est, apposite ad persuadendum. Lutz: *Artis Logicae Praecepta* (as note 98) p. 3, and 11 other references between 1605 and 1696, see Rother: *Philosophie in Basel* (as note 96) p. 702, n. 98.

Basle scholars who were committed Ramists did not agree with the definition of the purpose of rhetoric as persuasion (*ad persuadendum*), among them notably the young Lutz himself,[129] Felix Platter[130] and Samuel Grynaeus – the latter stating categorically: *certus, proprius et adaequatus finis est bene dicere.*[131] The great majority of the theses, however, followed Cicero (*De oratore* I [138]), stating that the purpose of rhetoric was to persuade, *persuadere.*[132] In the 1660s and 1670s many of these dissertations cite Aristotle (*Ars rhetorica* I 2). In this sense the *causa efficiens* of rhetoric is *ars,* the *causa materialis* is *quaestio quavis,* the *causa formalis* is *bene dicere* and the *causa finalis* is *persuadere.* A characteristic instance of the attempt to reconcile Ramus with Cicero or Aristotle is provided by Nicolaus Agricola's thesis, which argues that there was a great quarrel among scholars whether the purpose of rhetoric was *bene dicere* or *persuadere;* he himself considered both correct.[133] A similar argument was advanced by Friedrich Bauhin, who stated succinctly more than eighty years later: *Bene dicimus, ut persuademus.*[134] There is an almost uniform acceptance of the four parts of rhetoric in line with Lutz's *Praecepta,*[135] and again, it was Lutz himself who had previously upheld the Ramist position.[136] He refers to "that war between Brutus and M. Tullius and the wrestling of Ramus with Quintilianus" concerning the difference between rhetoric and dialectic.[137] For Lutz there is a great difference between the *praecepta* of dialectic and those of rhetoric: dialectic was *ars bene disserendi,* rhetoric by contrast was

129 *Positiones philosophicae controversae,* Basle 1613, thes. rhet. 2. *Dissertationum philosophicarum ἑπτὰς τρίνα,* Basle 1614, thes. rhet. 9.

130 *Decades quatuor thesium e philosophia in genere, eiusque parte instrumentali depromptarum,* Basle 1632, rhet. 2.

131 *Theses philosophicae miscellaneae,* Basle 1625, thes. rhet. 2.

132 There are 22 references between 1603 and 1696, see Rother: *Philosophie in Basel* (as note 96) p. 702, n. 100.

133 *Miscellanea philosophica ex philosophia deprompta,* Basle 1605, thes. rhet. 3.

134 *Positiones philosophicae,* Basle 1689, thes. 14.

135 Lutz: *Artis Rhetorica Praecepta* (as note 107) p. 3, and 23 references between 1604 and 1696, see Rother: *Philosophie in Basel* (as note 96) p. 703, n. 103.

136 *Positiones philosophicae controversae,* Basle 1613, thes. rhet. 5.

137 Themata de discrimine dialecticae atque rhetoricae, in: *Agon philosophicus, logicus partim, partim etiam physicus,* Basle 1613, thes. 1: [...] bellum illud Bruti cum M. Tullio et lucta P. Rami cum Quintiliano.

ars bene dicendi or, as Melanchthon had put it: *ars, quae docet viam ac rationem recte et ornate dicendi* (a reference to Melanchthon's *Elementorum rhetorices libri duo,* Basle 1563 and 1567). In his view dialectic must be understood as *ars rationis* and rhetoric as *ars orationis,* both are therefore to be seen as clearly differentiated.[138] Furthermore, he stresses firstly, that the doctrine of *inventio, dispositio* and *memoria* had been undeservedly attributed to rhetoric and secondly, that *bene dicere* rather than *persuadere* was the ultimate purpose of rhetoric, as indeed Ramus had argued against Cicero.[139] In 1632 Felix Platter,[140] too, and Johannes Rudolf Fäsch[141] in 1669, excluded (like Ramus) *inventio* and *dispositio* from rhetoric. Jakob Rüdin devotes himself to the problem of the dual, i.e. logical and rhetorical, *inventio* and *dispositio* which arises from Lutz's system of the *Praecepta,* presenting arguments in defence of Lutz's position.[142]

A further problem which arises in respect to the philosophy of Ramus is the relationship of logic and metaphysics, which does not figure at all in his system of knowledge. Heinrich Kisselbach addressed the problem in Basle.[143] He claimed that Ramus had attempted to eliminate metaphysics by contending that it was the same discipline as logic, as both were concerned with "being and non-being". In contrast, Kisselbach maintained that metaphysics and logic were indeed in conformity as far as their subject matter was concerned, i.e. "being", but not in their *modus considerandi.* He argued that metaphysics viewed "being" *universaliter* and was concerned with the transcendental, whereas logic used the most universal metaphysical terms and was concerned with those subjects of the special branches of

138 Ibid. thes. 3 and 4.

139 Ibid. Porismaton.

140 *Decades quatuor thesium e philosophia in genere, eiusque parte instrumentali depromptarum,* Basle 1632, thes. rhet. 4f.

141 *Theses subsequentes ex praecipuis philosophiae partibus selectae,* Basle 1669, thes. rhet. 2: Inventio et dispositio non sunt partes rhetoricae.

142 *Theses miscellaneae,* Basle 1665, rhet. 5: [...] omnino differunt, unde peccant, qui non nisi elocutionem et actionem partes rhetoricae faciunt. – In a similar way also Felix Platter: *Quaestiones ex amplissimo philosophiae campo collectae,* Basle 1646, thes. 2. Felix Platter: *Theses miscellaneae,* Basle 1648, thes. rhet. 2. Johannes Jakob Staehelin: *Theses ex rhetorica depromptae,* Basle 1678, thes. 7.

143 *Omnivarii ex fragrante et patenti philosophiae horto flosculi,* Basle 1645, thes. 26: Utrum metaphysica sit specie distincta a logica?

knowledge that had to be defined, divided, proved and ordered. In the same sense Jakob Rüdin argued that metaphysics must be accepted as a discipline which differs from logic.[144]

b) Bern and Lausanne

The extant theses and textbooks used at the *Hohe Schule* of Bern, where there existed two chairs of philosophy from 1598 onwards, and at the *Haute École* of Lausanne, where philosophy was taught by only one professor – compared with the nine chairs of philosophy at the University of Basle – also suggest a considerable influence of the philosophy and method of Ramus.

In Bern the teaching of logic and rhetoric was purely Ramist. Here were published in 1601 an edition of the *Rhetorica,* in 1616 of the *Rhetorica* and the *Dialectica* and in 1619 again an edition of the *Rhetorica* – always *in usum Scholarum.*[145] It is probable that the whole of the Ramist system was taught in Bern, for it is apparent that the Ramist textbooks, even though they were not all printed in Bern, were bound together in the following manner: 1. *Rami Grammatica* (Hanau 1612); 2. *Talaei Rhetorica* (Bern 1616); 3. *Rami Dialectica* (Bern 1616); 4. *Rami Geometria* (Hanau 1612); 5. *Rami Arithmetica* (Hanau 1611).[146] Firstly, mention must be made of

144 *Thesium seges ex Synopsi Logicae Burgersdicianae a clarissimo Adriaano Heereboord adornata, Rhetorica quin et contracta, Grammaticaque Latina celeberrimi Gerardi Joannis Vossii,* Basle 1668, thes. met. 1.

145 *Audomari Talaei Rhetorica e P. Rami* [...] *Praelectionibus observata. Recens in usum Scholarum Bernensium cum Scholiis Ioann. Piscatoris distinctius excusa,* Bern: Ioannes le Preux 1601. – *Audomari Talaei Rhetorica e P. Rami* [...] *Praelectionibus observata. Recens in usum Scholarum hac forma distinctius & emendatius excusa,* Bern: Abraham Weerlin 1616, 94 pp. in 12°. – *Dialecticae Libri Duo. Recens in usum Scholarum hac forma distinctius & emendatius excusi,* Bern: Abraham Weerlin 1616, 84 pp. in 12°. – *Audomari Talaei Rhetorica e P. Rami* [...] *Praelectionibus observata. Recens in usum Scholarum hac forma distinctius & emendatius excusa,* Bern: Thomas Villerianus, impensis M. Iohannis Willii et Iohannis Benthemii 1619.

146 So the copy in the *Universitätsbibliothek Basel.* The *Arithmetica* was also published in Bern: *Arithmeticae libri duo; nunc primum hac manuali forma* [...] *editi,* Bern: Abraham Weerlin 1617, 81 pp. in 12°.

Markus Rütimeyer (1580-1647), a theologian who, after studies in Bern, Herborn, Marburg, Heidelberg, Leiden, Franeker and Paris became professor of philosophy in Bern in 1613, but who four years later transferred to a pastorate. In 1617 and 1618 he published his *Idea Methodica Seu Analysis Logica Generalis Et Specialis Praeceptorum & exemplorum Dialecticae Petri Rami.*[147] Ramism in Bern did not, however, end with Rütimeyer's departure from the *Hohe Schule.* This is confirmed by the printed theses of Johannes Nicolaus (died 1676), who taught philosophy from 1658 to 1662 and later Hebrew and theology. Between 1659 and 1661 he published a five-part *Disputatio exegetica* on the first book of the *Dialectica* by Ramus.[148] Nicolaus represents a notion of philosophy which relies on the Cartesian postulate of evidence, defining philosophy as *habitus evidenter discursivus de rebus.*[149] In 1670 there appeared new editions of the *Rhetorica* by Talaeus and of the *Dialectica* by Ramus.[150] Even at the end of the seventeenth century Ramism was still present at Bern, although only in conjunction with Cartesianism which was represented by Samuel Leemann (1651-1709). Having studied in Bern, Lausanne, Geneva, Puylaurens, Saumur, Oxford, Utrecht, Leiden and Groningen, he first taught philosophy in Bern from 1684 to 1698, and then Hebrew and theology. His

[147] *Idea Methodica Seu Analysis Logica Generalis Et Specialis Praeceptorum & exemplorum Dialecticae Petri Rami perpetuis tabulis synopticis; textuque e regione respondentibus adornata, studio Marci Rutimeieri* [...], *accessit eiusdem Oratio panegyrica continens Scholarum Scholarumque fundatorum et conservatorum encomium,* Bern: Abraham Weerlin 1617, (8), 183, (15) pp. in 4°. Another edition: *Idea Methodica Seu Analysis Logica Generalis Et Specialis Praeceptorum & exemplorum Dialecticae, tabellis synopticis adornata, studio Marci Rutimeieri; accessit eiusdem oratio panegyrica,* Bern: Abraham Weerlin 1618.

[148] *Disputatio exegetica in cap. I. et II. lib. I. Dialecticae Rameae,* Bern 1660, (12 pp.). [...] *in cap. III. IV. et V,* Bern 1659, (11 pp.). [...] *in cap. VI. VII. VIII. et IX*, Bern 1659, (8 pp.). [...] *in cap. X. et XI,* Bern 1660, (11 pp.). [...] *in cap. XII. XIII. XIV. XV. XVI. et XVII,* Bern 1661 (16 pp.).

[149] *Theses miscellae,* Bern 1659, I 1. *Theses de philosophiae natura et partibus,* Bern 1660, cap. II.

[150] *Audomari Talaei Rhetorica e P. Rami* [...] *Praelectionibus observata. Recens in usum Scholarum hac forma distinctius & emendatius excusa,* Bern: Georg Sonnleitner 1670. *Dialectica* [...] *distinctius et emendatius excusa,* Bern: Georg Sonnleitner 1670, 77, (18) pp. in 8°.

1695 textbook of logic[151] and his theses printed between 1690 and 1697[152] are indebted to Cartesian philosophy. In his *De logica correctrice et directrice* logic is no longer defined as *ars bene disserendi* but as the "art which directs the mind in attaining the knowledge of truth".[153] He argues that it is necessary to shed the errors and prejudices of childhood and to doubt everything. Ramist logic is embedded in a system of concept, judgement, conclusion and method. The first part of this system is the *ars percipiendi,* which incorporated the Cartesian proposition that the clear and distinct perception is the criterion of truth, and the Cartesian doctrine of the innate ideas. A component of the *ars percipiendi* is the *ars inveniendi,* which is presented *juxta Ramum.* In the *ars judicandi* and the *ars ratiocinandi,* Leemann also follows Ramus, whereas in the fourth part, the *ars disponendi,* which is dedicated to method, he discusses the four rules of Cartesian logic, based on the *Discours de la méthode,* on Johannes Clauberg and on the *Ars cogitandi* of Port-Royal.

At the *Haute École* of Lausanne Jean-Rodolphe Le Fèvre (Faber, Fabri, ca. 1580-1650), who taught philosophy there from 1620 to 1624, was a representative of Aristoteleo-Ramism, similar to that at the University of Basle. In his theses,[154] especially those on physics, he reveals himself as an Aristotelian, whereas in his theses on logic he follows Ramus. His Aristoteleo-Ramist eclecticism is also evident in the title of the lost *Cur-*

151 *Compendium Logicum per Quaestiones & Responsiones Juniorum captui accomodatum in usum institutionis privatae,* Bern: Andreas Hügenet 1695, 68 pp. in 8°.

152 *Disputationum logicarum I. de logica correctrice & directrice,* Bern 1690, (20 pp.). *II. de arte percipiendi prior,* Bern 1694, (27 pp.). *III.* [...] *altera,* Bern 1694, (40 pp.). *IV.* [...] *posterior, in specie agens de variis argumentorum inveniendorum locis juxta Ramum,* Bern 1695, (32 pp.). *V. de arte judicandi prior,* Bern 1695, (36 pp.). *VI.* [...] *posterior,* Bern 1695, (20 pp.). *VII. de arte ratiocinandi prior,* Bern 1697, (32 pp.). *VIII.* [...] *posterior,* Bern 1697, (28 pp.). *IX. de methodo sive arte disponendi,* Bern 1697, (24 pp.).

153 Logica est ars regens mentem in cognitione veritatis. *De logica correctrice* (as note 152) thes. 1.

154 *Theses de iudicio,* Bern 1621, 15 pp. [...] *de mundo,* Geneva [ca. 1621-1623? copy damaged], (11 pp.). [...] *de coelo,* Bern 1621, 16 pp. [...] *de elementis physicis,* Bern 1621, 16 pp. [...] *de corpore perfecte mixto inanimato et animato,* Geneva 1622, 15 pp. [...] *de corpore vegetante, sentiente, et intelligente,* Geneva 1622, (16 pp.). [...] *miscellaneae,* Bern 1623, 16 pp.

sus Logicus, Totius Organi Stagyritani & Ramaei praecepta & usum demonstrans.[155] Georg Müller (ca. 1603-1684), born in the Palatinate, who thanks to his long span of teaching, from 1628 to 1650 as professor of philosophy and subsequently until 1684 as professor of theology, formed generations of students, continued this tradition of Aristoteleo-Ramism, as his theses on logic of 1629 and 1643 show.[156] As in Bern, some sort of Cartesio-Ramism seems to have been taught in Lausanne in the later seventeenth century, a trend represented by Jérémie Sterky, who was professor of philosophy in Lausanne from 1685 until 1700, later of theology and thereafter pastor in Berlin and professor of theology at the University of Frankfort on the Oder. In his *Institutiones logicae,*[157] subdivided into *De inventione* and *De judicio,* he offers – as he announces in Cartesian language – "a clear and manifest explanation of Ramist doctrines", adducing older as well as more recent philosophers.[158] His handbooks reveal Sterky to be an eclectic, who in metaphysics not only "accurately discusses the principles of old and new philosophers, but also, in so far as this is possible, reconciles their views one with the other",[159] or who in physics "compares them with each other",[160] drawing on the most varied sources from Aristotle, Copernicus, Ramus and Descartes to Pascal, Boyle and Christiaan Huygens.

155 Mentioned in: *Cursus Physicus,* Geneva: Pierre Aubert 1625, (40), 496, (32) pp. in 8°: p. 9 in the unpaginated preface.

156 *Theses de natura logices,* Bern 1629, 24 pp. *Apodixis logica,* Bern 1643, (19 pp.).

157 *Institutiones Philosophicae* [...] *Pars Prima In qua Praeter Generaliora Philosophiae prolegomena continentur Institutiones Logicae,* Bern: Andreas Hügenet 1694, 12, 417 pp. in 4°.

158 Ibid. p. 22.

159 *Institutiones Metaphysicae* [...] *In Quibus Non modo Veterum Recentiorumque Philosophorum principia accurate discutiuntur, verum etiam, qua fieri poterit, eorundem placita inter se conciliantur,* Geneva: Philippe Gamonet 1695, (18), 173, (7) pp. in 4°.

160 *Institutiones Physicae* [...] *In Quibus Veterum & Recentiorum Philosophorum principia accurate discutiuntur, eorumque placita inter se comparantur,* Geneva: Philippe Gamonet 1696, (8), 389, (2) pp. in 4°.

5. Conclusion

The influence of Petrus Ramus' philosophy and method on scholars and on higher education in Protestant Switzerland can be subdivided into different phases. Two scholars are representative of the first phase, which begins with Ramus' stay in Basle: Theodor Zwinger (1533-1588) and Johannes Thomas Freigius (1543-1583). Both were, each in his own way, intent on the application of the Ramist method, and both promoted, with different intensity, the diffusion of Ramus' works and thought. Freigius, the biographer of Ramus, editor of many of his works and himself an author, who wished to apply the Ramist method to all knowledge, considered himself Ramus' heir and successor. His resolute Ramism proved to be a serious obstacle to his academic career. Zwinger, who was personally more closely connected with Ramus than Freigius was, did not, however, share Ramus' radical critique of Aristotle, and even adapted Aristotle's works for educational purposes. He was the type of polyhistor who applied the Ramist method productively to particular areas of knowledge.

The second phase of influence is characterised by a growing distance from genuine Ramism leading to an eclectic levelling. The Swiss protagonist in this development was Ludwig Lutz (1577-1642), who combined the Aristotelian *Organum* with Ramist dialectic and who decidedly dissented from Ramus in the question of the relationship between logic and rhetoric. In the form of Lutz' levelling textbook-Ramism, the influence of Ramus extended even into the eighteenth century. This second, Aristoteleo-Ramist phase was – with some local variations – followed by a Cartesio-Ramist phase. In Basle, Cartesio-Ramism more or less abandoned Ramus in the field of logic, as the lectures after 1660 were based on the handbooks of the Aristoteleo-Ramist eclectic Franco Burgersdijk (1590-1635) and the Cartesian Adriaan Heereboord (1614-1661).

In Bern the phase of genuine Ramism represented by Markus Rütimeyer (1580-1647) was followed immediately by the Cartesio-Ramist phase, first noticeable in the late 1650s in Johannes Nicolaus (died 1676), then much more clearly in Samuel Leemann (1651-1709). Lausanne was, at least until the middle of the seventeenth century, under the influence of Aristoteleo-Ramism represented by Jean-Rodolphe Le Fèvre (ca. 1580-1650) and Georg Müller (ca. 1603-1684). In the eclectic compilation which comprised Aristotle, Descartes and experimental science, assembled by Jérémy Sterky who taught at Lausanne since 1685, Ramism was only of secondary importance.

In conclusion, it is possible to state that Ramus survived the seventeenth century in Switzerland, but in the process of reception the contours of his method, which had claimed to achieve a systematic penetration of all knowledge, became more and more diffuse in different eclecticisms.

Notes on Ramism in the Netherlands

THEO VERBEEK

Before discussing Ramism in the Netherlands we should be clear what we are talking about. I propose to see Ramism as a program rather than a doctrine, which is threefold: in education (expected it to be more systematic and problem orientated), in logic (reducing it to an *ars disserendi*) and in philosophy generally (where the study of nature, whatever that is, is expected to replace the study of Aristotle's texts). As such, Ramism expanded upon earlier trends – which makes it difficult to identify the program, except when one encounters a self-professed Ramist. On the whole, though, Ramism has never been in the Netherlands a cause for acrimonious disputes, as it was in France and sometimes Germany. In fact, the memory of Ramus' murder served as a reminder "not to be too quick with one's judgment or condemnation of someone's person or doctrine".[1]

In this paper I shall discuss two universities: Leiden, founded in 1575, and Franeker, founded ten years later. In Leiden two very famous Ramists held the chair of mathematics: Rodolphus Snellius van Royen (1546-1613) and his son Willebrord (1580-1626).[2] Rodolphus in particular distinguished himself as a Ramist propagandist through his commentaries and lectures on

1 Abraham Heidanus: *Consideratien over eenige saecken onlanghs voorgevallen in de Universiteyt binnen Leyden,* Amsterdam: Van Someren 1676, pp. 26-27.

2 Monographs are sadly lacking; for a biography of both Rodolphus and Willebrord Snellius, see the articles of Cornelis de Waard in *Nieuw Nederlandsch Biographisch Woordenboek,* vol. 7, cols. 1152-1163. Some source material was published (incompletely) by J. A. Vollgraff: Brieven van R. en W. Snellius, in: *Leidsch Jaarboekje* 11 (1914) pp. 103-111 (quoting from a correspondence with Aemilius van Rozendael, kept in Utrecht University Library). A copy of Ramus' *Optica* (Kassel 1606), with annotations in Willebrord Snellius' hand, is kept in Leiden University Library.

the Frenchman's dialectic, physics and arithmetic, which were published in Frankfort and Herborn.[3] Snellius had come to know Ramus during his student days in Marburg, which Ramus visited in 1569 and 1570. From 1581 he was professor extraordinary of mathematics in Leiden, where his most famous students were Prince Maurice of Nassau and Isaac Beeckman (1588-1637), Descartes' first Dutch friend – although at the time few people could have known him.[4] Among the Beeckman papers one finds a list of authors whom Snellius recommended him to read in mathematics, which include Ramus' geometry and arithmetic (including 'practical arithmetic'), as well as Euclid, Heron, Boethius, Ptolemy, Copernicus, Hermes Trismegistus (for astrology), Clavius, Comandinus, Regiomontanus, Vitello, Stevin, Oronce

3 *In aureum Philippi Melanchthonis de anima vel potius de hominis physiologia libellum, commentationes utilissimae cum lectissimis aliorum annotationibus sedulo collatae. Accesserunt D. Rudolphi Goclenii* [...] *theses quaedam ac disputationes de praecipuis materiis psychologicis selectissimae,* Frankfort: Saurius 1596. *In Petri Rami arithmeticam,* Frankfort 1596. *In Dialecticam Petri Rami Praelectiones,* Frankfort: Fischer 1596. *Commentarius doctissimus in dialecticam Petri Rami, forma dialogi conscriptus: in quo artis praecepta cum exemplis analytice explicantur* [...], Herborn: Corvinus 1587 and many other editions. *Snellio-Ramaeum philosophiae Syntagma, tomis aliquot separatis distinctum: Quibus cont. 1. Generales sincerioris philosophiae Ramaeae informationes, 2. Dialectica, 3. Rhetorica, 4. Arithmetica, 5. Geometria, 6. Sphaera, seu astronomia, 7. Physica, 8. Psychologia, 9. Ethica,* Frankfort: Fischer 1596. *Rudolphi Snellii Veteraquinatis partitiones physicae: methodi Rameae legibus informatae, exceptae olim ex dictantis eius ore in Schola Marpurgensi. Nunc primum in lucem editae,* Hanau: Antonius 1594. *In Audomari Talaei Rhetoricam commentationes,* Frankfort: Saurius and Fischer 1596, also known as *Dialogismus rhetoricus,* Leiden 1600. *De ratione discendi et exercendi logicam,* Herborn 1599. 2nd ed. 1609. *Rudolphi Snellii in P. Rami Geometriam praelectiones, cum lectissimis Lazari Schoeneri [et] Johannes Thomae Freigii explicationibus,* Frankfort: Fischer 1596. Many of these books were bound together, in various combinations.

4 For the most recent survey of Beeckman's life and works, see the article by Klaas van Berkel in: *Grundriss der Geschichte der Philosophie. Begründet von Friedrich Ueberweg. Die Philosophie des 17. Jahrhunderts,* vol. 2/2, Basle: Schwabe 1993, pp. 631-636. Apart from his medical thesis on tertian fever (Caen 1618), which is not completely known (*Journal tenu par Isaac Beeckman de 1604 à 1634,* ed. Cornelis de Waard, 4 vols., The Hague: Nijhoff 1939-1953, vol. 4, pp. 42-44), the only thing published at the time was a posthumous volume edited by Beeckman's younger brother Abraham: *Mathematico-physicarum meditationum, quaestionum, solutionum centuria,* Utrecht: Petrus Daniel Sloot 1644.

Finé and Pappus.[5] Snellius also advised the government about the Leiden *Duitsche School voor Mathematiek* (where courses were given in Dutch – *Duitsch* – rather than Latin), which was created by Prince Maurice as a school for engineers in the army. His son Willebrord (1580-1626) succeeded him in the chair of mathematics and published a commentary on Ramus' arithmetic.[6]

At Franeker, the best known Ramist was undoubtedly William Ames (1576-1633), who came to the Netherlands as an army preacher in 1610, was appointed professor of theology in Franeker in 1622 and ended up as a minister of the Church of Rotterdam.[7] Ames was opposed to the use of metaphysics in theology (*Adversus metaphysicam,* 1629), on the grounds that the use of metaphysical concepts would rob theology of its practical character. In his *Technometria* (published posthumously in 1643) he integrated the six fundamental *artes* of human life (theology, dialectic, grammar, rhetoric, mathematics and physics) into a single *pansophia.*[8]

5 *Journal* (as note 4) vol. 4, pp. 17-19. Beeckman adds the following note: Hi fuerunt auctores quos Snellius pater olim a me rogatus, mihi indicavit ad Mathesin exercendam, cum prius jussisset me dividere artem mathematicam in suas partes, quod feci uti videre est in prima columna; quae sequuntur ipse scripsit. Neque praeter ea mihi quicquam auxilii tulit, non quod denegaverit sed quod ausus non essem rogare. Ideoque necessarium fuit pro labore, quicquid teneo, ex iis libris haurire. p. 19.

6 *Petri Rami arithmeticae Libri II cum commentariis,* Leiden: Plantin Raphelengius 1613. His other publications include: *De re nummaria,* Leiden: Plantin Raphelengius 1613. *Cyclometricus, de circuli dimensione secundum logistarum abacos,* Leiden: Elzevier 1621. On Willebrord, see Gilbertus Jacchaeus: *Oratio in obitum W. Snellii,* Leiden: Jean Maire 1626.

7 *Medulla theologica,* Amsterdam: Janssonius 1628. 1659. *Disputatio theologica adversus metaphysicam,* Leiden: J. Livius 1632. *Demonstratio logicae verae,* Leiden: Livius 1632. *Philosophemata,* Amsterdam: Janssonius 1643. 2nd ed. 1651. *Technometria,* 1643. An English translation of this work (by Lee W. Gibbs) was published by the University of Pennsylvania Press (Philadelphia 1979). Parts of Ames' *Demonstratio logicae verae* were published in combination with Ramus' *Dialectica: P. Rami Dialecticae libri duo / quibus loco commentarii perpetui post certa capita subjicitur Guilielmi Amesii Demonstratio logicae verae, simul cum synopsi ejusdem, qua uno intuitu exhibetur tota ars bene disserendi,* Cambridge: J. Hayes 1672. A Dutch translation of Ames' *Medulla theologica* is due to Spinoza's friend Lodewijk Meyer (1638-1681): *Mergh der Ghodgeleerdtheidt,* Amsterdam: Abraham Hoopwater 1656. 1670.

8 On Ames, see K. L. Sprunger: *The learned Doctor William Ames: Dutch backgrounds of English and American Puritanism,* Urbana: University of Illinois Press 1972.

The Arts Faculty at Franeker had something of an anti-Aristotelian tradition. For example, some of the disputations of the first professor of philosophy (and rhetoric), Lollius Adama (1544-1609), certainly exhibit Ramist influence – notwithstanding the traditional depiction of Adama as an Aristotelian.[9] Henricus De Veno (died 1613), known as a follower of the Academy and in fact teaching a mixture of Platonic and Stoic ideas, was even temporarily suspended in 1609 and readmitted two years later on the condition that he would observe the *tranquillitas academica,* not become too familiar with students and read on Aristotle's physics and ethics every afternoon – which provides some clues as for the reasons of his suspension.[10] Let us say that the attitude towards Aristotle was rather relaxed. With Frederik Stellingwerff (died 1623), who had no official position but gave private lessons in philosophy, this detached attitude towards the teachings of Aristotle reaches a combative stage: He assailed not only Aristotle himself (whom he calls *metaphysicorum papa* and *fumosarum opinionum pontifex*), but also his followers, singling out Keckermann, Crellius, Pacius and Zabarella. Anti-Aristotelianism seems also to have characterized the teaching of Johannes Hachtingius (1594-1630), who in 1626 published a *Dialectica Petri Rami* – which, like his other works, failed to survive.[11] Finally, Johannes Phocylides Holwarda (1618-1651), a natural philosopher rather than a logician, presented an atomistic physics which, contrary to all expectations – given that Gassendi is cited as an astronomer –

9 Sybrand Galama: *Het wijsgerig onderwijs aan de Hogeschool te Franeker 1585-1811,* Ph. D. thesis, Leiden University 1954, pp. 39-47. 241-242. Paul Dibon: *L'enseignement philosophique aux universités néerlandaises à l'époque pré-cartésienne (1575-1650),* Ph. D. thesis, Leiden University 1954, pp. 131-134. Ferd. Sassen: *Geschiedenis van de wijsbegeerte in Nederland tot het eind der negentiende eeuw,* Amsterdam, Brussels: Elzevier 1959, p. 127.

10 Galama: *Wijsgerig onderwijs* (as note 9) p. 78; on De Veno in general, see also p. 317, and Dibon: *Enseignement* (as note 9) pp. 134-136; Sassen: *Geschiedenis* (as note 9) p. 127.

11 Galama: *Wijsgerig onderwijs* (as note 9) pp. 86-91. 290-291. Dibon: *Enseignement* (as note 9) pp. 137-138. Sassen: *Geschiedenis* (as note 9) p. 129. Since the completion of this article I located a copy of an oration by Hachtingius in the *Sächsische Landesbibliothek* of Dresden: *Oratio de successu quem Deus dedit exercitui Belgarum anno 1629,* Leeuwarden 1629. It is probably the only work of him to survive.

owes nothing to the famous French philosopher but goes back directly to Epicure and Lucretius.[12]

Less known figures should be cited as well. The Leiden professor of medicine Johannes Heurnius (1543-1601) followed Ramus' teaching in Paris in the mid-sixteenth century but details of his own teaching are not known, while his works are in the line of the reformed Galenism of Jean Fernel.[13] Cyprianus Regneri ab Oosterga (1614-1687), from 1641 professor of law in Utrecht, published a professedly Ramist book on logic for lawyers: *Demonstratio logicae verae juridica, variis canonibus et exemplis ad usum illustrata* (Leiden: Livius 1638). In Deventer Henricus Gutberleth (1572-1635), a professor of logic and ethics, seems to have caused trouble by comparing what he called the perfection of the logic of Ramus with the imperfection of the *Organon,* in spite of the fact that he had been asked to make a *systema ex logica Aristotelis.*[14] In fact, if those who appointed him had known his earlier works there could hardly have been any doubt in their minds about Gutberleth's Ramism, for already in 1614 he had published a *Discursus logicus ad methodicam Rameae dialecticae tractationem* for the *Hohe Schule* in Herborn (reprinted in 1637).

For various reasons, however, it is difficult to assess the significance of all this. First of all, Ramus himself was a radical representative of a broad movement of didactic reform, which in the Low Countries goes back to the late Middle Ages and which produced people like Agricola and Erasmus.[15]

12 Galama: *Wijsgerig onderwijs* (as note 9) pp. 91-100. 291-293. Dibon: *Enseignement* (as note 9) pp. 148-150. Sassen: *Geschiedenis* (as note 9) pp. 129-131. Apart from a work on natural philosophy (*Philosophia naturalis, seu physica vetus-nova,* Franeker: Alberti 1651) Holwarda is mainly known for his books on astronomy, some of which were in Dutch (*Friesche sterren-konst,* Pt. 1, Harlingen: Hendrik Jansz 1650).

13 Heurnius was born in Utrecht, studied medicine, philosophy and mathematics in Louvain, Paris, Padua and Pavia, and in 1581 was appointed professor of medicine in Leiden. He was a highly esteemed practitioner and teacher; cf. G. A. Lindeboom: *Dutch medical biography,* Amsterdam: Rodopi 1984.

14 A. A. M. de Haan: Geschiedenis van het wijsgerig onderwijs te Deventer, in: H. W. Blom, H. A. Krop, M. R. Wielema, eds.: *Deventer denkers: De geschiedenis van het wijsgerig onderwijs te Deventer,* Hilversum: Verloren 1993, pp. 29-122 (on pp. 39-45).

15 Rien n'est plus malaisé que de circonscrire l'influence de Ramus dans la seconde moitié du 16e siècle. Nous aurons à plus d'une reprise l'occasion de rappeler que l'influence de l'auteur des célèbres *Animadversiones aristotelicae* (1551) ne doit

Moreover, apart from Snellius perhaps, none of the figures cited above were Ramists in a dogmatic sense of the word. Even Stellingwerff, who generally sides with the *Veromanduus,* does not hesitate to criticize him. Hachtingius' works are lost but the fact that, apart from his *Dialectica Petri Rami,* which must have been a collection of disputations, he also wrote a *Logicae Aristotelico-Rameae Compendium* and a *Fasciculus praeceptorum Aristotelico-Rameorum totius Logicae substantiam continens,* rather suggests an attempt to achieve a synthesis of Aristotle and Ramus. In fact, the picture is dominated by eclecticism, also prevalent among so-called Aristotelians, most of whom followed Keckermann. Finally, all met with resistance. The Leiden petition of 1582, in which students complain about the *via compendiosa* of their teachers, was written in part in protest against Snellius, if only because it explicitly mentions the *Physica* of Cornelius Valerius which he used as a textbook.[16] After the Synod of Dordrecht (1618-1619) orthodox theology took an explicitly anti-Ramist turn by allying itself with scholastic philosophy.[17] But violent opposition to Aristotle

pas éclipser aux Pays-Bas l'influence – aussi, sinon plus marquante – d'humanistes comme R. Agricola, Laurent Valla et Louis Vivès [...]. Dibon: *Enseignement* (as note 9) p. 10.

16 P. C. Molhuysen: *Bronnen tot de Geschiedenis van de Rijksuniversiteit te Leiden,* vol. 1, pp. 103*-104*; cf. Ferd. Sassen: *Het oudste wijsgeerig onderwijs te Leiden,* Amsterdam: Noord-Hollandsche Uitgeversmaatschappij 1941, pp. 8-14; Dibon: *Enseignement* (as note 9) pp. 12-15. Snellius published several commentaries on Valerius: *In Ethicam Cornelii Valerii,* Frankfort 1565. *In Sphaeram Cornelii Valerii Praelectiones,* Frankfort 1596. *In Physicam Cornelii Valerii annotationes,* Frankfort 1596 (also containing Snellius' own *Pneumatologia* and a series of notes by Goclenius). Like Snellius, Cornelius Valerius (or, in Dutch, Wouters, 1512-1578) was born in Oudewater. Educated in Utrecht at the famous Hieronymus School, he moved to Louvain where he became a professor of Latin in 1557; on him, see *Nieuw Nederlandsch Biographisch Woordenboek,* vol. 5, cols. 992-993. Until the end of the seventeenth century there are numerous editions of his works. His letters and Latin poems were published by H. de Vocht: *Epistolae et Carmina*, Louvain: Uystpruyst 1957.

17 Cf. Gysbertus Voetius: De philosophia scolastica, in: *Disputationes theologicae selectae* I, pp. 12-29; see also John E. Platt: *Reformed Thought and Scholasticism,* Leiden: Brill 1982. This movement culminated with *Philosophia prima reformata* (1657) and *Theologia naturalis reformata* (1656) by Paulus Voetius, a son of Gysbertus. Reservations about Suárez' position become apparent in Jacobus Revius' *Suarez repurgatus* (Leiden 1643).

remained extremely rare anyway. Apart from the outbursts of Stellingwerff, which probably never disturbed anyone outside Franeker, it was limited to two works that would remain extremely popular until mid-century: *Exercitationes philosophicae* (Leiden 1620) by David Gorlaeus (born 1591/92), and *Philosophia naturalis reformata* (Dublin 1642) by two brothers, Gerard and Arnold Bootius from Utrecht. Both works may owe something to the Ramist tradition. Gorlaeus, who was an 'intimate friend' of Stellingwerff and apparently incited him to publish his views, studied in Leiden while Snellius was professor there.[18] And in their preface to *Philosophia naturalis reformata* the Bootii declare that under the impulse of their education in Utrecht they became anti-Aristotelians.[19] Since at that time the University of Utrecht was not yet founded they must be referring to the Hieronymus School, a secondary school, where philosophy teaching was usually limited to logic and dialectic. This brings us to the place of logic and philosophy in the curriculum.

The original teaching program or *Hypotyposis* of Leiden University (1575), designed by the first professor of theology, Guillaume Feugueray, prescribes a course of study which integrates the *trivium* (*schola trivialis* or as it was called 'Latin school') and the academy and which, within the academy, makes a distinction between the *artes* (the old *quadrivium*) and the higher faculties.[20] The *trivium* (three years) was to be devoted to a practically oriented (*exemplis magis et usu quam praeceptis docemus*) study of languages (Latin and to a lesser extent Greek), illustrated by texts: the *Bu-*

18 On Gorlaeus, see F. M. Jaeger: *Historische studiën: Bijdragen tot de kennis van de geschiedenis der natuurwetenschappen in de Nederlanden gedurende de 16e en 17e eeuw,* Groningen: Wolters 1919. For the most recent survey, see the article by Wolfgang Uwe Eckart in: *Grundriss der Geschichte der Philosophie. Begründet von Friedrich Ueberweg. Die Philosophie des 17. Jahrhunderts,* vol. 4/2, Basle: Schwabe 2001, pp. 927. 935-936. 985. A second book, also posthumous, *Idea physicae,* Utrecht 1651, is even more rare than the first.

19 On the Bootii there is nothing at all, apart from an entry in the *Dictionary of National Biography* (they served as physicians in the English army in Ireland). A manuscript of *Philosophia naturalis* with notes, probably by Arnold Bootius, is in the *Bibliothèque Nationale de France* (Fonds Latin: 12.975).

20 It is published in Molhuysen: *Bronnen* (as note 16) vol. 1, p. 26; see also L. Knappert: Een academisch leerplan van den jare 1575, in: *Leidsch Jaarboekje* 6 (1909) pp. 73-79; Dibon: *Enseignement* (as note 9) pp. 8-11.

colica, the *Batrachomuomachia* and Cicero's correspondence in the first year, more important works in the second and third year. The stage of the *artes,* which comprises four years, would be devoted to rhetoric in the first year, the practical study of the rules of dialectic (*scholasticae dialecticae disciplinae regulis contenti non sumus: earum regularum usum etiam volumus*) in the second year, mathematics in the third year and physics and ethics in the fourth year. The authors and works prescribed by Feugueray were Virgil's *Georgica,* Seneca's *Quaestiones naturales* and Lucretius' *De natura rerum*. It is only then that the student enters the 'higher faculties' of theology, medicine and law, which took seven years to complete.

The fact that no lip service is paid to Aristotle, that there is no mention of metaphysics, that there is much emphasis on use and practice, and that Virgil's *Georgica* are considered a suitable textbook in physics certainly rings a Ramist bell, but Dibon is undoubtedly right to conclude that both Ramus and the program are inspired by the same anti-scholastic spirit of humanism.[21] Moreover, it is not clear to what extent Feugueray's program has been implemented. The first part of his proposal at any rate, which aimed at an integration of the *schola trivialis* and the academy proved unacceptable to the Leiden Magistrate. The strict curriculum never became a reality either. Students could and did enter the university at any point and often skipped the faculty of arts – a subject of many complaints both in Leiden and Franeker, and later in Utrecht and Groningen.[22] On the other hand, the faculty of arts often acquired independent status, particularly in Leiden which from the end of the sixteenth century became world famous as a center of literary and historical studies. In any case, philosophy never was really important until the scholastic reform of orthodox theology and later the rise of Cartesianism. Professors of philosophy were among the lowest paid in the university and the number of students that took a degree in philosophy was never very high.[23]

21 Dibon: *Enseignement* (as note 9) p. 10.

22 For Franeker, see Galama: *Wijsgerig onderwijs* (as note 9) pp. 29-32; for Leiden, H. Wansink: *Politieke wetenschappen aan de Leidse Universiteit 1575-±1650,* Utrecht: H&S 1981, p. 26.

23 For some figures see Wansink: *Politieke wetenschappen* (as note 22) pp. 24-32.

For a full view of the teaching of philosophy we should also look at the *schola trivialis,* a form of secondary teaching where traditionally logic and dialectic were taught alongside (Latin) grammar and rhetoric.[24] Until the first quarter of the century there must have been little agreement on that front, for we encounter numerous complaints – especially from theologians – that students when reaching the university displayed an alarming divergence in their knowledge of logic. There is little precise that we know. In Harderwijk secondary and higher teaching were integrated at the *Gymnasium Illustre*.[25] The official program stipulated that logic should be taught from "the *Dialectices Elementa* of Petrus Ramus" and that in the higher forms a comparison should be instituted between the logic of Ramus and "the Peripatetic logic of Molinaeus".[26] In 1625 the States of Holland issued a *schoolordre,* intended to affect certain unity in pre-university teaching. Schools were required to have six forms (but until fairly recently pupils could enter the school at any stage); logic was to be taught from the *tertia* onwards, occupying up to four hours a week in the *prima;* it should be taught from a textbook commissioned from the Leiden professor of phi-

24 At least until 1650; after that date there is a growing tendency to replace the teaching of logic and dialectic by that of Latin grammar; cf. E. J. Kuiper: *De Hollandse 'Schoolordre' van 1625,* Groningen: Wolters 1958, pp. 140. 157. 199. 202 etc. This book is also the source of much of the following.

25 The *Schola Illustris* or *Gymnasium Illustre* counted as a form of higher education which, however, since it was created by a municipal authority (instead of the sovereign, that is, the States), did not have the right to confer academic degrees. Some of them were eventually converted into universities, like Harderwijk (1648) and Utrecht (1636). Their official use was to keep often very young boys at home instead of sending them to the university, but, in fact, they were a means for the more wealthy townships to affirm their identity. On Harderwijk, see H. Bouman: *Geschiedenis van de voormalige Geldersche Hoogeschool en hare hoogleeraren,* 2 vols., Utrecht 1844-1847; A. A. M. de Haan: *Het wijsgerig onderwijs aan het Gymnasium Illustre en de Hogeschool te Harderwijk 1599-1811,* Ph. D. thesis, Leiden University 1960.

26 De Haan: *Wijsgerig onderwijs* (as note 25) pp. 10-11. *Elementa logica* (Leiden: Raphelengius 1598) of the Leiden professor Petrus Molinaeus (Du Moulin, 1568-1658) was an immensely popular book, being reprinted several times both in the Netherlands (7th ed., Amsterdam: Janssonius 1620; *editio ultima,* Amsterdam 1645) and abroad (Bremen: Villiers 1622). I suppose one can call it semi-Ramist, insofar as it pays not only much attention to judgment and invention but also to the syllogism; cf. Dibon: *Enseignement* (as note 9) pp. 52-54.

losophy Franco Burgersdijk (1590-1635), *Institutiones logicae* (1626).[27] In his preface Burgersdijk discerns three schools in logic: the Aristotelians, whom he considered to be too dogmatic, though he admits that the basis of the teaching of logic remains the *Organon;* the Ramists, whom he praises for their insistence on method and practical exercise, but finds too narrow in their conception of logic; and a third school, to which he attaches the name of Bartholomäus Keckermann (1572/73-1609), whom he praises for his attempt to ally Aristotle and Ramus but does not find suitable for use in schools because he is too verbose.[28] In fact, his own textbook is based on Keckermann: it gives a full and extensive exposition on the syllogism but there are also chapters on invention and method. Apparently, it fulfilled a need even outside academic circles for in 1646 it was translated into Dutch.[29] Until the end of the seventeenth century it ran into many editions, not only in the Low Countries but also in England. In a companion to Burgersdijk's *Institutiones,* which was first published in 1650, the Cartesian Adriaan Heereboord (1614-1661) declares that he had always been using it for his lectures ever since his appointment in Leiden (1641).[30]

With Heereboord, and even Burgersdijk, we leave the period in which Ramism formed a meaningful category, although the fact that we find an edition of Ramus' *Dialectica* as late as 1664 is telling.[31] Still, the fact that Heereboord, a Cartesian, was concerned at all with the teaching of logic raises the question of the relation between Cartesianism and Ramism. There are two aspects to the issue: one concerns Descartes and the other concerns the Cartesians. I do not think that it is helpful to situate Descartes

27 J. B. M. van Rijen: Burgersdijk, logician or textbook writer? in: E. P. Bos, H. A. Krop, eds.: *Franco Burgersdijk (1590-1635),* Amsterdam: Rodopi 1993, pp. 9-28.

28 Cf. William and Martha Kneale: *The Development of Logic,* Oxford: Clarendon Press 1962, p. 305.

29 *Institutio logica, dat is, reden-konstigh onder-wys, be-ghreepen in twe boeken,* Amsterdam: Gerrit Willemsz 1646. In the same year a Dutch translation was published of Burgersdijk's *Synopsis: Frank Bughersdijks Kort be-ghrijp van 't reden-konstigh onderwijs, oft reden-konstigh ruigh be-werp,* Amsterdam: Gerrit Willemsz 1646, repr. 1648.

30 Adriaan Heereboord: *Ermeneia logica seu Explicatio tum per notas tum per exempla Synopseos Logicae Burgersdicianae,* Leiden: David van Lodesteyn, Severinus Matthaei 1650.

31 Petrus Ramus: *Dialectices Libri II,* Tiel: G. van Duyn 1664, edited with commentaries and annotations by J. van Aelhuysen.

in a Ramist perspective – as has been done recently.[32] Even if we grant the rather obvious point that Descartes rejected Aristotle and Aristotelian logic and that, if anything, he preferred a logic of invention, the idea that an *ars disserendi* could ever be serviceable in a context of science and the search for truth is, I believe, completely alien to Descartes' mind. At any rate, an assessment of the possible influence, if any, of Ramus on Descartes should be based on precise indications, not on the use of similar terms.[33] Here are some points of reference.

First of all, the presence of Ramism as an organized or at least recognizable force does not seem to have promoted the cause of Cartesianism. In Franeker, for example, where Ramism was current shortly before the publication of the *Discours de la méthode* – Hachtingius died in 1630 – Cartesianism was very slow in being accepted. The first professor of philosophy showing Cartesian sympathies was Johannes Greidanus (1630/31-1668).[34] His is a very hesitant form of Cartesianism, which consciously attempts a reconciliation with Aristotelian ideas; indeed, Greidanus explicitly sees himself as a member of what he calls the *secta electiva,* which is not Cartesian.[35] Moreover, he based his teaching of logic on

32 André Robinet: *Aux sources de l'esprit cartésien: L'axe La Ramée-Descartes: De la Dialectique de 1555 aux Regulae,* Paris: Vrin 1996; see also his contribution to *Descartes: Objecter et répondre,* ed. J.-M. Beyssade, J.-L. Marion, Paris: PUF 1994, pp. 65-88 (Originalité des Réponses: lumière naturelle et dialectiques).

33 The relation with Beeckman, which is mentioned by Robinet: *Aux sources* (as note 32) pp. 177-181, might be a promising venue, given the fact that Beeckman was a student of Snellius. But thus far no study has been made of Snellius' physics (consigned to his commentaries on Valerius), nor indeed of Beeckman's.

34 Galama: *Wijsgerig onderwijs* (as note 9) pp. 100-105. 288-289. The first openly 'Cartesian' professor was Johannes Wubbena (died 1681), appointed in 1664 and suspended (not because of his ideas but of his dissolute life) in 1678. He left no works: Galama, pp. 105-107. 319; C. Louise Thijssen-Schoute: *Nederlands cartesianisme,* Amsterdam: Noord-Hollandsche Uitgeversmaatschappij 1954, repr. Utrecht: H&S 1989, pp. 502-505. He was succeeded by Johannes Schotanus a Sterringa (1643-1699) who provided a poetic paraphrase of Descartes' *Meditationes* as well as defending Descartes against the critique of Pierre-Daniel Huet; cf. Galama: *Wijsgerig onderwijs* (as note 9) pp. 116ff. 302-305; Thijssen-Schoute: pp. 521-523.

35 *Institutiones physicae,* Leeuwarden: H. Rintjes 1664, quoted by Thijssen-Schoute: *Nederlands cartesianisme* (as note 34) p. 504.

Burgersdijk,[36] while his *Idea logicae novantiquae* (Franeker 1659) was based on the logic of Clauberg.[37] Second, notwithstanding the modern tendency to associate Descartes primarily with method, this was not the case of his first followers, all of whom totally ignored what Descartes had to say about the rules of method.[38] Heereboord, for example, who was a professor of logic and ethics, never mentioned it. In fact, the first to deal with them in a systematic way was Johannes Clauberg (1622-1665), who in 1654 published a *Logica vetus et nova* (Amsterdam 1654) which, apart from the reprint in *Opera omnia philosophica* (2 vols., Amsterdam 1691), was republished twice in the Netherlands.[39] Still, a case can be made that the two earliest followers of Descartes in the Netherlands, Henricus Reneri (1593-1639) and Henricus Regius (de Roy, 1598-1679), were to a certain extent influenced by the Ramist tradition.

Little is known of Reneri, except some correspondence (which however is mainly personal), an inaugural lecture and a few disputations.[40] In this

36 A set of lecture notes on logic by a student, Willem Siccama, is kept in the library of the *Fries Genootschap* (Leeuwarden); cf. Galama: *Wijsgerig onderwijs* (as note 9) p. 289; Thijssen-Schoute: *Nederlands cartesianisme* (as note 34) p. 504.

37 This is said on the authority of Dibon, who was allowed to make a study of the only known copy, which survives in a private collection: *Enseignement* (as note 9) pp. 159-160. In 1660 disputations organized by Greidanus on this logic were heavily criticized for being *novitatum plenae* and eventually forbidden by the administration of the university; Galama: *Wijsgerig onderwijs* (as note 9) p. 102; Thijssen-Schoute: *Nederlands cartesianisme* (as note 34) p. 503.

38 Naturally, one should take into account the fact that the *Regulae ad directionem ingenii* were not published until 1684; cf. Descartes: *Regulae ad directionem ingenii,* texte critique établi par Giovanni Crapulli avec la version hollandaise du XVIIème siècle, The Hague: Nijhoff 1966.

39 Amsterdam 1654; Franeker 1692. For a more complete bibliographical survey see now: Theo Verbeek, ed.: *Johannes Clauberg (1622-1665) and Cartesian Philosophy in the Seventeenth Century,* Dordrecht: Kluwer 1999, pp. 187-194.

40 The correspondence can be found in the University Library of Leiden (letters to Rivet and De Wilhem), the Utrecht Municipal Archives (letters to Cornelis Booth) and the Bodleian (letters to Isaac Vossius). The inaugural lecture is published as part of a collective volume on the inauguration of the Utrecht Illustrious School, the precursor of the university: *Illustris Gymnasii Ultrajectini inauguratio,* Utrecht: Aegidius Roman 1634. For the disputations, see Paul Dibon: *Le fonds néerlandais de la Bibliothèque académique de Herborn,* Amsterdam: Noord-Hollandsche Uitgeversmaatschappij 1957, p. 106.

context, the inaugural lecture (1634) is the most relevant because it contains a program for the teaching of logic. This should be practical throughout. Beginners will study it from passages of famous rhetoricians, poets or historians, which provide the most interesting applications of the *ars disserendi.* More advanced students must follow lessons and do exercises on *loci communes:* They must learn to gather them from correspondence, speeches and historical works, to classify them and to make indices and mark them (from 1 to 5) according to their importance. Students must also carry out *analytical exercises,* that is, summarize the thought of an author in the form of a list or a scheme and they must engage in *methodic exercises,* the importance of which Reneri illustrates by referring to the treatise on method by Jacobus Acontius (died 1566).[41] These consist of compositions and resolutions. A very Ramist sounding program, although, again, little is known about the execution.[42]

Regius in turn, who was not a logician and, strictly speaking, not even a philosopher but a physician, followed a perfectly Ramist method of exposition. In his last letter to him Descartes blames Regius for not doing enough to prove his position and for presenting his views "through definitions and divisions" and "by proceeding from the general to the particular".[43] The criticism is, on the whole, justified. Regius began every discussion by presenting a division, say between A and B, then defines A and discussed the various ingredients and implications of A, and finally returning to B to deal with it in the same manner. Descartes' second way of characterizing Regius' method, as "proceeding from the general to the more particular", is also accurate. In the *Physiologia,* for example, Regius starts with a definition of

41 *De methodo e Opuscoli religiosi e filosofici e Stratagemata Satani,* ed. G. Radetti, Florence 1944.

42 Descartes' own *Discours de la méthode* would not be written until one year later, almost under the eyes of Reneri for that matter, in Utrecht, where Descartes settled in the spring of 1635; see Theo Verbeek: "*Une université pas encore corrompue ...*": *Descartes en de eerste jaren van de Universiteit Utrecht / Descartes et les premières années de l'Université d'Utrecht,* Utrecht: Utrecht University 1992.

43 Descartes to Regius, July 1645, *Œuvres de Descartes,* ed. Adam/Tannery, vol. 4, pp. 248-249 (CSM III, p. 254). This may be one of the few points on which Descartes agreed with Gysbertus Voetius; cf. René Descartes et Martin Schoock: *La querelle d'Utrecht,* textes établis et traduits avec une introduction et des notes par Theo Verbeek, Paris: Les impressions nouvelles 1988, p. 90.

medicine and the different parts of medicine, which leads him to define health as the principal object of physiological knowledge, and that in turn is the introduction to a discussion on perceptible and imperceptible parts.[44] However, the very words used by Descartes are an echo of Ramus. It is important, therefore, to note that Descartes rejected this method, which in fact cannot be used in any of the stages of the mature Cartesian project.

By way of conclusion I would like to draw attention to a passage in a letter of Joannes De Raey (1622-1702), professor of philosophy at Leiden, and later in Amsterdam, written to a friend, another Cartesian, the theologian Christoph Wittich (1625-1687). It is a private letter, written in Dutch on 12 August 1680, and published by De Raey himself, in Latin, in his *Cogitata de interpretatione* (Amsterdam 1692).[45] In it De Raey reviewed his career, mainly in order to show that his aversion of Spinoza was consistent with his earlier commitments. According to De Raey he had started his academic career by emphasizing the peculiar nature of philosophy: Philosophy is contemplation of nature and as such presupposes a radical break with common sense. But he also discovered that there is a difference between Aristotle himself and Aristotelianism, implying that the original Aristotle is closer to Descartes than one might think.[46] But then he proceeded to discuss logic in a passage that deserves to be quoted in full:

> I had spent my youth in the study of logic, never to regret it. In the school I had learned the dialectic of Petrus Ramus, which until the present day I believe to have been to my advantage. In my academic studies I had examined several other logicians and compared some of them with Ramus. But then I started to doubt whether that technique, which goes by the double name of logic and dialectic, was really as useful and necessary as I always thought. Indeed, I realized on the one hand that many of those who find this technique important are not made wiser by it, but on the other hand I believed that without the knowledge of logic I could not do my work. I proposed these doubts to Mr. Clauberg when he asked me, by letter, my opinion on his *Logic,* as he often did for that matter in serious concerns. I already knew that among the heathens Epicure rejected the art of logic and that among the Christians Augustine has written somewhere that the art of arguing and of using one's reason

44 *Physiologia,* Utrecht: Aegidius Roman 1641, I, i, 1-5, p. 1.

45 *Cogitata,* pp. 645-661.

46 See Theo Verbeek: Tradition and novelty: Descartes and some Cartesians, in: Tom Sorell, ed.: *The Rise of modern Philosophy,* Oxford: The Clarendon Press 1992, pp. 167-196.

> [*ars disserendi atque ratione utendi*] is learned not by teaching and rules but by practice, as we become artisans by doing. A bit later, whilst I was still pondering and doubting, Clauberg's *Logic* was published. I use the book, lecture on it in my private classes but without the success I had hoped for. I wonder about the reason and I realize – and how happy I am that I did so and how useful the discovery was – that the mistakes and all the doubts that one can have on the necessity of logic, and all the confusion in the thoughts of the learned (which is the cause of so many evils in science) comes only from this, that they never understood that logic can and must be twofold, as we have shown on the ground that in one man, understanding can be twofold and actually is, whenever we have found a piece of true philosophy – philosophy, I say, which is the object of this second way of understanding. I also understood clearly that the mixture of these two ways of understanding in one logic and the confusion that springs therefrom of the respective language, notions, and concepts must cause a similar confusion in other disciplines. Accordingly [I understood] that one logic must be separated from the other, insofar as the one, which is the best known, is comprised in the dialectic of Petrus Ramus and the other (of which we shall show elsewhere that Plato was aware of it) in the method of the right use of reason of René Descartes, of which he has given us an example in the *Meditations on First Philosophy* and in the First Part of his *Principles.*[47]

This passage, like the letter as a whole, raises numerous historical and philosophical questions, only some of which can be discussed here. First of all, the letter confirms that in the years 1630-1640 (when De Raey attended secondary school) Ramus' *Dialectica* still was the basis of logical teaching in the *schola trivialis.* Second, it shows that there are different ways of being a Cartesian; indeed, if we reconstruct De Raey's objection to Clauberg's *Logic* it is that Clauberg conceived of logic and method as a *medicina mentis* – a medicine which, De Raey would say, most people do not need because the way they think and talk is perfectly appropriate for their purposes.[48] Third, the letter signifies a definitive break between philosophy and dialectic: Dialectic is suitable for daily life, whereas the method of philosophy is, well, suitable for philosophy, that is, for the contemplation of nature. Accordingly, De Raey reintroduces a much older model, which, via Agricola and Cicero's *De oratore,* ultimately goes back to Aristotle, except

47 *Cogitata,* p. 659.

48 In a note De Raey says that Clauberg's *Logic* was not free of "this way of mixing things", that is, of confusing the two types of understanding but he speaks forgivingly of it because, he says, "Clauberg followed my lessons only in 1648 and 1649 when I did not insist as much on this separation". *Cogitata,* p. 259.

that what in Aristotle is called dialectic is replaced by (Cartesian) method, while that which Aristotle terms the art of persuasion is replaced by what De Raey calls dialectic, that is, the dialectic of Ramus.[49]

It is difficult to draw any definite conclusion on the basis of such diverse materials. So much is clear, apart from secondary teaching, Ramism never was present as a very strong or even recognizable movement. The only figures that could have created such a movement – in philosophy Snellius and in theology Ames – never gained enough influence. The strong association of scholasticism and orthodox theology, which was very strong between 1620 and 1660, made this impossible. We may also conclude that there is no positive connection between Ramism and Cartesianism, in spite of the fact that the anti-Aristotelian rhetorics of the Ramist tradition sometimes creates the impression of a relation on a more profound level. Indeed, Descartes' own insistence on the practical results of his philosophy does ring a vaguely Ramist bell. The particular evolution of Dutch Cartesianism was on the contrary into a speculative direction, precisely to avoid all confusion with practical concerns.

49 Aristotle: *Rhetorica* I, i, 1354a1-30.

The Influence of Ramism on the Academies of Bremen and Danzig: A Comparison

THOMAS ELSMANN

The State and University Library of Bremen whose institutional origin owes much to the *Bibliotheca Bremensis* of 1600 but which in terms of holdings goes much further back,[1] contains *inter alia* a collection of some 3800 letters written in the sixteenth and seventeenth centuries, partly originals, partly copies. Some pieces are of considerable interest for our topic, first and foremost two letters dated 1570. On 9 January Petrus Ramus wrote from Heidelberg to the rector of the Bremen academy (or school),[2] the Flemish scholar Johannes Molanus (ca. 1510-1583).[3] Ramus began his letter with a eulogy of the Bremen educational institution: should he be asked to name distinguished schools in Germany, he would most certainly mention that of Bremen. Should the war in France continue, he proposed to stop at Bremen to visit Molanus on his planned journey to England; if not he would invite Molanus to Paris. On 26 February of the same year Molanus replied from Bremen.[4] In addition to

1 Extensively on this Thomas Elsmann: Bremen, Staats- und Universitätsbibliothek, in: *Handbuch der historischen Buchbestände in Deutschland,* ed. Bernhard Fabian, vol. 1: *Schleswig-Holstein, Hamburg, Bremen,* Hildesheim, Zurich, New York 1996, pp. 289-308.

2 Bremen, Staats- und Universitätsbibliothek: msa 0007/203 (copy with corrections by Molanus in his own hand; further copy: msa 0012/170).

3 On Molanus see Thomas Elsmann: Albert Rizäus Hardenberg und Johannes Molanus in Bremen. Zwei Humanisten im konfessionellen Zeitalter, in: *Wessel Gansfort (1419-1489) and Northern Humanism,* ed. Fokke Akkerman, Gerda C. Huisman, Arjo J. Vanderjagt, Leiden, New York, Cologne 1993 [Brill's Studies in Intellectual History, vol. 40] pp. 193-209, esp. pp. 204-209.

4 Bremen, Staats- und Universitätsbibliothek: msa 0007/204 (copy with correction in Molanus' own hand; further copy: msa 0012/171).

profuse thanks for Ramus' praise we find a short appreciation of Ramus' works and a remark which is of special interest to us here. Molanus said that he had been using Ramist dialectic in his teaching for the past six years. This suggests 1564 as the latest date of the incorporation of Ramist teaching in German schools. Molanus' contacts with Ramism, however, seem to have been established a few years earlier: it is very likely that this occurred when Molanus was teaching at the *Gymnasium* in Duisburg.

To come across such evidence is in most cases a happy accident. Where one has to depend on such chance finds, however, it will also be clear that the reception of a philosophical-methodical re-orientation, in this instance a totally new approach, its impact and significance are extremely difficult to verify. Yet, it is only through micro-analysis (which follows below) that one is able to draw a differentiated, meaningful picture. Walter Ong[5] and Joseph Freedman[6] have already provided preliminary investigations in two areas by examining the publications of Ramus' works and their reception (or rejection) in European schools, academies and universities from 1570 to 1630. The key findings of these studies is, first, a steady rise in the number of Ramist imprints after 1580, which reaches its peak around the turn of the century. After 1630 one witnesses a sudden decline. Second, it becomes clear that Ramism was integrated into the teaching of schools, academies and academic *Gymnasia* predominantly in Reformed, but also in Lutheran, regions. The evidence for the presence of Ramist works in individual schools, however, does not mean that there was a consistent and thorough Ramist type of instruction in these schools. Such matters are difficult to ascertain since the tradition is intermittent, and in view of the erroneous impression that educational institutions staunchly adhered to a specific philosophical-methodological outlook in line with their confessional standing. Herborn, founded in 1584, seems to have been the exception in Germany. Although an equilibrium between Ramism and Melanchthonian-Aristotelianism appears to have been instituted there, in actual fact the academy was committed to Ramism.[7]

5 Walter J. Ong: *Ramus and Talon Inventory,* Cambridge, Mass. 1958.

6 Joseph S. Freedman: The Diffusion of the Writings of Petrus Ramus in Central Europe, c. 1570 - c. 1630, in: *Renaissance Quarterly* 46 (1993) pp. 98-152.

7 See Gerhard Menk: *Die Hohe Schule Herborn in ihrer Frühzeit (1584-1660). Ein Beitrag zum Hochschulwesen des deutschen Kalvinismus im Zeitalter der Gegen-*

The early Herborn records relate essentially to the last decade of the sixteenth century, while evidence from Korbach and Marburg goes back to the 1570s,[8] which help elucidating the significance of the Bremen findings.[9] In the following pages I shall present the most important stages in the reception of parts of the Ramist corpus in the Bremen school system – against the background of the confessional and intellectual developments in this city – before attempting a comparison with concurrent developments in the Hanseatic sister city of Danzig.

Johannes Molanus, Ramus' correspondent, was born around 1519 in Nieuwekerk near Ypres. After studying at Louvain and an enforced exile to Emden, on account of his conversion to Protestantism, Molanus reached Bremen around 1553. He became a teacher at the school which had been founded in 1528 in the former Dominican cloister; he was allowed to keep a private college for pupils as well. Molanus settled in a community which was torn by intensely acrimonious confessional conflicts.[10] Following an early and rigorously introduced Reformation, the community split between followers of Luther, sympathisers of Melanchthon, and converts to the Reformed tradition. Molanus adopted an anti-Lutheran persuasion but without feeling the need of articulating his position publicly. Yet, his association with the Cathedral preacher Albert Rizaeus Hardenberg (ca. 1510-1574),[11]

reformation, Wiesbaden 1981 [Veröffentlichungen der Historischen Kommission für Nassau, vol. 30] pp. 203ff.

8 See Arnd Friedrich: *Die Gelehrtenschulen in Marburg, Kassel und Korbach zwischen Melanchthonismus und Ramismus in der zweiten Hälfte des 16. Jahrhunderts,* Darmstadt 1983 [Quellen und Forschungen zur hessischen Geschichte, vol. 47].

9 On Bremen see also: Thomas Elsmann: Das Bremer Gymnasium Illustre und seine Vorläufer in ihrer Bedeutung für den Ramismus in Deutschland (1560-1630), in: *Northern Humanism in European Context, 1469-1625: from the 'Adwert Academy' to Ubbo Emmius,* ed. Fokke Akkerman, Arjo J. Vanderjagt, Adrie H. van der Laan, Leiden, Boston 1999 [Brill's Studies in Intellectual History, vol. 94] pp. 99-108.

10 See Herbert Schwarzwälder: *Geschichte der Freien Hansestadt Bremen,* vol. 1: *Von den Anfängen bis zur Franzosenzeit (1810),* expanded and improved edition, Bremen 1995, pp. 231-267. Ortwin Rudloff: Bremen, in: *Theologische Realenzyklopädie,* vol. 7, Berlin, New York 1981, pp. 153-168, esp. pp. 156-159.

11 See Wilhelm H. Neuser: Albert Rizaeus Hardenberg (ca. 1510-1574), in: *Theologische Realenzyklopädie,* vol. 14, Berlin, New York 1985, pp. 442-444. Willem Janse: *Albert Hardenberg als Theologe. Profil eines Bucer-Schülers,* Leiden, New York, Cologne 1994.

the leader of the opposition within the town, proved fatal in view of the latter's increasing isolation, brought about by a Lutheran reaction politically supported from outside. Such reaction rendered Molanus' position precarious, despite the fact that he was not a theologian. Molanus was saved by a timely invitation in March or April 1559 to teach at the newly established *Gymnasium* at Duisburg; the invitation had been mediated by his future father-in-law, the geographer Gerhard Mercator (1512-1594). We have fairly good evidence for his time in Duisburg in the form of letters; but Molanus made very few references to the content of the instruction there. In any case, his sojourn in Duisburg remained a brief episode, partly because the Duisburg school foundered and partly because a Philippist, crypto-Calvinist orientation had gained the upper hand in Bremen. The result was the renewed invitation to Molanus to settle in Bremen and his appointment to the position of rector of the Bremen school in 1563.

Molanus' letter to Ramus allows us to conclude that the "reformation" of the Bremen school was guided, from the start, by Ramist principles. It meant nothing less than turning away from the Aristotelian-Melanchthonian education which, from its Wittenberg power-base, dominated most of the schools and universities at the time. The redirection did not signify rejection! The extant lecture *indices* contain the first mention of Ramus as a guide for the analysis of the speeches of Cicero in 1566 in the *Secunda* class.[12] They also attest to the reception of the *Dialectic*[13] in 1568 and in 1576/77 the reception of the *Rhetoric* of Talaeus,[14] for the benefit of a school with five, six or seven classes in which the so-called *Prima* – that is an academic-

12 *Brevis Index Lectionum, et Exercitiorum quibus (si Deus volet) his aestivis mensibus naviter incumbemus in Schola huius clarissimae Reipub. 1566 Bremae,* Bremen: [Arend Wessel] 1566 (Wolfenbüttel, Herzog August Bibliothek). On a possible effect of the Ramus reception in Bremen on Groningen see Thomas Elsmann: Ubbo Emmius: Inhalt und Ziel des Unterrichts. Eine Analyse auf dem Hintergrund seiner Bildungsbiographie, in: *Ubbo Emmius. Ein Ostfriesischer Gelehrter in Groningen,* ed. Willem J. Kuppers, Groningen, Emden 1994, pp. 186-191. 197.

13 *Brevis Artium et Lectionum Index, quibus Deo praeside, aestate veniente anni 68. In Schola Bremensi naviter intendemus* [Bremen: Arend Wessel] 1568 (Bremen, Staats- und Universitätsbibliothek).

14 *Hibernae lectiones et operae temporibus divisae. Anno novissimi temporis 1576. Adhuc regente Scholam cum Deo Johannes Molano* (Bremen, Staats- und Universitätsbibliothek: msa 0007/269; further copy: msa 0012/258).

propaedeutic class – was missing. Simultaneously, one finds manuals by Melanchthon[15] – for example grammars that relate to the method of learning languages – as well as works by Aristotle, for example the *Ethics* and *De anima* with Melanchthon's commentary.

The case of Molanus is an early instance of a general trend: the essentially selective and eclectic reception of Ramus' works. We are not dealing with two mutually exclusive alternatives – either Aristotle/Melanchthon *or* Ramus – but with a trend towards conjunction and co-existence: Aristotle/ Melanchthon *and* Ramus. Molanus variously confirmed his preference for Ramus in the areas of dialectic and rhetoric in his letters; he significantly speaks of *noster Ramus* in a letter to Gerhard Mercator.[16] In many respects he simply considered the Melanchthonian equivalents out-of-date, as too much influenced by an outmoded tradition, while the Ramist method was clearer, methodologically more plausible and therefore to be preferred in instruction. He also argued that Ramist empiricism equipped much better for confessional confrontations, especially where the contradictions within Lutheranism were concerned, that is, in the debates over the Eucharist.[17] Other letters confirm that Molanus recommended Ramus' *Dialectic* to his students and that in the 1560s and 1570s the Bremen rector apparently prepared written theses or commentaries on the Ramist method at the request of other scholars – though these failed to survive.[18]

15 On Melanchthon as author of school manuals most recently: *Melanchthon und das Lehrbuch des 16. Jahrhunderts,* ed. Jürgen Leonhardt, Rostock 1997 [Rostocker Studien zur Kulturwissenschaft, vol. 1].

16 Staats- und Universitätsbibliothek Bremen: msa 0007/179 (copy with corrections in Molanus' own hand; further copy: msa 0012/124): Molanus to Gerhard Mercator, Bremen, 14 October [15]67.

17 See Jürgen Moltmann: *Christoph Pezel (1539-1604) und der Calvinismus in Bremen,* Bremen 1958 [Hospitium Ecclesiae, vol. 2] pp. 33ff.

18 The following letters by Molanus and his correspondents contain references to Ramus: Staats- und Universitätsbibliothek Bremen: msa 0012/139: Molanus to Caspar Olevianus, Bremen, 7 April 1568; msa 0007/219 (copy with corrections by Molanus; further copy: msa 0012/195): Jakob Barlaeus to Molanus, Heidelberg, 9 September 1572; msa 0007/240 (copy with corrections by Molanus; further copy: msa 0012/226): Molanus to Gerhard Schroeder, Bremen, 14 January 1575; msa 0007/251 (copy with corrections by Molanus; further copy: msa 0012/237): Molanus to Jakob Commelinus, [Bremen], 30 August 1575; msa 0007/254 (copy with corrections by Molanus; further copy: msa 0012/240): Molanus to Jakob Comme-

In 1584, one year after the death of Molanus, the *schola Bremensis* was extended to include the previously lacking academic-propaedeutic class, the *Prima,* which contained the academic subjects. It appears that Ramus played at first a more marginal role among the successors of Molanus. In 1589 Melanchthon dominated rhetoric and dialectic,[19] in 1590, however, the Talaeian rhetoric, for one, secured its place again.[20] In 1596/97 rector Nathan Chytraeus (1543-1598), who had arrived from Rostock, adopted a comparable procedure;[21] he favoured Melanchthon for dialectics and Talaeus for rhetoric. This vacillation in the philosophical-methodological orientation is not unusual; it demonstrates the great freedom that was accorded to individual rectors.

The actual *epigone* of Molanus, in respect of methodology, was Andreas Widmarius (1552-1621), who arrived in Bremen at the beginning of the seventeenth century. He was the father of Abdias Widmarius (1591-1668), and was later to become professor at Groningen. Widmarius created the basic structure of the *schola Bremensis*, which, in its content, remained binding for decades. However, it was overlaid by a complicated, and in part inexplicable, canon of requirements. Consequently, Widmarius' tenure as

linus, Bremen, 20 September 1575; msa 0007/271 (copy with corrections by Molanus; further copy: msa 0012/260): Molanus to Jakob Mylius, Bremen, 8 November 1576; msa 0012/272 (copy with corrections by Molanus; further copy: msa 0012/261): Molanus to Jakob Mylius, Bremen, 13 December 1576; msa 0007/314 (copy with corrections by Molanus; further copy: msa 0012/310: Molanus to N.N., [Bremen, 22 January 1578]; msa 0007/315 (copy with corrections by Molanus; further copy: msa 0012/311): Molanus to N.N., Bremen, 5 February [1578]; msa 0007/323 (copy with corrections by Molanus; further copy: msa 0012/323): Molanus to N.N., Bremen, 16 August 1579; msa 0007/325 (copy with corrections by Molanus; further copy: msa 0012/325): Molanus to Reiner Lange, Bremen, 14 January 1579; msa 0010/142 (autograph): Sebastian Trescho to Johannes Hildebrand, Braunschweig, 24 March 1583 (Trescho asks here for Molanus' theses on Ramus).

19 *Index Artium & Lectionum, quibus (Deo preside) Aestivis mensibus Anni 89. in Schola Bremensi naviter incumbemus.* Bremen: Bernhard Peters 1589 (Bremen, Staats- und Universitätsbibliothek).

20 *Index Artium et Lectionum, quibus (Deo praeside) Aestivis mensibus Anni 90. in Schola Bremensi naviter incumbemus,* Bremen: Arend Wessel Erben 1590 (Bremen, Staats- und Universitätsbibliothek).

21 *Lectiones Hiberniae Scholae Bremanae. Anno Christi MDXCVI. exeunte. & XCVII ineunte* [Bremen:] Arend Wessel Erben 1596 (Rostock, Universitätsbibliothek).

rector was described as an unhappy one by his contemporaries. It was said that he made excessive demands on his pupils and that he, a non-theologian (actually the last in a long line of non-theologians), tried to keep the school free from ecclesiastical intervention. Widmarius constructed a clear separation in terms of the content of instruction between *Prima* and the subdivided *Secunda,* on the one hand, and the other classes, on the other. While in the classes below *Prima* and *Secunda* only Ramus and Talaeus were used to teach dialectic and rhetoric,[22] for the advanced classes other authors were prescribed. Thus, rhetoric was taught on the basis of Aristotle, Melanchthon and Talaeus. Widmarius pursued a fairly independent course in the *Prima.* Aristotle was the guide in ethics and Euclid in geometry – complemented by Ramus' treatise and commentary on Euclid. At the same time dialectic was taught on the basis of Aristotle, Melanchthon and Ramus; rhetoric, as in the *Secunda,* was communicated on the basis of Aristotle, Melanchthon and Talaeus. This separation governed also the guidelines in respect of content for the period after 1610 – Ramus and Talaeus were accepted as the indispensable authors on account of their clarity and didactic plausibility.

In 1610 under the aegis of the new rector, Matthias Martinius (1572-1630) the *Prima* was replaced by a *schola publica* with four faculties. This turned the Bremen school into a Reformed university-type *Gymnasium Illustre*. Although he had been influenced by the Herborn Ramists, Martinius insisted on the joining of Ramus and Aristotle for the Arts Faculty [23] – though the manuals by Ramus and Talaeus dominated the subordinate *Paedagogeum,* the *schola privata.* This applied to dialectic and rhetoric and, in a less pronounced form, also to the Latin and Greek grammars used in Bremen schools, where in the *Praefationes* Melanchthon and Ramus

22 See *Scholae Bremanae: Favente Deo opt. Maximo, Rectore Andrea Wedemejero, Inde ab autumno anni praesentis M.D.C. deinceps administrandae,* Bremen: Johann Wessel 1600 (Bremen, Staats- und Universitätsbibliothek); *Elenchus Gymnasii Bremani: Ad singularis cujusdam, quae Senatus Bremani pietas est, liberalitas: universorumque studiorum & laborum rationem reddendam: Repetitus inde ab Autumno anni M.D.CVI. exemplisque peculiarium exercitationem declaratus. Leges quoque ejusdem Gymnasii ac Scholae loco postremo sunt addita,* Bremen: Johann Wessel 1606 (Bremen, Staats- und Universitätsbibliothek).

23 Bremen, Staatsarchiv der Freien Hansestadt Bremen: 2-T.5.a.1.b.

were cited as guides. In 1619 and 1625, respectively, the *Dialectic* and *Rhetoric* by Ramus and Talaeus were printed in Bremen,[24] while an *index lectionum* of 1636/37 confirms once more that both authors were used in the *Secunda*.[25] However, already in the following year a change becomes evident. In the only known copy of the 1638/39 *index* the entries relating to Ramus and Talaeus are deleted by hand.[26] Talaeian rhetoric and Ramist dialectic were replaced by combinations of textbooks and florilegia compiled by members of staff who were no longer bound to any methodological-philosophical approach. In the methodological approach of the *Paedagogeum* Comenius replaced Ramus in Bremen.

If we draw a conclusion in respect of the Ramus reception in Bremen, some general and some specific points can be made. Bremen reflects the previously observed tendency that Ramus loses his attraction fairly significantly in the 1630s. The beginning of the Ramus reception coincides with the confessional transformation of the community to the Reformed Faith. It is also clear that Ramus and Talaeus were considered innovators in matters of didactic and their works, published in school editions, were increasingly used as preliminary propaedeutics. A distinctive Bremen characteristic is undoubtedly the early reception of Ramism, which is quite obviously attributable to Johannes Molanus. His activities and those of his successors as rectors also confirm a far-reaching freedom in the determination of the content of instruction. Since at no time did either church or secular authorities seek to determine the actual form of instruction in Bremen, the rectors had a completely free hand there. In respect of the reception of Ramism they developed quite distinctive individual selective models. Another point

24 See Ong: *Ramus and Talon Inventory* (as note 5) nos. 150. 413. 680. To date the following have not been found: Markus Rütimeyer: *Examen Dialecticae Rameae,* Bremen: [Johann Wessel] 1613 and Bremen: [Johann Wessel] 1618. The edition of 1613 is listed in Ong under no. 397. Both imprints are listed in Martin Lipenius: *Bibliotheca realis philosophica,* tom. 2, Frankfort on the Main: Johannes Friderici, Egidius Vogel 1682, p. 1283.

25 *Index Lectionum et Exercitationum, in Illustri Schola Bremensi, Deo volente, proxima hieme anni MDCXXXVI. Habendarum,* Bremen: Berthold de Villiers 1636 (Bremen, Staats- und Universitätsbibliothek).

26 *Index Lectionum et Exercitationum, in Illustri Schola Bremensi, Deo volente, proxima hieme anni MDCXXXVIII. Habendarum,* Bremen: Berthold de Villiers 1638 (Bremen, Staatsarchiv der Freien Hansestadt Bremen).

specific to Bremen is the late printing of the Ramist dialectic and the Talaeian rhetoric. The reason for this state of affairs is undoubtedly the late development of the Bremen printing presses.[27] We know that for the first decades of the seventeenth century Ramus and Talaeus imprints from Frankfort and Hanau were used in the school and the *Gymnasium Illustre* respectively. It was presumably the rapidly increasing numbers of pupils and then the rapid decline of the German book trade occasioned by the Thirty Years' War that prompted the printing of textbooks 'at home'.

Let us now turn to a comparison with Danzig. The confessional conditions in the context of the historical development present the following picture:[28] In 1522 Danzig was seized by Protestantism; the new movement was simultaneously anti-aristocratic and also directed against the Polish royal dynasty; even though the dynasty maintained prudent relationship with the city. When the Royal Privilege of Religion of 1557 sanctioned the emancipation of the new faith it simultaneously recognised the *status quo* which had existed in any case. Despite its close personal tie with Luther in the early stages, the Danzig Church did not follow the Lutheran Church slavishly. Rather, it opted for a more moderate stand with definite Philippist influences. In addition, there remained a strong Catholic congregation with ties to Poland.[29] Yet while Bremen had developed its Reformed Church via a transitional stage of Philippism, championed by an intellectual-administrative elite, a similar course met with determined resistance in Danzig. The city had been open to Reformed ideas in the 1570s and despite adherence to the *Confessio Augustana,* Calvinist pastors, preachers, and teachers were appointed. The years between 1580 and 1612 witnessed the most virulent

27 Thomas Elsmann: Humanismus, Schule, Buchdruck und Antikenrezeption. Anmerkungen zur Bremer Entwicklung bis 1648, in: *Stadt und Literatur im deutschen Sprachraum der Frühen Neuzeit,* ed. Klaus Garber, Stefan Anders, Thomas Elsmann, vol. 1, Tübingen 1998 [Frühe Neuzeit, vol. 39] pp. 203-238.

28 See Heinz Neumeyer*: Kirchengeschichte von Danzig und Westpreußen in evangelischer Sicht,* vol. 1: *Von den Anfängen der christlichen Mission bis zum Ende des 18. Jahrhunderts,* Leer 1971, pp. 75-119; *Historia Gdańska,* red. Edmund Cieslak, vol. 2, Gdańsk 1982, pp. 233ff.

29 On the religious minorities in Danzig see Hans Hecker: Religiöse Minderheiten und Politik in Danzig, in: *Kirchen und Bekenntnisgruppen im Osten des Deutschen Reiches. Ihre Beziehungen zu Staat und Gesellschaft,* ed. Bernart Jähnig, Silke Spieler, Bonn 1991, pp. 221-244.

controversies between Lutheranism and Calvinism in the city.[30] Although most clergymen, the academically trained Council, the teachers of the academic *Gymnasium* (to be described below) and those of the parish schools, adhered to Calvinism, most members of the community did not. Lutheranism was the preferred denomination of the middling bourgeoisie and the merchants as well as the lower class.

The conflict grew more vehement when it became intermingled with social and political issues. The confessional barriers had solidified between the Reformed city churches (St. Peter and Paul, Trinity and St. Elizabeth) and the Lutheran churches; this meant that a rapid Calvinisation had become impossible. Such failure permitted a slow resurgence of Lutheranism, which culminated in 1612 in a mandate applied for by the organised citizenry and granted by King Sigismund III Vasa. It decreed that all those professing the Reformed Faith had to relinquish public office. They subsequently lost all influence, though were not persecuted. As their religious centres they retained control of the churches of St. Peter and Paul and St. Elizabeth. They also retained as their intellectual centres the parish schools connected with these churches and until 1631 – with some restrictions – the academic *Gymnasium*. This *Gymnasium* was founded in 1580,[31] but it originated in the so-called *particular,* that was established in 1558 – a school with four classes, following Melanchthonian principles. Twenty years later its form was changed. When the school had been closed as a result of the occupation of the city by Stephan Báthory, a new creation was undertaken in 1580 with a form of organisation which resembled that of Bremen in its sub-divisions. Two academic classes followed upon three or four preparatory ones. Besides the school of Toruń (1568) and the university of Königsberg (1544) the *Gymnasium* in Danzig became the most im-

30 Detailed treatment in Michael G. Müller: *Zweite Reformation und städtische Autonomie im Königlichen Preußen. Danzig, Elbing und Thorn in der Epoche der Konfessionalisierung (1557-1660),* Berlin 1997 [Publikationen der Historischen Kommission zu Berlin].

31 See *Historia Gdańska* (as note 28) pp. 359ff. 707ff.; Theodor Hirsch: *Geschichte des academischen Gymnasiums in Danzig,* Danzig 1837; *Gdańska gimnazjum Akademickie 1558-1958,* Gdynia 1959; further details can be verified in Włodzimierz Zientara: *Gottfried Lengnich. Ein Danziger Historiker in der Zeit der Aufklärung,* Part 1, Toruń 1995, *passim.*

portant educational institution in the region. It is of some significance that the Bremen rector Matthias Martinius, in a memorandum of 1609, described the Danzig *Gymnasium* as exemplary but counted it among the Reformed.[32] This demonstrates with some clarity the inner-urban confessional conflicts as they affected the *Gymnasium*.[33]

In addition to a number of Reformed professors, the major exponent of this development was the rector, Jakob Fabricius (1551-1629), who was, however, only appointed after the Bremen superintendent, Christoph Pezel (1539-1604) had refused the position in Danzig. Fabricius in his active years (until 1613) ensured that the *Gymnasium* had an adequate structure; he also managed to bring about a steady rise in the number of pupils. Yet, due to a protracted illness Fabricius failed to protect the *Gymnasium* against, largely confessionally motivated, attacks. Three extensive memoranda by Fabricius of the years 1616, 1618 and 1628 bear witness to such confrontations. The memorandum of 1618 is germane to our topic.[34] This is a report in which Fabricius discusses the instruction provided in the two academic classes, primarily the philosophical section.[35] Since the beginning of the seventeenth century a marked discontinuity, as far as the school's

32 On the Calvinist academic schools see Notker Hammerstein: Schule, Hochschule und Res publica litteraria, in: *Res publica litteraria. Die Institutionen der Gelehrsamkeit in der Frühen Neuzeit,* ed. Sebastian Neumeister, Conrad Wiedemann, vol. 1, Wiesbaden 1987 [Wolfenbütteler Arbeiten zur Barockforschung, vol. 14] pp. 92-110; Gerhard Menk: Die kalvinistischen Hochschulen und ihre Städte im konfessionellen Zeitalter, in: *Stadt und Universität,* ed. Heinz Duchhardt, Cologne, Weimar, Vienna 1993 [Städteforschung; Reihe A, Darstellungen, vol. 33] pp. 83-106.

33 Concerning the close link between the institutions of Bremen and Danzig which are revealed in the increasing number of Danzig students in Bremen, especially since the resurgence of Lutheranism see Thomas Elsmann: Danziger akademisches Schrifttum im Umfeld des Bremer Gymnasium Illustre, in: *Hospitium Ecclesiae* 19 (1993) pp. 101-116.

34 Archiwum Panstwowe Gdańsk 300,42/301,45ff.

35 On the teaching of philosophy at the Danzig *Gymnasium* see Izydora Dąmbska: Logika w gimnazjum akademickim gdańskim w pierwszej połowie XVII wieku [Logic at the Danzig academic *Gymnasium* in the first half of the seventeenth century], in: *Rocznik Gdański* 15/16 (1956-1957) pp. 199-223; Bronisław Nadolski: *Nauczanie filozofii gdańskim gimnazjum akademickim za profesury Bartholomieja Keckermanna* [Philosophical instruction at the Danzig academic *Gymnasium* under Bartholomäus Keckermann], in: *Gdańskie Zeszyty humanist* 5 (1962) pp. 321-331.

staff was concerned, had taken place; the decisive development was the appointment of Bartholomäus Keckermann (1572/73-1609) to the chair of philosophy in 1602. Keckermann,[36] intellectually formed by the university of Heidelberg, was considered one of the most prominent philosophers of the German-Reformed confessional region. He personally symbolises the link between the Eastern Calvinism with that of Southern Germany and its intellectual powerhouse. Keckermann wrote a number of so-called *Systems* on ethics, logic, physics, metaphysics etc. – textbooks in the form of compendia which remained influential in Germany until the time of the Thirty Years' War. Following Keckermann's death in 1609, there followed in rapid succession the Scot Andrew Aidius (died 1630), Johann Rodenborch (ca. 1573-1617) and finally the Lutheran Andreas Hojer (died after 1630), all of whom were Aristotelians: at least that is what their writings and the *indices* of their lectures suggest.[37] In 1618 Fabricius was faced with the need to defend the school against accusations from outside, which concerned primarily allegations that the school disseminated heterodox – i.e. Calvinist – beliefs.

The culmination point was reached with Keckermann's manuals which were used, although not uninterruptedly, in the instruction at the Danzig *Gymnasium*. It seems certain that only the *Systema logicae* and the *Systema physicum* were retained as textbooks after his death. According to Fabricius, however, Keckermann's logic was accepted only with considerable reservations by Lutherans, both within and outside the school. Thus Andreas Hojer had written a refutation of the two Lutheran logicians Christoph Scheibler (1589-1653) and Jacob Martini (1570-1649), in which he had accused Keckermann of errors the latter had supposedly committed because of his false faith. Fabricius' survey of 1618 demonstrated – apart from Keckermann's manuals – a philosophical-methodological orientation indebted to Aristotle and Melanchthon. Thus Keckermann was barely

36 See Wilhelm Kühlmann: Bartholomaeus Keckermann, in: *Literatur-Lexikon. Autoren und Werke deutscher Sprache,* ed. Walther Killy, vol. 6, Gütersloh, Munich 1990, pp. 265-266; Wilhelm Schmidt-Biggemann: *Topica Universalis. Eine Modellgeschichte humanistischer und barocker Wissenschaft,* Hamburg 1983, pp. 89-100 *et passim.*

37 Archiwum Panstwowe Gdańsk 300,42/164.

within the bounds of acceptability. In all this the name of Petrus Ramus has been missing so far. Yet he, too, is mentioned by Fabricius and that in a rather revealing context. Fabricius does not simply outline the actual and the envisaged curriculum, he also lists the authors whose instruction manuals could under no circumstances be tolerated by the urban authorities, namely Petrus Ramus for dialectic and the Heidelberg professor of physics Fortunatus Crellius (died 1605) and his manuals on logic which had seen several new editions since the late sixteenth century. Fabricius argued that these had been introduced in *Reformatis Ecclesijs* and merely caused theological controversies. Here we are confronted by a simple identification of Ramism with the Reformed Faith which is the direct result of the controversial situation in Danzig and which, in its expression of bitterness, is used primarily as a rhetorical device. It was quite evidently far less to do with Ramus, Crellius and also Keckermann as personalities but was basically concerned with their confessional persuasions.

If we consider the whole span from the foundation of the *Gymnasium* in 1580 until the year 1630 with the appointment of the decidedly Lutheran rector Johannes Bolsack (1600-1674), we can see that, with the exceptions of the years Keckermann was in office, a sustained Aristotelian-Melanchthonian orientation in the instruction manuals: these were partly written by Danzig teachers like Andreas Hojer, partly by outsiders like Samuel Heiland (1533-1592) for ethics, Sebald Hauenreuter (1508-1589) for physics and Lucas Lossius (1508-1582) for dialectic and rhetoric in the form of erotematically written compendia on the dialectic and rhetoric of Melanchthon. Peter Ramus and his followers, as far as we can judge from the available sources, played no consequential role in the instruction of the *Gymnasium*. Their names were mentioned, but only in a negative sense. Keckermann as well as Aidius, Hojer and the influential Heinrich Nicolai (1605-1660), appointed 1630, were critics of the Ramist method, although from totally different motivations and against utterly different confessional backgrounds.

A characteristic of the Danzig school system was the fact that six or seven parish schools acted as academic-propaedeutic feeders; some of these had already come into existence in the late Middle Ages. Their main emphasis was undoubtedly on the provision of basic humanist propaedeutic instruction, concentrating in their advanced classes on Latin and Greek.

The individual institutions were largely dependent on their rectors for their proper functioning; there was as yet no uniform orientation in their teaching, not least because the schools were confessionally aligned to either Lutheranism or the Reformed Faith. The few existing *indices lectionum* bear witness to an orientation that resembles those of the lesser classes of the *Gymnasium:* dialectic and rhetoric were taught on the basis of Melanchthon's manuals or their revision by Lossius. The practicability of Ramus and Talaeus, so frequently attested, nevertheless did not induce teachers to make use of these authors, with one notable exception. This was Johannes Martini (1558-1629), rector of St. Mary's since 1613 and, besides Keckermann and Fabricius, one of the most renowned representatives of the Reformed Church in Danzig at the beginning of the seventeenth century.[38] In 1614 he presented his *Institutiones Rhetoricae ex Philippo Melanchthone et Audomaro Talaeo conscriptae* and his *Compendium Rhetoricae*. The former was intended for instruction at St. Mary's school. It is impossible to find out whether and if and for how long they were used, since no curricula for this period survived.[39] There were, however, six printed editions before 1643.

The Danzig findings, by contrast with those of Bremen, do not support the thesis of an active and positive reception of Ramus; they do, however, clarify a number of points which explain his rejection in a confessional age. The most prominent of these is the ever-present rhetorical simplistic equation of Ramism and Calvinism in Danzig. The results of our enquiry may be different for each community but they also demonstrate, through the very particularities of each, the importance of such micro-analysis if we wish to gain an understanding of the actual influence of Ramism on education.

38 For his biographical data see Ephraim Praetorius: *Athenae Gedanenses,* Leipzig 1713, pp. 174-175; *Allgemeine Deutsche Biographie,* vol. 20 (1884) p. 510; *Altpreußische Biographie,* vol. 2, Marburg 1967, p. 422.

39 Archiwum Panstwowe Gdańsk 300,42/194; for the school consult also Theodor Hirsch: *Die Oberpfarrkirche von St. Marien in Danzig in ihren Beziehungen zum kirchlichen Leben Danzigs überhaupt,* vol. 2, Danzig 1843.

Melanchthon's Opinion of Ramus and the Utilization of Their Writings in Central Europe[1]

JOSEPH S. FREEDMAN

Philipp Melanchthon (1497-1560) and Petrus Ramus (1515-1572) can be considered as two of the sixteenth century's most significant educators.[2] To what extent were the writings of these two authors utilized

1 I am indebted to Dr. Heinz Scheible (Heidelberg) for calling my attention to the letter by Melanchthon as cited in note 4 below. I also wish to thank Robert Kolb (Concordia Seminary, Saint Louis, Missouri) for reading the draft of this article and offering some helpful suggestions. I have published a short assessment of the passage transcribed in note 6 below in: Philipp Melanchthon's Views Concerning Petrus Ramus as Expressed in a Private Letter Written in 1543: A Brief Assessment, in: Barbara Bauer, ed.: *Philipp Melanchthon und die Marburger Professoren,* Marburg: Universitätsbibliothek 1999, vol. 2, pp. 841-848. That publication contains a complete transcription and German translation of that letter. The articles contained in notes 50, 58, 68, and 69 have been reprinted in *Philosophy and the Arts in Central Europe, 1500-1700: Teaching and Texts at European Schools and Universities,* Aldershot, UK, Brookfield, VT: Ashgate 1999 [Variorum Collected Studies Series, CS 626].

2 Concerning Ramus' life, career, and writings see Walter J. Ong: *Ramus, Method, and the Decay of Dialogue,* Cambridge, MA: Harvard University Press 1958; Idem: *Ramus and Talon Inventory,* Cambridge, MA: Harvard University Press 1958. Concerning Melanchthon refer to the article cited in the following note as well as to the articles by Heinz Scheible: Melanchthon, Philipp (1497-1560), in: *Theologische Realenzyklopädie [TRE]* 22, Berlin, New York: Walter de Gruyter 1992, pp. 371-410; Idem: Philipp Melanchthon. Der bedeutendste Sohn der Stadt Bretten, in: Alfons Schäfer, ed.: *Geschichte der Stadt Bretten von den Anfängen bis zur Zerstörung 1689,* Bretten 1977, pp. 256-282. Especially valuable for locating many of the sixteenth-century editions of works by Melanchthon, Ramus, and other authors associated with them is the *Verzeichnis der im deutschen Sprachbereich erschienenen Drucke des 16. Jahrhunderts (VD 16),* ed. Bayerische Staats-

in Central European schools and universities during the sixteenth and the early seventeenth centuries? Were Melanchthon and Ramus regarded as complementary or as contrary authorities? The search for answers to these two questions requires examination of the ways in which writings on the arts by Ramus (i.e., logic, rhetoric, grammar, arithmetic, and geometry) and Melanchthon (i.e., logic, rhetoric, grammar, physics, the soul, and ethics) were utilized during that period. This article will attempt to provide such answers through discussion of the following ten points:

1. the demand for Ramus' writings on logic as well as other arts disciplines;
2. the demand for Melanchthon's writings on logic and the other arts;
3. adoption of, and commentaries on, Melanchthon's writings on the arts;
4. adoption of, and commentaries on, Ramus' writings on the arts;
5. polemical writings against Ramus' writings on the arts;
6. the lack of extant polemical writings against Melanchthon's writings on the arts;
7. the frequency with which sixteenth- and seventeenth-century writings on the arts authored by both Lutherans and Calvinists utilized works by Ramus in combination with works by Melanchthon;
8. the eclectic and independent manner in which Melanchthon, Ramus, Aristotle, and other authors were utilized by late sixteenth- and early seventeenth-century writings on the arts;
9. revised versions of Ramus' and Melanchthon's writings on the arts;
10. differences between individual commentaries on Ramus' and Melanchthon's writings on the arts.

In 1569, Ramus praised in glowing terms not only Melanchthon's efforts to foster mathematical studies, but also Melanchthon personally:

> Just as Plato revived the study of mathematics in Greece through the great power of his eloquence and erudition, so Melanchthon found [mathematical studies] already greatly encouraged in most academies of Germany, with the exception of Wittenberg. Whereupon, through the force of much and varied instruction and through the example of his pious and upright life which, at least in my opinion, no doctor or professor in that country has ever attained, he wondrously ignited [those studies], with

bibliothek in München and Herzog August Bibliothek in Wolfenbüttel, Stuttgart: Anton Hiersemann 1983ff. In referring to this bibliography, the numbers concerning each individual entry – which are prefaced by the first letter of the author or title of that entry – will be cited.

the result that Wittenberg became superior not only in theology and eloquence, in which fame it especially excels, but also in the studies of the mathematical discipline.[3]

Far less complimentary was the manner in which Petrus Ramus is mentioned in a letter written by Philipp Melanchthon on June 11, 1543.[4] This autograph letter is contained on both sides of a single leaf. Melanchthon, who resided in Bonn at the time, addressed his letter to Cunradus Embecanus – care of Johannes Gymnicus – in nearby Cologne.[5] The final nine lines concern Ramus and focus on the first edition of the latter's textbook on logic, which had been published earlier that year:

3 Robert S. Westman: The Melanchthon Circle, Rheticus, and the Wittenberg Interpretation of the Copernican Theory, in: *Isis* 66 (1975) pp. 165-193 (172). With several minor alterations, I have adopted Westman's translation of the Latin language passage given in Petrus Ramus: *Scholarum mathematicarum, libri unus et triginta,* Basileae: Per Eusebium Episcopium et Nicolai fratris haeredes 1569, p. 66 [Basle UB: K c, IV 3; Leiden University Library: Thys. 1521.2; Yale University Library, Beinecke Library: Q 33. L 37]. The Latin text reads as follows: "Plato pro summa eloquentiae & eruditionis authoritate mathematicum studium in Graecia excitavit; Melanchthon in plerisque Germaniae academiis jam mirabiliter excitatum reperit, sed Vitembergae perexiguum: itaque pro variae ac multiplicis doctrinae, pro vitae innocentioris & sanctioris authoritate, quantam meo quidem iudicio, nemo doctor vel professor, in patria consecutus est unquam, mirabiliter inflammavit: ut Viteberga non solum theologiae & eloquentiae, quibus laudibus tum praecipue praestabat, sed mathematicae disciplinae studiis antecelleret." Westman (pp. 170-172) provides well argued and informative discussion of Melanchthon's career as an educator, including discussion of the careers of some of Melanchthon's disciples.

4 Philippus Melanchthon to D. Cunradus Embecanus apud D. Gymnicum Coloniae die xi. Junii (Bonn 1543); the original copy of this letter is owned by the Folger Shakespeare Library in Washington, D.C. and has the call number F. c. 2. It is also referred to in Heinz Scheible, ed.: *Melanchthons Briefwechsel* 3: *Regesten 2336-3420 (1540-1543),* Stuttgart - Bad Cannstatt: Frommann-Holzboog 1979, p. 400 (no. 3259); when making this reference, Scheible is able to provide sufficient evidence to show that it was written in the year 1543. On the basis of those letters in which Melanchthon mentions his place of residence, he was in Bonn from no later than May 5, 1543 until at least June 15, 1543 barring one brief absence. See Scheible: *Melanchthons Briefwechsel* 3, pp. 390-402 (nos. 3235-3262).

5 This address appears in lines 12 through 20 of the verso side of this leaf as cited in note 4. Concerning Johannes Gymnicus (ca. 1480-1544) see *Neue Deutsche Biographie* 7, Berlin: Duncker und Humblot 1965, p. 362; unfortunately, research done in conjunction with this article did not result in any information concerning the identity of Cunradus Embecanus (Einbecanus).

> Concerning Petrus Ramus' textbook on dialectic, I do not know how I should respond. It is a good effort, but contains many foolish statements. And he certainly sings to himself using long flutes (he certainly expresses himself using lengthy, irrelevant utterances) in a typically French manner. But nothing of utility is offered to the reader. Perhaps the first edition will sell well. But after that there will be no demand for this work.[6]

1543 saw the first publication of two major works by Ramus on the subject matter of logic: the *Remarks on Aristotle (Aristotelicae animadversiones)* and *The Structure of Dialectic (Dialecticae partitiones).*[7] The latter work was republished in September of the same year with the title *Training in Dialectic (Dialecticae institutiones).*[8] Melanchthon was not the only academician to criticize Ramus' logic in 1543. Two of Ramus' colleagues at the University of Paris – infuriated by Ramus' virulent criticism of Aristotle's logic – also attacked Ramus that year.[9]

Yet Melanchthon's prognosis that there would be little demand for Ramus' *Dialectic* proved wrong. In fact, the latter's most popular writings were those on logic. Ramus' *Remarks on Aristotle* was printed at least ten times by the year 1557; his *Teaching of Dialectic* went through at least

6 These nine lines appear on lines 20 and 21 of the recto side of the letter and on lines 1 through 7 of the verso side as cited in note 4 above. I have transcribed the original text here as follows: "De dialectica Petri Rami nescio quid respondeam. Magno conatu, magnas nugas dicit. Et plane μακροῖς αὐλοῖς sese canit valde gallice. Sed utilitatis nihil adfert lectori. Fortassis editio prima vendibilis esset. Sed semel perlectus liber non rursus expetitur." When utilizing the term αὐλοῖς, Melanchthon also probably intended to refer to ἀΰλοις (i.e., immaterial/irrelevant things). In using the term *plane* Melanchthon also my have wished to allude to the term *planus* (i.e., impostor, charlatan) as well.

7 Petrus Ramus: *Dialecticae partitiones, ad celeberrimam & illustrissimam Lutetiae Parisiorum Academiam,* Parisiis: Excudebat Jacobus Bogardus 1543 [Saint Louis University, Pius XII Library: Saint Louis Room, BC 60. L 23 1543]; this is the work which Melanchthon was criticizing (see the previous note); also see Ong: *Ramus, Method* (see note 2) pp. 172-175; Idem: *Ramus and Talon* (see note 2) pp. 46-49 and the following note.

8 Petrus Ramus: *Dialecticae Institutiones. Aristotelicae animadversiones,* Faksimile-Neudruck der Ausgaben Paris 1543 mit einer Einleitung von Wilhelm Risse, Stuttgart - Bad Cannstatt: Friedrich Frommann, Günther Holzboog 1964.

9 See Ong: *Ramus and Talon* (see note 2) pp. 493-496 as well as *Peter Ramus' Attack on Cicero. Text and Translation of Ramus' Brutinae Quaestiones,* ed. James J. Murphy, trans. Carole Newlands, Davis, CA: Hermogenes Press 1992, pp. ix-x. xxxiv.

sixteen imprints – and six revisions – through the same period.[10] Additional revised versions of the latter work were published with the titles *Dialectic (dialectique; dialectica)* and *Dialectic in Two Books (Dialecticae libri duo)* beginning in the year 1555, and printed more than sixty times over the next century.[11] In addition, by 1650 there appeared two Dutch, fourteen English, and five German versions of that latter work.[12] In a similar fashion, Ramus' works on rhetoric – some of which were written in collaboration with his friend Omer Talon – went through well over eighty editions between 1545 and 1650,[13] while his writings on Latin, Greek, and French grammar enjoyed a total of more than fifty editions by 1650.[14] Ramus' writings on arithmetic and geometry enjoyed smaller print-runs, but they, too, were quite popular.[15]

In the past forty years there has been a substantial amount of scholarship on Ramus; this has included bibliographies, general studies, and scholarship devoted especially to Ramus' writings on logic, rhetoric, grammar, and mathematics.[16] This has been matched by a similar magnitude of scholarship on Philipp Melanchthon, including some very useful bibliographies.[17] The overwhelming portion of the secondary literature on Melanchthon has been focused upon his theological works and the confessional impact of his

10 Ong: *Ramus and Talon* (see note 2) pp. 48-56.

11 Ibid. pp. 178-269.

12 Idem.

13 Ibid. pp. 82-137; for the purposes of this article, Talon's works will be subsumed under those of Ramus.

14 Ibid. pp. 311-332. 337-350.

15 Ibid. pp. 68-69. 333-337. 368-374.

16 The following two bibliographical articles of literature pertaining Petrus Ramus are well organized and very informative: Peter Sharratt: The Present State of Ramus Studies, in: *Studi francesci* 47-48, Anno 16, Fascicolo 2-3 (Maggio-Dicembre 1972) pp. 201-213; Idem: Recent Work on Peter Ramus (1970-1986), in: *Rhetorica* 5, no. 1 (Winter 1987) pp. 7-58; also refer back to note 2.

17 The following two bibliographies are especially useful: Ralph Keen: *A Checklist of Melanchthon Imprints Through 1560,* St. Louis: Center for Reformation Research 1988 [Sixteenth Century Bibliography 27]; Wilhelm Hammer: *Die Melanchthonforschung im Wandel der Jahrhunderte,* 3 vols., Gütersloh: Verlagshaus Gerd Mohn 1967-1981. There also is a good, chronologically arranged bibliography of Melanchthon's writings in Karl Hartfelder: *Philipp Melanchthon als Praeceptor Germaniae,* Berlin: A. Hofmann & Comp. 1889 [Monumenta Germaniae Paedagogica 7] pp. 567-620.

writings.[18] Only a handful of studies have examined Melanchthon's pedagogy.[19] However, his writings on the arts were used extensively in the late sixteenth and early seventeenth centuries, contributing in large part to his reputation as the *Praeceptor noster* and the *Praeceptor Germaniae.*[20] Yet these very writings have generally been ignored by modern scholars.[21]

Beginning in the year 1525, neither Martin Luther nor Philipp Melanchthon were required to lecture on specific subjects pending consultation with their colleagues at the University of Wittenberg.[22] From then

18 This is evident from the literature cited in Scheible: Melanchthon *TRE* (see note 2) pp. 397-410 as well as from Peter Fraenkel, Martin Greschat: *Zwanzig Jahre Melanchthonstudium. Sechs Literaturberichte 1945-1965,* Genève: Librairie Droz 1967 [Travaux d'humanisme et renaissance 93].

19 Hermann-Adolf Stempel: *Melanchthons pädagogisches Wirken,* Bielefeld: Luther-Verlag 1979 [Untersuchungen zur Kirchengeschichte 11]; Hartfelder: *Melanchthon als Praeceptor Germaniae* (see note 17). Hartfelder also edited a large set of short pedagogical writings by Melanchthon: *Melanchthoniana Paedagogica. Eine Ergänzung zu den Werken Melanchthons im Corpus Reformatorum,* Leipzig: Druck und Verlag von B. G. Teubner 1892.

20 Hartfelder: *Melanchthon als Praeceptor Germaniae* (see note 17) notes (p. viii) that Andreas Osiander refers to Melanchthon as the *Praeceptor Germaniae* in the year 1551; Melanchthon is referred to as *Praeceptor noster* in a passage cited in note 61 below.

21 Some aspects of Melanchthon's logic and rhetoric are discussed in Peter Mack: *Renaissance Argument. Valla and Agricola in the Traditions of Rhetoric and Dialectic,* Leiden, New York, Cologne: E. J. Brill 1993 [Brill's Studies in Intellectual History 43] pp. 320-333. Melanchthon's rhetoric is the subject of the following recent monograph (which also includes a reprint and abridged translation of the 1542 edition of Melanchthon's *Elementorum rhetorices libri duo*): Joachim Knape: *Philipp Melanchthons 'Rhetorik',* Tübingen: Max Niemeyer 1993 [Rhetorik-Forschungen 6]. Some discussion of Melanchthon's rhetoric is given in Debora K. Shuger: *Sacred Rhetoric. The Christian Grand Style in the English Renaissance,* Princeton, NJ: Princeton University Press 1988. Some relevant secondary literature is cited in Scheible: Melanchthon *TRE* (see note 2) pp. 398-400. Hartfelder: *Melanchthon als Praeceptor Germaniae* (see note 17) still has the only general survey of Melanchthon's writings which have appeared to date. Melanchthon's mathematics and physics are the subject matter of Wilhelm Bernhardt: *Philipp Melanchthon als Mathematiker und Physiker,* Wittenberg: Im Verlage des Vereins für Heimathkunde 1865; also see Westman: Melanchthon Circle (see note 3).

22 See Scheible: Melanchthon *TRE* (see note 2) p. 374; Idem: Aristoteles und die Wittenberger Universitätsreform, in: Michael Beyer, Günther Wartenberg, eds.: *Humanismus und Wittenberger Reformation. Festgabe anläßlich des 500. Geburtstages des Praeceptor Germaniae Philipp Melanchthon,* Leipzig: Evangelische Verlagsanstalt 1996,

on until his death in 1560, Melanchthon was permitted to teach any subject he wished within the philosophy and theology faculties. It is known that shortly before his death Melanchthon was lecturing on logic, Greek drama (Euripides), introductory Greek, New Testament, ethics, and history.[23]

Thus, Melanchthon's interest in, and concern for, the arts did not end with the early years of the Reformation.[24] He placed substantial emphasis on his instruction in arts disciplines throughout his academic career.[25] He wrote textbooks on logic, rhetoric, Latin grammar, Greek grammar, ethics, physics, and the soul (*de anima*) – as well as prefaces to a number of works on astronomy and other mathematical disciplines.[26] He also published commentaries on, and editions of, many ancient Greek and Latin authors. These included commentaries on Aristotle's *Ethics* and *Politics,* on Cicero's *De amicitia* and *De officiis,* on Cicero's letters and on many of his orations,

pp. 123-144 (141); Walter Friedensburg: *Urkundenbuch der Universität Wittenberg,* Teil 1, Magdeburg: Selbstverlag der Historischen Kommission 1926, pp. 133. 167.

23 Hammer: *Melanchthonforschung* (see note 17) vol. 1, p. 166 (no. 218).

24 The assumption that Melanchthon's "humanism" (i.e., his writings on many arts subjects) served only as a prelude to his work as a theologian and reformer seems to be implied in much of the secondary literature pertaining to Melanchthon. For example, see Wilhelm Maurer: *Der junge Melanchthon zwischen Humanismus und Reformation,* Band 1: *Der Humanist,* Band 2: *Der Theologe,* Göttingen: Vandenhoeck & Ruprecht 1967-1968.

25 For example, several of Luther's letters draw attention to the fact that Melanchthon apparently preferred teaching Greek to teaching theology. See Luther's letter to Elector Frederic of Saxony dated March 23, 1524 as well as his letter to Elector John of Saxony dated February 9, 1526. Martin Luther: *Werke. Kritische Gesamtausgabe. Briefwechsel,* Weimar: Hermann Böhlaus Nachfolger 1933, vol. 3, pp. 258-259. vol. 4, pp. 29-30; English translations of these two letters are in *Luther's Correspondence and other Contemporary Letters,* translated and edited by Preserved Smith, Charles M. Jacobs, Philadelphia, PA: The Lutheran Publication Society 1918, vol. 2, pp. 223-224. 363-364. I wish to thank Rebecca Peterson (University of North Texas) for this reference. Also refer to the following edition of texts: Johannes Haussleiter: *Melanchthon-Kompendium. Eine unbekannte Sammlung ethischer, politischer und philosophischer Lehrsätze Melanchthons in Luthers Werken,* Greifswald: Druck und Verlag von Julius Abel 1902.

26 The following collection of astronomical and astrological writings also contains a number of prefaces and translations by Melanchthon: *Artis divinatricis quam astrologiam seu iudiciam vocant, encomia & patrocinia,* Parisiis: Excudebat Christianus Wechelus 1549 [University of Illinois: x520.9 / Ar 78].

and on works by Homer, Ovid, Virgil as well as editions of the corpus of Aristotle's works.[27]

Walter Ong has noted that Ramus continually revised his works throughout his career.[28] Melanchthon's works went through a similar process, though the ways in which his textbooks on the arts – that is, his textbooks on grammar, rhetoric, logic, ethics, and physics – evolved has not been studied in any detail. And bibliographical information concerning the further publication of these works after the year 1560 is still rudimentary.

In the case of Melanchthon's logic, Karl Gottlieb Bretschneider has distinguished between three principal stages of its development.[29] The first took the shape of the relatively short *Compendium of Dialectic (Compendiaria dialectices ratio),* which first appeared in print in 1520 and was reprinted over twenty times by 1545.[30] The second consisted of a series of works which began with the publication in 1528 of his *Dialectic in Four Books (Dialectices libri quatuor)*, followed by the *Dialectic (Dialectica)* in 1532, his *Dialectic in Three Books (Dialecticae libri tres)* in 1534, and his *Precepts of Dialectic (Dialecticae praeceptiones)* in 1544.[31] The third and final form of Melanchthon's logic was a lengthy textbook first published in 1547 with the title of *Questions on Dialectic (Erotemata dialectices)*; this latter textbook was reprinted at least twenty

27 See Keen: *Checklist* (see note 17) pp. 142-149. 187-207; Scheible: Melanchthon *TRE* (see note 2) p. 387.

28 A detailed presentation of the evolution of Ramus' logic is given in Nelly Bruyère: *Méthode et dialectique dans l'œuvre de La Ramée. Renaissance et âge classique,* Paris: Librairie philosophique J. Vrin 1984 [De Pétrarque à Descartes 45].

29 Carolus Gottlieb Bretschneider, ed.: *Philippi Melanthonis Opera quae supersunt omnia,* vol. 13, Halis Saxonum: Apud C. A. Schwetschke et filium 1846 [Corpus Reformatorum 13] cols. 509-510. Also see Ong: *Ramus, Method* (see note 2) p. 236.

30 Keen: *Checklist* (see note 17) pp. 18-19; Philippus Melanchthon: *Compendiaria dialectices ratio,* Wittenbergae: Apud Melchiorem Lottherum iuniorem 1520 [University of Illinois: 160 / M482c]. The University of Illinois also owns copies of the Leipzig 1521 (Lipsiae: Apud Melchiorem Lottherum iuniorem) [Stonehill 497], Basle 1522 (Basileae: Per Adamum Petri, Januario 1522) [160 / M482c / 1522a] and Cologne 1522 (Apud inclytam Coloniam: Jo. Soter excudebat, 1522 mense Aug.) [x160 / M482c / 1522 (1)] editions of this same work. Ong translates the title of this work into English as *A Brief Notion of Dialectic,* see Ong: *Ramus, Method* (see note 2) p. 236.

31 Keen: *Checklist* (see note 17) pp. 42-44.

times by 1560 and saw many additional reprintings thereafter into the seventeenth century.[32]

Melanchthon's writings on rhetoric and grammar went through a similar evolution. The first publication of his *Three Books on Rhetoric (De rhetorica libri tres)* in 1519 was followed by his *Principles of Rhetoric (Institutiones rhetoricae)* in 1521 and his substantially longer *Elements of Rhetoric in Two Books (Elementorum rhetorices libri duo)* in 1531.[33] The first two works were reprinted numerous times in the decade following their original appearance; the third continued to be republished long after Melanchthon's death in 1560.[34] His textbooks on Greek grammar and Latin grammar were first published in 1518 and 1525, respectively.[35] Joachim Camerarius' edition of the former was first published in 1545 and continued to be reprinted well into the seventeenth century.[36] Jacob Micyllus' edition of the latter work first appeared in 1540 and was superseded by Joachim Camerarius' edition of it in the year 1550.[37] Camerarius' edition of Melanchthon's textbook on Latin grammar frequently reappeared in print through the sixteenth century and beyond.[38]

32 Ibid. pp. 44-45. Also refer to the list given in *VD 16* (see note 2) M 3242-3270.

33 See Bretschneider: *Melanthonis Opera* 13 (see note 29) cols. 413-416; Keen: *Checklist* (see note 17) pp. 50-54. Knape: *Melanchthons Rhetorik* (see note 21) discusses these three versions of Melanchthon's rhetoric.

34 *VD 16* (see note 2) M 3101-3136. M 3514-3527. M 4179-4186.

35 Keen: *Checklist* (see note 17) p. 66; Henricus Ernestus Bindseil, ed.: *Philippi Melanthonis Opera quae supersunt omnia,* vol. 20, Halis Saxonum: Apud C. A. Schwetschke et filium 1854 [Corpus Reformatorum 20] cols. 5-6. Keen (p. 68) also lists an edition of Melanchthon's Latin grammar from the year 1517, while Bindseil 20, cols. 193-194 regards the 1525 edition to be the first.

36 Bindseil: *Melanthonis Opera* 20 (see note 35) cols. 9-12. Concerning Joachim Camerarius (1500-1574) see Frank Baron, ed.: *Joachim Camerarius (1500-1574),* Munich: Wilhelm Fink Verlag 1978 [Humanistische Bibliothek, Reihe 1: Abhandlungen, Band 24]; ADB 3, pp. 720-724; Jöcher 1, pp. 1593-1594.

37 Bindseil: *Melanthonis Opera* 20 (see note 35) cols. 197-214. Concerning Jacob Micyllus (1503-1558) refer to ADB 21, pp. 704-707; Jöcher 3, pp. 524-525; Jöcher-Adelung 4, pp. 1702-1703; J. Classen: *Jacob Micyllus ... als Schulmann, Dichter und Gelehrter,* 1. Abt., Frankfort on the Main: Joh. Christ. Hermann'scher Verlag 1858.

38 An edition from the year 1524 is mentioned in Keen: *Checklist* (see note 17) p. 99; 1538 is the year of the earliest editions mentioned in *VD 16* (see note 2) M 3961 and in Henricus Ernestus Bindseil, ed.: *Philippi Melanthonis Opera quae supersunt omnia,* vol. 16, Halis Saxonum: Apud C. A. Schwetschke et filium 1850 [Cor-

Melanchthon's textbook on ethics was first published as the *Epitome of Moral Philosophy (Moralis philosophiae epitome)* in the year 1524. His *Commentary on the Soul (Commentarius de anima),* and his *Introduction to Physics (Initia doctrinae physicae)* first appeared in print in the years 1540 and 1549, respectively.[39] Melanchthon's textbooks on ethics, the soul, and physics each had been republished over a dozen times by his death in 1560 and continued to appear in print frequently into the early seventeenth century.[40]

Melanchthon's writings on grammar, rhetoric, logic, ethics, physics, and the soul were also the source of numerous commentaries which were widely used in Central Europe during the late sixteenth and early seventeenth centuries. Commentaries by two of Melanchthon's disciples – Lucas Lossius (1508-1582) and Victorinus Strigel (1524-1569) – were especially influential. Lossius spent his career as a teacher at the Johanneum school in Lüneburg, where he published numerous works arising out of his instruction there.[41] His commentary on Melanchthon's logic and rhetoric was first published in 1550 and went through numerous editions continuing into the

pus Reformatorum 16] cols. 9-12. See Philippus Melanchthon: *Philosophiae moralis epitome ... Nunquam antea excusa* (Argentorati): (Apud Cratonem Mylium) (mense Martio) 1538 [University of Illinois: 170 / M48p].

39 See Bretschneider: *Melanthonis Opera* 13 (see note 29) cols. 1-4. Melanchthon's *Commentary on the Soul* was republished – in revised form – in 1553 with the title *Book on the Soul (Liber de anima)*; the 1553 edition has been republished in *Melanchthons Werke* 3: *Humanistische Schriften,* ed. Richard Nürnberger, Gütersloh: Verlagshaus Gerd Mohn 1961, pp. 303-372. Also refer to the following first editions: Philipp(us) Mel(ancthon): *De anima commentarius,* Parisiis: Apud Jacobum Kerver 1540 [University of Illinois: uncat.]; Philipp(us) Melan[c]th(on): *Liber de anima recognitus,* Witebergae: Excudebat haeredes Petri Seitzii 1553 [University of Illinois: uncat.]. Also see Philipp(us) Melan[c]t(hon): *Initia doctrina physicae, dictata in Academia Wittebergensi,* Excusa Witebergae: Per Johannem Lufft 1549 [University of Illinois: uncat.].

40 *VD 16* (see note 2) M 2748-2794. M 3469-3486 for lists of sixteenth century editions of both versions his *de anima* as well as of his textbook on physics and ethics. Bindseil: *Melanthonis Opera* 16 (see note 38) cols. 1-20 gives a detailed history of the evolution of Melanchthon's writings on ethics, which included a commentary on Aristotle's ethics first published in the year 1529 (ibid. cols. 3-4).

41 Concerning Lossius, see ADB 19, pp. 220-221; concerning Strigel refer to ADB 36, pp. 590-594; Charles H. Lohr: Renaissance Latin Aristotle Commentaries, Authors So-Z, in: *Renaissance Quarterly* 35, no. 2 (Summer 1982) pp. 164-256 (177-178).

early seventeenth century.[42] He also published several commentaries on Melanchthon's Latin grammar. Victorinus Strigel taught at the Universities of Jena, Leipzig, and Heidelberg.[43] He published commentaries on Melanchthon's rhetoric and on his logic; his commentaries on Melanchthon's ethics and on his treatise on the soul both first appeared posthumously.[44]

Many additional adaptations of and commentaries on Melanchthon's works on the arts were published in Central Europe during and after his lifetime.[45] A tabular summary of Melanchthon's rhetoric by Georg Major

42 Lucas Lossius: *Erotemata dialecticae et rhetoricae Philippi Melanchthonis, et praeceptionum Erasmi Roterodami, de utraq: copia verborum et rerum, iam primum ad usum scholarum <quas vocant triviales> breviter selecta et contracta,* Francofurti: Apud Petrum Brub. 1550 [Munich UB: 8° Melan. 106:2]. A list of sixteenth-century editions is given in *VD 16* (see note 2) L 2740-2763; two seventeenth-century editions can be cited here: Lucas Lossius: *Erotemata dialecticae et rhetoricae Philippi Melanchthonis et praeceptorum Erasmi Roterodami, de utraque copia verborum et rerum, jam primum ad usum Scholarum <quas vocant triviales> breviter selecta et contracta,* Lipsiae: Impensis Jacobi Apelii excudebat Laurentius Cober 1618 [Wolfenbüttel HAB: O 78. 8° Helmst.], and: Witebergae: Sumptibus haeredum Selfischianorum excudebat Augustus Boreck 1620 [Wolfenbüttel HAB: O 79. 8° Helmst.].

43 Lucas Lossius: *Grammatica Philippi Melanthonis, ad usum puerorum in breves interrogationes contracta ... Erotemata grammaticae* (Excusum Witebergae): (Apud honestum virum Gerogium Rhaw) 1544 (mense Octobri) [Wolfenbüttel HAB: P 865. 8° Helmst.]; Lucas Lossius: *Exemplorum in grammaticis Philippi Melanthonis latinis, eorum, quae sunt in etymologia, completio, & brevis interpretatio,* Franc.: Apud haeredes Chr. Egen. 1560 [Munich SB: L. lat. 483 (1, 2)]. Hammer: *Melanchthonforschung* (see note 17) vol. 1, p. 102 (no. 113c) notes that the following work by Lossius is – despite its title – an independent work: Lucas Lossius: *Annotationes in grammaticen D. Philippi Melanthonis latinam,* Franc.: Apud Chr. Egenolphum (1552f.) [Nuremberg StB: Strob. 77. 8°]. Hammer also cites further imprints of this latter work from the years 1555, 1557, 1559, 1565, 1568, 1571, 1577, and 1581. See Hammer: vol. 1, pp. 115-116 (no. 129a). 152-153 (no. 193a). 302 (no. 421a). 338-339 (no. 469a). 362 (no. 508a); vol. 3, pp. 73 (no. 129a). 85-86 (no. 143a). 104 (no. 193a). 146 (no. A356a). 158 (no. A390a). 171 no. (421a). 192 (no. 469a). 206 (no. 508a) as well as *VD 16* (see note 2): L 2699-2706.

44 Lohr: Aristotle Commentaries (see note 41) p. 178; Hammer: *Melanchthonforschung* (see note 17) vol. 1, pp. 360 (no. 505). 369 (no. 517). 415 (no. 602); vol. 3, p. 230 (no. 505); *VD 16* (see note 2): S 9603-9606.

45 In addition to the examples given in notes 46 through 49 below, see to the following: Johannes Stigelius: *De anima commentarii ... Philippi Melanchthonis, explicatio, ...*

(1502-1574) was printed together with tabular summaries of works by Petrus Mosellanus and Desiderius Erasmus.[46] Commentaries on Melanchthon's Latin grammar were prepared for use in individual grades at schools in Saxony.[47] A compendium of "Philippic" logic and rhetoric – including numerous dichotomous tables – was prepared by Thomas Sagittarius (1577-1621) for use of students at Breslau schools.[48] And Gregor Richter (1560-1624) published an alphabetically-arranged compendium of

nunc primum in gratiam studiosorum in lucem edita, Mulhusii Duringorum: Excudebat Georgius Hantzsch 1570 [Washington, D.C., Folger Library: BR / 336 / D3 / Y7 / Cage]; Jacobus Sommerfeldt: *Praxis dialectices Dn. Philippi Melanthonis,* Francofurti: Imprimebat Andreas Eichorn 1596 [Wolfenbüttel HAB: O 81. 8° Helmst.]; Martinus Crusius: *Elementorum rhetorices libri duo ... quaestionibus et scholijs explicati in Academia Tybingensi,* Basileae: Ex officina Oporiana (1564 mense Martio) [University of Illinois: Baldwin 2142]; Henricus Decimator: *Annotationes in grammaticam D. Philippi Melancht. latinam, tam docentibus quam discentibus profuturae: non tantum ad explicationem huius libri sed etiam aliorum. In usum & gratiam scholarum collectae & conscriptae,* Magdeburgi: Excudebat Paulus Donatus impensis Ambrosii Kirchneri 1598 [San Marino, The Huntington Library: 384463].

46 *Tabulae de schematibus et tropis, Petri Mosellani. Item. Rhetorices Phil. Melanchthonis tabulae* [by Georg Major]. *Item libelli Erasmi de duplici copia, scilicet rerum & verborum, tabulae,* Norimbergae: Apud Johan. Petreium 1540 [University of Illinois: x870 / M852t / 1540]. The University of Illinois owns editions of this work printed in 1532 (Lipsiae: Excudebat Nicolaus Faber [x871 / c7rh.l (5)], in 1539 (Augustae: Per Alexandrum Weissenborn) [uncat.], and in 1566 (Argentorati): (Excudebat Christianus Mylius) [uncat.]. Also refer to note 65 and the corresponding passage in the text of this article.

47 *Quaestiones de primis rudimentis grammaticae, ex Philippo Melanthone, in usum secundae classis scholarum in Electoratu Saxoniae collectae,* Lipsiae: (Hans Steinmann) (1579) [Wolfenbüttel HAB: Alv.: Ca 140 (3)]; *Quaestiones de primis rudimentis grammaticae, ex Philippo Melanthone, in usum tertiae et quartae classis scholarum in Electoratu Saxoniae collectae,* Lipsiae: (Hans Steinmann) 1579 [Wolfenbüttel HAB: Alv.: Ca 140 (4)].

48 (Thomas Sagittarius): *Dialecticae Philippeae compendium ... (Rhetoricae Philippeae compendium pro tyronibus scholarum Vratislaviensium mandato ejusdem senatus conscriptum),* Breslae: Impensis haeredum Joannis Eyringii, & Johannis Perferti bibliopol. Vratisl. 1617 [Washington, D.C., Folger Library: PA / 2313 / D 46 / 1617 / Cage]. Concerning the career of Thomas Sagittarius see ADB 30, p. 173; Jöcher 4, p. 29; Zedler 33, p. 591. Also see Michael Neander*: Compendium dialecticae ac rhetoricae Philippi Melancht[h]onis,* Islebii: Urbanus Gubisius excudebat 1580 [Wolfenbüttel HAB: O 72. 8° Helmst.].

opinions by Melanchthon and his disciples with regard to recommended authors and writings.[49]

These publications were matched by a large number of Central European adaptations of Petrus Ramus' writings and commentaries on his works. Although use of his writings in Central Europe can be traced back to the 1560s, they appear to have had their greatest impact during the last three decades of the sixteenth century and the first three decades of the seventeenth.[50] Johannes Thomas Freigius (1543-1583) and Friedrich Beurhaus (1536-1609) were among the more prominent Central European commentators of Ramus. The former taught at the University of Freiburg (Breisgau), at the Altdorf Academy, and in Basle during his relatively short career, while Beurhaus taught at a school in Dortmund for over forty years.[51]

Ramus' writings were also discussed within a large body of polemical literature that circulated in late sixteenth- and early seventeenth-century Europe. Some examples can be given here. Walter Ong has already documented Ramus' polemical correspondence with Jakob Schegk (1511-1587), a professor at the University of Tübingen.[52] In the year 1570, Zacharias Ursinus (1534-1583), a professor of theology at the University of Heidelberg, a former student and a follower of Melanchthon, wrote a short treatise in which he ar-

49 Gregorius Richterus: *CRISEIS Melanchthoniae, de eligendis optimis quibusque in omni disciplinarum genere auctoribus, tam docentibus quam discentibus bonas artes,* Islebiae: Typis & impensis Henningi Grossij bibliopolae Lipsensis 1597 [University of Illinois: uncat.]; as noted in Hammer: *Melanchthonforschung* (see note 17) vol. 1, p. 424 (no. 619), this work was orginally published with the title *Iudicia florentis scholae Melanchthonis de lectione bonorum autorum in omnia genere disciplinarum,* Witebergae: Typis M. Simonis Gronenbergii sumptum impendente Henning Großio bibliopola Lipsensi 1592 [Wolfenbüttel HAB: 78.2 Gram]. Concerning Gregor Richter (1560-1624), a Lutheran minister in Görlitz and an opponent of Jakob Böhme, see ADB 28, pp. 459-460; Jöcher-Adelung 6, pp. 2076-2077.

50 See Ong: *Ramus, Method* (see note 2) p. 296 as discussed in Joseph S. Freedman: The Diffusion of the Writings of Petrus Ramus in Central Europe, c. 1570 - c. 1630, in: *Renaissance Quarterly* 46, no. 1 (Spring 1993) pp. 98-152.

51 Concerning Freigius see ADB 7, pp. 341-343; Jöcher 2, pp. 737-738; Zedler 9, pp. 1804-1805; Ong: *Ramus, Method* (see note 2); Wolfgang Rother: Ramus and Ramism in Switzerland (see the present volume) pp. 16-21; with regard to Beurhaus see ADB 2, pp. 584-585; Jöcher 1, p. 1062; Zedler 3, p. 1581.

52 Ong: *Ramus and Talon* (see note 2) pp. 374-380; concerning the career of Jakob Schegk (Degen) see ADB 5, pp. 21-22; Jöcher 4, pp. 235-236; Ong: *Ramus, Method* (see note 2).

gued against using Ramus' logic and rhetoric in school instruction.[53] It appeared in print in 1586 and was followed by a rebuttal in the following year.[54] The publication of a number of polemical treatises accompanied an unsuccessful attempt to introduce the use of Ramus' logic at the University of Helmstedt during the final years of the sixteenth century.[55] Several decades later, Georg Gutke (1589-1634), the rector of the *Gymnasium des Grauen Klosters* in Berlin, and Jakob Grosse (1592-1652) exchanged a series of polemical writings concerning the pedagogical value of Ramus.[56]

53 Zacharias Ursinus: *Bedencken ob P. Rami Dialectica und Rhetorica in die Schulen einzuführen an Pfaltzgraff Friedrich Churfürsten den III. gestellet ... im 1570. Jar / als Ramus solches in der Universitet zu Heidelberg zuthun fürhett,* Gedruckt zu Franckfurt an der Oder: Durch Andream Eichorn 1586 [Wolfenbüttel HAB: 50.10 Pol. (13)]. Concerning Ursinus refer to the following: *Lexikon für Theologie und Kirche* 10, Freiburg: Verlag Herder 1965, p. 571; ADB 39, pp. 369-372; Jöcher 4, pp. 1741-1742.

54 *Bedencken ob P. Rami Dialectica und Rhetorica in die Schulen einzuführen / Gestelt durch Zachariam Ursinum ... Sampt einem andern Bedencken von obgesetzter Fragen gemeltem des Herr Ursini bedencken entgegensetzt. Durch Fridericum Trinaeum Vilnesem. Mit angehengtem kurtzen Sendschreiben und von dieser Uneinigkeit woher sie entstanden und wie sie bis ahero von beider Part gelehrten getrieben sey / etc. Gethan zum Bericht an einen guten Freund durch Casparum Pfaffradium,* Gedruckt zu Erffurd: Durch Esaiam Melchlern in vorlegung Otto von Rißwick 1587 [Bamberg SB: Ph.o.194].

55 Two of the printed works arising out of this exchange at Helmstedt are: Cornèlius Martini: *Ad Conradi Hoddaei ... teretismata quibus ... primae contra Ramistas disputationi de subiecto & fine logicae obstrepit, responsio autoris* (Helmaestadii): (Excudebat Jacobus Lucius) (1597) [Berlin, Pr. SB: Nl 3600]; Antonius Notholdius: *Magistri Cornelii Martini ... Programma cum responsione studiosorum Rameae philosophiae, in Academia Julia: qua studium rameum a reprehensione M. Cornelii vindicatur, & Melanchthoni, nec non optimae institutionis legibus consonum probabur in lucem edita,* Lemgoviae: Apud haeredes Conradi Gronthenii 1597 [Berlin, Pr. SB: Nl 3530 (3)].

56 Jacobus Grossen: *Dissertatio pro P. Rami logica, quibusdam porismatibus nuper contra Ramum, disputatione aliqua, propositis opposita ... Anno 1621. 20. Septemb.,* Magdeburgi: Typis Andreae Bezelij [Hildesheim StB: And. 3328 (4 an)]; Georgius Gutkius: *Amicum cum amico dissidium, hoc est discursus pro logica Peripatetica in scholas reducenda: oppositis duabus pro P. Rami logica dissertationibus, anno 1621, a quodem habitis ... An dialectica seu logica Rami digna sit, quae juventuti commendetur, inculcetur? In posteriore porismata ... examinantur,* Berlini: Typis Rungianis sumptibus Martini Guthij 1622 [Gdańsk, Bibl. PAN: Fa 22132.8° (1 an)]; Jacobus Grossen: *Apologeticum responsum calumniis quibusdam*

Numerous works – written by Lutherans, Calvinists, and Roman Catholics – were published in opposition to Melanchthon's theological views.[57] Yet treatises attacking Melanchthon's writings on the arts disciplines – i.e., his textbooks on physics, the soul, ethics, logic, rhetoric, and grammar – have yet to be discovered. This is in sharp contrast with the frequent publication in late sixteenth- and early seventeenth-century Central Europe of works devoted, entirely or partially, to attacking Ramus' writings on the arts.

At least three reasons can be ventured in accounting for such discrepancy. First, Ramus habitually attacked Aristotle, Cicero, and Quintilian in his writings. The extent to which scholars active during the sixteenth and seventeenth centuries subscribed to the opinions of these ancient authors is debatable.[58] Yet it was generally agreed that it was indecorous to criticize them as openly and violently as Ramus chose to do.[59] Melanchthon, on the

oppositum ..., Rintelii: Typis Petri Lucii typogr. acad. 1623 [Hildesheim StB: And. 3328 (5 an)]; Georgius Gutkius: *Vindiciarium pro discursu logico, repositarum responso apologetico M. Jacobi Grossen &c.,* Berolini: Typis Georgii Rungianis sumptibus Johannis Kallij 1623 [Gdańsk, Bibl. PAN: Fa 22132.8° (2 an)]. Concerning the career of Gutke see Julius Heidemann: *Geschichte des Grauen Klosters zu Berlin,* Berlin: Weidmannsche Buchhandlung 1874, pp. 140-144; a brief biography of Jakob Grosse is given in Jöcher 2, p. 1199.

57 "Am wenigsten konnte sich Melanchthons eigentümliche Theologie durchsetzen. Nachdem man einmal seine Abweichungen von Luther erkannt hatte, war für die lutherische Orthodoxie nur noch die stillschweigende oder auch ausdrückliche Verdammung des Praeceptor Germaniae möglich. Und reformiert war Melanchthon nun wirklich nicht, wenn auch vielen seiner Schüler schließlich doch nur der Übertritt in gemäßigt reformierte Kirchen wie die kurpfälzische übrigblieb. So geriet Melanchthons Theologie zwischen die konfessionellen Fronten." Scheible: Melanchthon Bretten (see note 2) p. 278. Hammer also cites numerous examples of theological writings by both Protestants and Roman Catholics – which are directed in whole or in part against Melanchthon; for example, see Hammer: *Melanchthonforschung* (see note 17) vol. 1, pp. 38-40 (nos. 27-29). 103-104 (no. 115); vol. 3, pp. 49 (no. 97). 66 (no. 115).

58 The following two articles raise this question with respect to the opinions of Aristotle and Cicero: Joseph S. Freedman: Aristotle and the Content of Philosophy Instruction at Central European Schools and Universities during the Reformation Era (1500-1650), in: *Proceedings of the American Philosophical Society* 137, no. 2 (June 1993) pp. 213-253; Idem: Cicero in Sixteenth- and Seventeenth-Century Rhetoric Instruction, in: *Rhetorica* 4, no. 3 (Summer 1986) pp. 227-254.

59 See Murphy: *Ramus' Attack* (see note 9) pp. vii-viii and *Arguments in Rhetoric against Quintilian,* Translation and Text of Peter Ramus' *Rhetoricae Distinctiones*

other hand, not only published commentaries on writings by Aristotle, Cicero, and Quintilian but he also refrained from openly differing from them.[60] Melanchthon was also praised by some late sixteenth-century authors for his pedagogical ability to explain Aristotle's writings.[61]

Second, in late sixteenth- and early seventeenth-century Central Europe Ramus' writings were frequently criticized for relying too heavily on definitions and dichotomies of concepts with little substantive treatment of such concepts.[62] In actual fact, however, the value placed on definitions and dichotomies within academic writings everywhere in Europe cannot be associated solely with Petrus Ramus or with the influence of his writings. Definition theory and classification theory were normally discussed as part of the subject matter of sixteenth- and seventeenth-century logic textbooks.[63] Printed works – both by individuals who supported Ramus and by

in Quintilianum (1549), ed. James J. Murphy, trans. Carole Newlands, Dekalb, IL: Northern Illinois University Press 1992, pp. 1-2. 7. 19-22. 41-44. It should also be noted that sixteenth- and seventeenth-century authors sometimes felt obliged to defend themselves against claims of displaying unwarranted novelty. Refer to the discussion of this point given in Joseph S. Freedman: *European Academic Philosophy in the Late Sixteenth and Early Seventeenth Centuries. The Life, Significance, and Philosophy of Clemens Timpler (1563/4-1624),* 2 vols., Hildesheim, Zurich, New York: Georg Olms Verlag 1988 [Studien und Materialen zur Geschichte der Philosophie 27] pp. 437-439. 727-728.

60 Melanchthon's positive yet critical attitude towards Aristotle is well described in Scheible: Melanchthon *TRE* (see note 2) p. 374.

61 The following examples can be given here: "Erasmus Rudinger Tomo 4. Int. B. 5. b. 9. Aprilis 1559. Explicabimus libellum D. PHILIPPI MELANTH. Praeceptoris nostri, V. CL. quem ex doctrina Aristotelis de Physica collegit, quae in hoc erudite explicatur, cum Aristotelica legere minime sit expedita res, etiam non vulgaris doctrinae: & in tanto numero philosophantium & versantium scripta Aristotelis, qui vere illa & utiliter, atq[ue] erudite explicare poßint, non est valde magnus numerus." Richter: *Criseis Melanchthoniae* (see note 49) p. 657 (also see ibid. pp. 657-660); similar comments with respect to Melanchthon's ability to explain Aristotle's writings on logic are given in Matthaeus Dresserus: *Oratio de dialectica Philippi Melanchthonis,* Lipsiae 1580, sig. A5r-B6r [Saint Louis / Concordia Seminary: Rare Book / LB / 2361 / P 898 (1)].

62 Criticisms by Bartholomäus Keckermann of this effect are discussed in Freedman: Ramus (see note 50) pp. 125. 129-130.

63 Refer to the discussion given in Joseph S. Freedman: The Study of Sixteenth- and Seventeenth-Century Writings on Academic Philosophy: Some Methodological Considerations, in: Freedman: *Philosophy and Arts* (see note 1) pp. 2-7.

others who did not – which totally or partially consisted of dichotomous tables appeared throughout the sixteenth and seventeenth centuries.[64]

Georg Major's synopsis of Melanchthon's rhetoric textbook – which was printed at least fifty times between 1526 and 1561, consisted almost entirely of dichotomous charts.[65] Victorinus Strigel praised Melanchthon's ability to formulate definitions.[66] And early editions of Ramus' writings did not contain the dichotomous tables that later came to be associated with him and his followers.[67] Nonetheless, it appears to have been the writings of Ramus – and not those of Melanchthon – that were most frequently associated with an over-dependence on definitions and dichotomies.

A third reason why Melanchthon's works on the arts escaped the kind of criticism that was leveled at similar works by Ramus, can be attributed to the fact that Melanchthon published on different arts disciplines than Ramus did. During the sixteenth and seventeenth centuries, arts disciplines could include any of the following: metaphysics, physics, mathematical disciplines (e.g., arithmetic, geometry, music, astronomy, astrology), ethics, family life (*oeconomica*), politics, logic, rhetoric, grammar, poetics, history, and mechanical arts (e.g., agriculture, fabric-making, navigation, weaponry).[68] Gram-

64 Numerous examples of such works can be given. For example, see Remigius Rufus Candidus: *Raemundi Lullii ... in rhetoricen isagoge perspicacibus ingeniis expectata* ([Parisiis]: (Sub praelo Ascensiano) (Ad 14 kalendas Decembris 1515) [San Marino, CA: The Huntington Library: 385891]; Caspar Macer: *Artis rhetoricae praeceptiones schematibus* (Augustae Vindelicorum): Philippus Ulhardus 1561 [University of Illinois: uncat.]; (Rodolphus Goclenius): *Resolutio systematis logici maioris (Bartholomaei Keckermanni) in tabellas pleniores, quam antehac fuerint,* Hanoviae: Apud Guilelmum Antonium 1610 [Passau SB: Hi [b] 21 (nr. 3)]; Johannes Stierius: *Praecepta ethicae, sive philosophiae moralis ex Aristotele, aliisque probatis auctoribus collecta, et ad juvandae memoriae causa tabulis synopticis inclusa ... Editio tertia,* Gothae: Excusa typis Petri Schmid 1643 [Wolfenbüttel HAB: O 71. 4° Helmst. (17)]; Kilianus Rudrauffius: *Cursus metaphysicus methodus per tabellas ... redditus,* Giessae Hassorum: Typis Müllerianis, sumptibus Anthonii Fabri 1685 [Wiesbaden LB: Bd 10793 (2)]. Also see the work cited in note 45 above.

65 *VD 16* (see note 2) M 2061-2111.

66 "Victorinus Strigel in Eth. lib. 1. pag. 36 PHILIPPUS fuit artifex in componendis definitionibus." Richter: *Criseis Melanchthoniae* (see note 49) p. 633.

67 Refer to notes 91 and 92 below.

68 For a discussion of this topic, and further references, see Joseph S. Freedman: Classifications of Philosophy, the Arts, and the Sciences in Sixteenth- and Seventeenth-

mar, rhetoric, and logic (i.e., the *trivium*), mathematical disciplines (and especially arithmetic and geometry) and religious instruction formed the core of the Protestant school curriculum in Central Europe from about 1570 to about 1630.[69] During most of the sixteenth century, physics (including discussion of the soul), astronomy, and ethics as well as the advanced study of logic and rhetoric were the principal arts subjects offered in Central European Protestant universities and in university-level instruction offered at many Central European academies.[70] Metaphysics, family life, and optics largely disappeared from the university curriculum in the early sixteenth century and only began to reappear in the early seventeenth.[71]

Melanchthon published textbooks on grammar, rhetoric, and logic as well as on physics, the soul, and ethics. Although his mathematical publications did not extend much beyond a number of prefaces within writings by other authors on astronomy and geometry, he supported the study of mathematics at the University of Wittenberg, and many Wittenberg professors and students went on to teach mathematics elsewhere.[72] Ramus published textbooks on grammar, rhetoric, logic as well as on arithmetic and geometry. Melanchthon's arts textbooks fit admirably into the university-level curriculum in sixteenth-century Central Europe and were suitable for the school curriculum as well.[73] Ramus' arts textbooks, on the other hand, fit

Century Europe, in: *The Modern Schoolman* 72, no. 1 (November 1994) pp. 37-65. Also see Freedman: Aristotle (see note 58) pp. 231-234. 252-253.

69 Detailed discussion of this point is given in Freedman: Ramus (see note 50); also see Joseph S. Freedman: Philosophy Instruction within the Institutional Framework of Central European Schools and Universities during the Reformation Era, in: *History of Universities* 5 (1985) pp. 117-166.

70 See Freedman: Philosophy Instruction (see note 69).

71 Idem; also see Freedman: Ramus (see note 50) p. 142.

72 Compare the discussion in Westman: Melanchthon Circle (see note 3) pp. 170-172 and *Artis divinatricis patrocinia* (see note 26).

73 Melanchthon's works were used to teach grammar, rhetoric, logic, and theology in Lutheran schools at Nuremberg and Flensburg in the years 1575 and 1586, respectively. See Paulus Sperlingius: *Scholae Flensburgensis administratio,* Witebergae: Impressa Matthaeo Welaco 1589, sig. M3r-M5r [Copenhagen, Royal Library: LN 1525]; Ordo lectionum in schola Sebaldina Norimbergae. Conscriptus et oblatus a Paulo Praetorio, rectore eiusdem scholae. Pridie Calendarium Januarij anno ineunte 1575 [Nuremberg SA: RST Nürnberg, A-Laden, Akten S I, L 132, no. 2]. Melanchthon's writings were used at the University of Rostock to teach theology, the soul, and logic in 1565, and theology, logic, history, and rhetoric in 1568, and at the

very well into the Central European school curriculum beginning in the year 1570 but did not cover the principal subjects taught at the university level.[74]

This appears to be one important reason why Melanchthon's writings were widely used at the university level while Ramus' were not. Much of the polemic directed against Ramus' writings on the arts arose in connection with attempts to introduce his works into the university-level arts curriculum in Central Europe.[75] This arts curriculum largely consisted of instruction in the philosophical disciplines. From the second half of the sixteenth century through most of the seventeenth, philosophy generally included theoretical philosophy (metaphysics, physics, and mathematics) and practical philosophy (ethics, family life, and politics); philological disciplines (logic, rhetoric, and grammar) were sometimes included as well.[76] Ramus did not publish textbooks on any of the disciplines belonging to theoretical or practical philosophy and, consequently, his writings – as opposed to those by Melanchthon – were criticized primarily because they did not fit into university-level philosophical instruction.[77]

Still, the writings of both Ramus and Melanchthon were often used in tandem within academic writings on logic, rhetoric, and grammar authored by Central European Protestant pedagogues. The decades between 1580 and 1620 saw the frequent publication and utilization of textbooks – often referred to as "Philippo-Ramist" – which sought to utilize the writings of Ramus and Melanchthon or excerpt/adapt them.[78] Such works usually looked for middle ground between – or at least sought to present side-by-side – the positions of

University of Jena to instruct rhetoric, logic, history, ethics, physics, the soul, and theology in 1564, and rhetoric, logic, ethics, and the soul in 1592; see Freedman: Philosophy Instruction (see note 69) pp. 156. 160.

74 This is one of the central arguments presented in Freedman: Ramus (see note 50).

75 Ibid. pp. 138-140.

76 Refer to Freedman: Classifications (see note 68); Freedman: Philosophy Instruction (see note 69) pp. 117. 125-129.

77 Freedman: Ramus (see note 50) p. 139.

78 Some examples of such works are cited in ibid. pp. 117 (1a. 2b. 4a. 4b). 146-148. The term "Philippo-Ramist" (*Philippo-Rameus*) was also current during that period; it is used as a subject-category in an early seventeenth-century bibliography of philosophical works. See Paulus Bolduanus: *Bibliotheca philosophica, sive: elenchus scriptorum philosophorum atque philologorum ... usque in annum ... 1614 descripserunt ... digesta,* Jenae: Apud Joannem Weidnerum impensis haeredum Thomas Schureri bibl. Lips. (Typis Johannis Weidneri) 1616, pp. 121-123 [Hannover LB: Ba = A 123].

both men and/or of their disciples on selected points of doctrine from the disciplines of logic, rhetoric, and grammar. During the same period, certain curricular plans for Central European schools mention both Melanchthon and Ramus; this was the case for both Lutheran and Calvinist schools.[79]

Sixteenth- and seventeenth-century authors generally were eclectic and independent when selecting and using authorities either for instruction or publication.[80] Thus Melanchthon's and Ramus' works were often utilized in conjunction with some additional authority or authorities. Some textbooks and curriculum plans purported to blend both Aristotle and Ramus.[81] In his textbook on "Melanchthonian" logic published in the year 1610, Andreas Cramer stated that he has joined "Ramist" method with "Aristotelian" order.[82]

79 The use of Melanchthon's and Ramus' writings at both Lutheran and Calvinist schools in Central Europe is documented in Freedman: Ramus (see note 50); his writings on the arts were printed in many European cities (e.g., Venice, Lyon, and Paris) for Roman Catholic as well as for Protestant readers. The use of Ramus' writings in Roman Catholic schools was also mentioned in ibid. pp. 124. 128. Caspar Macer, professor of rhetoric at the (Roman Catholic) University of Ingolstadt from 1559 to 1564, cites Ramus within his rhetoric textbook; see Macer: *Artis rhetoricae praeceptiones* (see note 64) sig. A2v-A4v; concerning Macer see Jöcher-Adelung 4, p. 309 and Horst Peter Gerber: *Die Professoren der philosophischen Fakultät der Universität Ingolstadt vor und nach ihrer Übernahme durch die Jesuiten. Ihre Schriften,* M. D. thesis, University of Erlangen-Nuremberg 1974, pp. 77-80.

80 For example, the eclectic use of Aristotle's writings by Clemens Timpler and Michael Coelius in the early seventeenth century is discussed in Freedman: Aristotle (see note 58) pp. 230-231. 250-252. The eclectic manner in which sixteenth and seventeeth century academics utilized Cicero's writings is highlighted in Freedman: Cicero (see note 58).

81 One such textbook was Amandus Polanus von Polansdorf: *Syntagma logicum Aristotelico-Ramaeum ad usum inprimis theologicum accommodatum. Cui synopsis totius logicae, vice tabularum praemissa est,* Basileae: Typis Conradi Waldkirchii 1605 [Amberg SB: Philos. 663]; one such curriculum plan was *Index lectionum, hac hyeme {aestate} habendarum in illustri schola Nassovica Herbornensi publica ... in utroque paedagogeo, Herbornensi et Sigenensi ... Inchoabuntur lectiones Calend. Octobr. Anno 1610 {Cal. Aprilis 1611}* [Wiesbaden SA: Abt. 95, Nr. 1750[2] [anno 1610 and 1611].

82 Cramer begins his introduction of this textbook on logic with the following sentence: "Quemadmodum in Logicis praelectionibus nostris methodi Rameae vestigia hactenus legimus: sic in posterum ad Aristotelis ordinem nos convertemus, &, cum eodem P. Rami placita convertemus." Andreas Cramerus: *Comparationum logicarum libri quatuor, dialecticae Philippeae examussim respondentes a ... Christiano Becmanno Hannoverano quondam rectore publice praelecti, hactenus suppreßi,*

Furthermore, Conradus Bergius published an "Aristotelian-Lullian-Ramist" logic in the year 1615 while Johann Heinrich Alsted (1588-1638) and Sigmundus Coccejus published a disputation in 1610 which was titled as a "Mauritian-Melanchthonian-Ramist" compendium of Latin grammar.[83] And Johannes Brunnemann published a logic textbook in the year 1639 which – to judge by its title – was based primarily on Aristotle and Philipp Melanchthon.[84]

But more importantly, most authors who published textbooks, disputations, and other academic writings during this period exhibited a large degree of independence, as far as the form and content of these writings are concerned. Works on the same subject matter (e.g., Aristotle's logic or Ramus' logic) could be organized in diverse forms; the individual points of doctrine contained therein (e.g., the concept of method) were frequently discussed eclectically without clear and total dependence on Aristotle, Cicero, Melanchthon, Ramus, or any other authority.[85]

Numerous examples of such eclectic writings can be given. Among Friedrich Beurhaus' publications were a synopsis on Ramus' rhetoric, a commentary on Ramus' logic, and a defense of Ramus' logic.[86] However, he also published a logic textbook which intended to find consensus between the individual opinions of Ramus and Melanchthon, as well as a commentary on Cicero's logic which also included material from Aristotle's and Boethius' logic.[87] In 1608 Severinus Sluterus (1571-1648) pub-

nunc in usum schola Quedlinburgenis editi (Magdaeburgi): (Typis Andreae Bezelii sumtibus Levini Braunß bibliopolae) (1610) p. 1 [Hannover LB: P-A 118].

83 These works are cited in Freedman: Ramus (see note 50) pp. 117 (no. 1b). 118 (no. 7). 145.

84 Johannes Brunnemannus: *Enchiridion logicum ex Aristotele & Philippo potissimum ita concinnatum,* Francofurti: Impensis Melchioris Closemanni bibliop. 1639. [Wolfenbüttel HAB: 20 Log.]. Also see Johannes Scharfius: *Manuale logicum, ex Aristotele & Philippo Melanchthone, communi Germaniae praeceptore, ita concinnatum ... Editio quinta,* Wittebergae: Impensis Balthasaris Mevii bibliop. typis Johannis Haken 1647 [University of Illinois: x160 / Sch 17m5].

85 Concerning the eclectic use of Aristotle's and Cicero's works see Freedman: Aristotle (see note 58); Idem: Cicero (see note 58). Some eclectic discussions of method utilizing writings by Ramus and Melanchthon are presented in Freedman: Ramus (see note 50) pp. 106-111.

86 See a short list of Beurhaus' writings in Freedman: Ramus (see note 50) p. 145 as well as Fredericus Beurhusius: *Ad P. Rami dialecticam variorum et maxime illustrium exemplorum naturalis artis progressu, inductio ... quae Paedagogiae logicae pars tertia,* Coloniae: Apud Maternum Cholinum 1583 [Darmstadt LHB: U 358].

87 See Freedman: Ramus (see note 50) p. 145.

lished a commentary on Ramus' logic in which he incorporated views of Aristotle; in 1610 he published a commentary on Aristotle's logic in which he compared views of Aristotle and Ramus on various points of doctrine.[88] And Rudolph Goclenius (1547-1628) edited a commentary on Ramus' logic by Christoph Cramer, in which opinions of Jacopo Zabarella, Jakob Schegk, Matthias Flacius Illyricus Jr., Friedrich Beurhaus, Johannes Piscator, Rodolphus Snellius and Goclenius himself, were also discussed.[89]

Any attempt to assess the influence of Philipp Melanchthon and Petrus Ramus must take the following variable into account: the form and content of the various writings by both men did not remain fixed in the decades following their initial appearances. Nelly Bruyere's study of Ramus' logic has traced the evolution of that work over the course of 25 years following the appearance of the first edition in the early part of 1543.[90] That first edition was in prose form, utilizing lengthy sentences while lacking chapters, sections, or other page breaks.[91] This fact – applicable to some of Ramus' other early writings as well – may have prompted Melanchthon's criticism.[92] However, later editions of his textbooks on logic and on other subjects generally saw the introduction of sections – usually in the form of books (*libri*) and chapters – as well as the occasional inclusion of tables of contents and/or subject indexes or dichotomous charts.[93]

88 Ibid. pp. 118 (no. 6a-6b). 150; a short biography of Severinus Sluterus is given in Jöcher 4, p. 640.

89 Ibid. pp. 118 (no. 8a). 150-151; concerning Rudolph Goclenius see the brief discussion given and literature cited in Freedman: *European Academic Philosophy* (see note 59) pp. 92. 524.

90 This study by Bruyère is cited in note 28.

91 See Ramus: *Dialecticae partitiones* (see note 7). Also refer to Murphy: *Arguments in Rhetoric* (see note 59) p. 17; Ong: *Ramus and Talon* (see note 2) pp. 48-49.

92 Ramus' writings attacking Cicero's and Quintilian's rhetoric (see notes 9 and 59) were also written in a similar form of prose. Yet it should be noted that Ramus was not the only academician active during the mid-sixteenth century who wrote in this manner; for example, see the following writings: (Jacobus Schegkius): *Erotemata in physica Aristotelis,* Tubingae 1538 [Mainz, Bibliothek des Priesterseminars: 7 / 69 (3)]; Sebastianus Foxius Morzillus: *Ethices philosophiae compendium, ex Platone, Aristotele, alijsque ... authoribus collectum,* Basileae: Per Joannem Oporinum (ex officina Joannis Oporini) (1554 mense Martio) [Berlin, Pr. SB: Np 2348].

93 This point can be documented by looking at the chronological span of Ramus editions on logic and on other subjects which are cited in Ong: *Ramus and Talon* (see note 2).

Walter Ong has noted that both Melanchthon and Ramus revised their respective textbooks on logic during the late 1540s in order to add sections on the concept of method.[94] Melanchthon's textbooks on other arts disciplines were either revised or expanded. Joachim Knape, for example, has discussed the early evolution of Melanchthon's textbook on rhetoric, but there is no recent scholarship on the evolution of Melanchthon's textbooks on other disciplines. In the case of many authors who cited writings by Melanchthon or Ramus, it is difficult, if not impossible, to determine which editions they actually used. Furthermore, the numerous commentaries on Melanchthon and Ramus which appeared during the late sixteenth and early seventeenth centuries differed from one another in form and content. Most of these commentaries have gone unstudied for centuries. Yet some of them went through many editions over two or more decades, and their form and content may have changed over time. Commentators on individual works by Melanchthon and Ramus were not necessarily in agreement on all points. And some Melanchthon commentaries essentially were – notwithstanding the appearance of his name on the title pages – independent works.[95]

The path that began with the initial comments by Ramus and Melanchthon concerning one another and which evolved into the utilization of their writings by authors of subsequent decades is not an easy one to follow. Melanchthon's criticism of Ramus apparently did not result in a controversy between the two men. Ramus, who spoke very highly of Melanchthon in 1569, most likely never learned what Melanchthon had said about him twenty-six years earlier. Melanchthon probably never made public his remark concerning Ramus, and he perhaps was embarassed or angered at how inaccurate his assessment of Ramus' professional future actually turned out to be.

This letter containing Melanchthon's criticism of Ramus provides an interesting, albeit brief, glimpse at a previously neglected aspect of Melanchthon's thinking. But assessing such criticism within a historical context does not lead to any exciting revelations. The subsequent relationship be-

94 Ong: *Ramus, Method* (see note 2) p. 236.

95 For example, the following "Melanchthon commentary" is essentially an independent work: Gregorius Bersmannus: *Erotemata dialectices Philippi Melanchthonis, collectaneis novarum quaestionum, ex Aristotelis aliorumque virorum doctiß. scholis, illustrata ac locupletata,* Servestae: Excudebat Bonaventura Faber 1593 [Wolfenbüttel HAB: O 75. 8° Helmst.]; the same is true of Lucas Lossius' *Annotationes in grammaticen D. Philippi Melanthonis latinam* (see note 43).

tween the writings of Melanchthon and Ramus can best be described both as quite complex and as generally harmonious.

Numerous editions of writings by Melanchthon and Ramus appeared well into the seventeenth century. During the late sixteenth and early seventeenth centuries, Central European teachers of the arts made eclectic use in their instruction of writings by Aristotle, Cicero, Melanchthon, Ramus, and other authors. In publications on grammar, rhetoric, and logic that arose from such instruction, both Calvinist and Lutheran authors found it useful to utilize the writings of Melanchthon and Ramus in concert with one another. Yet these authors generally did so in ways that were eclectic and multifaceted. Finally, the actual relationship between Melanchthon and Ramus became blurred as well as increasingly remote, as succeeding generations of authors used, interpreted, and reinterpreted the writings of the two men.

*****

Library and archive locations are given – together with call numbers – for all the materials cited in this article which originated before the year 1801. The following abbreviations are used for this purpose: Bibl. PAN = Library of the Polish Academy of Science; HAB = Herzog August Bibliothek; LB = Landesbibliothek; LHB = Landes- und Hochschulbibliothek; Pr. SB = Staatsbibliothek Preußischer Kulturbesitz; SA = Staatsarchiv; SB = Staatsbibliothek; StB = Stadtbibliothek; UB = Universitätsbibliothek; University of Illinois = University of Illinois, Urbana-Champaign, University Library / Special Collections.

The following *abbreviations* are used to refer to biographical encyclopedias cited within this article: ADB = *Allgemeine Deutsche Biographie,* 56 vols., Leipzig: Duncker und Humblot 1875-1912; Jöcher = Christian Gottlieb Jöcher: *Allgemeines Gelehrten-Lexicon,* 4 vols., Leipzig: J. F. Gleditsch 1750-1751; Jöcher-Adelung = Johann Christoph Adelung, Heinrich Wilhelm Rotermund: *Fortsetzungen und Ergänzungen zu Christian Gottlieb Jöchers allgemeinem Gelehrten-Lexico,* Leipzig et al. 1784-1897; Zedler = (Johann Heinrich Zedler): *Grosses vollständiges Universal-Lexicon aller Wissenschaften und Künste,* 64 vols., 4 supplementary vols., Halle, Leipzig: Johann Heinrich Zedler 1733-1754.

Ramus' Metaphysics and its Criticism by the Helmstedt Aristotelians

RICCARDO POZZO

The object of this paper is the incidence of the metaphysics of Petrus Ramus in the late sixteenth century. In order to develop the terms of the discussion properly, however, it is necessary to begin with a short exposition of Philipp Melanchthon's views on the relation between logic and metaphysics. In the decades before and after the *Formula of Concord* some German philosophers declared themselves to be Philippists, others Ramists, others Philippo-Ramists. It was not until the Peripatetics, guided by Melanchthon's pupil Johann Caselius, came back in full vigor at the *Academia Julia* in Helmstedt that clarity was to be made about the actual extent of Ramus' innovation in metaphysics and the necessity of its critical appraisal from the background of the Aristotelian tradition.

1. Melanchthon on Logic and Metaphysics

Philipp Melanchthon reformed the university curriculum in Protestant Germany. Like many other Renaissance philosophers, Melanchthon felt a profound need of harmonizing Plato and Aristotle, whereby the latter, for clarity, should be considered before the former, for depth.[1]

1 Cf. Melanchthon: *De philosophia* (1537), in: *Opera quae supersunt omnia,* vol. 11, ed. Karl Gottlieb Bretschneider, Halle: Schwetschke 1843, repr. New York, Frankfort: Johnson and Minerva 1963, p. 423: Id certamen diremit Theodorus Gaza, inquiens suum cuique locum tribuendum esse: ita lectionem Platonis multum profuturam esse, si quis in Aristotele recte institutus, postea Platonem legat. Nam cum afferet lector Aristotelicam methodum, facile quasi intra certas metas includet eas res, quae apud Platonem late dissipatae sunt.

Wilhelm Risse highlighted a development in the three editions of Melanchthon's dialectic, from a Ciceronianism laced with Neoplatonism to a progressive recognition of the primacy of Aristotle.[2] But Melanchthon's Aristotle, too, Wilhelm Maurer stressed, "should be interpreted Platonically – which meant, like in the Middle Ages Neoplatonically".[3] Melanchthon's main contribution, according to Cesare Vasoli, was to join together, in a new methodical unity, the Ciceronian doxastic topic, the Stoic distinction between *systema et ordo* and the Aristotelian theory of demonstration.[4] The rediscovery of the "metaphysical content" in Melanchthon's philosophy has been one of the merits of Walter Sparn's book on the *Wiederkehr der Metaphysik*. According to Sparn, Melanchthon's exhortation *ad res ipsas* implies a comprehensive metaphysical project aimed at retranslating Aristotelian realist metaphysics in a system of ideal relations held together by mental procedures.[5] More recently, Ulrich-Gottfried Leinsle interpreted Melanchthon's logic as "supported by metaphysical elements".[6]

In fact, Melanchthon, like most Renaissance philosophers, completely neglected the systematic role played by metaphysics, arguing that: (a) logic can take the place of metaphysics with regard to assessing principles, (b) metaphysics is not only useless with regard to the concept of God, it is pernicious, (c) the main advantage of metaphysics is to be seen for grammar, if one just thinks about the thirty-two definitions of logical concepts given by Aristotle in *Metaphysica* Δ or about the criteria for truth and certainty in Λ.[7] On the other side, there is according to Leinsle a "real reference" in Melanchthon's logic in as far as it is aimed at "comprehending the *ordo rerum,* and individuating those *genera rerum* which permit to distinguish among the sciences".[8] In other words, Melanchthon's logic is aimed at a

2 Wilhelm Risse: *Logik der Neuzeit, Bd. 1: 1400-1650*, Stuttgart - Bad Cannstatt: Frommann-Holzboog 1964, p. 83ff.

3 Wilhelm Maurer: *Der junge Melanchthon zwischen Humanismus und Reformation,* Göttingen: Vandenhoeck & Ruprecht 1967, p. 92.

4 Cesare Vasoli: *La dialettica e la retorica dell'Umanesimo*, Milan: Feltrinelli 1968, p. 250.

5 Walter Sparn: *Wiederkehr der Metaphysik,* Stuttgart: Calwer Verlag 1976, p. 109.

6 Ulrich-Gottfried Leinsle: *Das Ding und die Methode*, Augsburg: Maro 1985, p. 11.

7 Ibid.

8 Ibid. pp. 13-14.

conceptual comprehension of being and therefore of categories and universals as well. The appropriate tool for this purpose is the doctrine of the *loci communes*. The goal of logic consists in the individuation (*inventio*) and in the preparation for discussion (*iudicium*) of a system of epistemological relations (*loci*). With regard to the *loci communes,* Plato furnishes the material contents, and Aristotle the formal requirements. And it is no wonder that although Melanchthon, following Aristotle, considers the universals (*praedicabilia*) in a nominalist fashion as *gradus vocum, voces,* or as *tituli vocum communium,*[9] at the same time, he holds to the real reference of the categories.

Among the categories are to be taken *omnes voces et res,* so that eventually the whole *rerum natura* would be distinguished by means of the categories.[10] Of course, the real reference of the *loci* to the *rerum natura* necessarily presupposes the world-view of Plato's ontologically grounded dialectic. Plato's theory of knowledge in *Respublica* Z and E, and the theory or the *genera summa* in *Sophistes* – not Aristotle's *Organon,* not his *Metaphysica* – are decisive for Melanchthon's *loci communes.* Let us consider briefly Melanchthon's position about the problem of the universals. At a first glance, it seems as if Melanchthon would maintain the nominalistic approach aiming at reducing the universals to mental acts: "This form is not a thing outside the understanding, but rather the act of understanding itself, the act painting that image."[11] At closer look, however, this statement turns out to prefigure the notion of a subjectivity constituted by means of language, which fulfills the requirement of really referring to *rerum natura* as carrier and user of mental structures. We see here the emergence of the possibility of substituting either *system* or *habit* for logic.

9 Melanchthon: *Erotemata dialectices* (1547), in: *Opera quae supersunt omnia,* vol. 13, ed. Karl Gottlieb Bretschneider, Halle: Schwetschke 1846, repr. New York, Frankfort: Johnson and Minerva 1963, p. 518.

10 Ibid. pp. 526-527.

11 [...] non est quiddam extra intellectionem [...] est revera actus intelligendi, pingens illam imaginem in mente, quae ideo dicitur communis, quia applicari ad multa individua potest. Melanchthon: *De philosophia* (as note 1) p. 520.

2. Ramus on Logic and Metaphysics

Petrus Ramus developed his positions following a pattern similar to that of Melanchthon. In the several editions of his dialectic one can note a tension between the Platonism of his early years and the gradual shift to Aristotle in the later years. Ramus' aim was to renew Aristotelian logic by referring to Platonic principles in order to restore the "genuine" Aristotle, i.e., to restore Aristotelian logic out of its Platonic nucleus.

The dichotomy between Platonism and Aristotelianism has been variously assessed by scholars. Walter J. Ong stressed the importance of Ramus' Aristotelianism, considering his initial Platonism as not genuine.[12] Wilhelm Risse traced a progression from Platonism to Aristotelianism between the 1543 edition of the *Dialectica,* that of 1555 and that of 1572, involving a gradual but complete shift from an initial conception of a metaphysically grounded logic to an acceptance of logical formalism.[13] Cesare Vasoli interpreted Ramus' development as a continuous debate over the Aristotelian-Galenic theory of method with the aim of providing a new kind of epistemology that would take into account both Platonic-Stoic and Aristotelian-scholastic issues.[14] Wilhelm Schmidt-Biggemann drew attention to Ramus' role in setting up a new model of science centered on the idea of a *topica universalis.*[15] *Pace* Ong, Nelly Bruyère argued for the continuous presence of a Platonic foundation in the idea of a *signatura rerum* from the biblical *Urim-Thumim* and Plato's theory of ideas,[16] while Ulrich-Gottfried Leinsle pointed out several elements in Ramus indicating an implicit acceptance of Aristotelian-scholastic metaphysics.[17]

Leinsle begins by admitting that Ramus presented himself as an anti-metaphysician. In the famous quote from the preface of the *Scholae in liberales artes:* "Fourteen metaphysical books are fourteen heaps of tautolo-

12 Walter J. Ong: *Ramus, Method, and the Decay of Dialogue,* Cambridge: Harvard University Press 1958, p. 208.

13 Risse: *Logik der Neuzeit* (as note 2) p. 124.

14 Vasoli: *Dialettica e retorica* (as note 4) pp. 3-4.

15 Wilhelm Schmidt-Biggemann: *Topica universalis,* Hamburg: Meiner 1982, pp. 59ff.

16 Nelly Bruyère: *Méthode et dialectique dans l'œuvre de La Ramée,* Paris: Vrin 1984, pp. 110-111.

17 Leinsle: *Ding und Methode* (as note 6) p. 22.

gies, I declare […] the question is whether the whole of Aristotelian metaphysics were nothing else than logic obscured by several logical and certain theological sophisms. This is what I principally proposed myself to discuss and demonstrate."[18]

Nonetheless, it is surprising that Ramus actually wrote a book on Aristotle's metaphysics, the *Scholarum metaphysicarum libri quatuordecim* (1566). In fact, according to Ramus, scholars could manage without metaphysics, given that its contents are likely to be dealt with in the domains of logic and theology. Aristotle introduced metaphysics and called it *sophia*. The fact is, though, that the true *sophia* belongs to the realm of dialectic. One may ask what warranted Ramus' dissolution of metaphysics into dialectic? According to Leinsle, Ramist dialectic *is* metaphysics. Dialectic for Ramus is "*sapientia katholikotate* [...] insofar as only the dialectic teaches about the causes of being, which are common to all arguments and judgments [...]. Dialectic deals truly with the essence common to all beings; the single substances belong in the single disciplines, among which is also theology. Dialectic appears thus to take the place of the classical *prima philosophia* [...] as a normative epistemology."[19]

Ramus does not need any metaphysics because his logic is already metaphysics. This very simple point was highlighted at the end of the sixteenth century by the Helmstedt Aristotelians. They saw that the function taken over by Ramus' dialectic as a unitary fundamental-science presupposes the dismissal of the Aristotelian distinction between *intentio prima* and *intentio secunda.* According to Leinsle, Ramus shows a "rare mixture of nominalism and Platonism", the former referred to the formal structure of *praedicabilia* and *praedicamenta,* the latter to the topical approach, "which presupposes a grounding in the *dialectica naturalis,* in that faculty, i.e., in that light given by God to the human mind, which enables him to

18 Quatuordecim metaphysicos libros, quatuordecim logicarum tautologiarum cumulos esse statuo [...] quaeraturque utrum tota Aristotelis metaphysica nil aliud sit, quam logica logicis plurimis et theologicis quibusdam sophismatis obscurata: Id enim est, quod praecipue disserendum, demonstrandumque mihi proposui. *Scholae in liberales artes,* Basle: Wechel 1569, repr. ed. Walter J. Ong, Hildesheim: Olms 1970, praef., sig. Nn2r.

19 Leinsle: *Ding und Methode* (as note 6) p. 22.

overcome the difficulties of his everyday life".[20] And this holds true not only in the explicitly Platonic first edition of Ramus' logic but applies to the 1555 edition as well, where Ramus "underlines the naturalness of his unitary method, thus anchoring it to a revisited understanding of the Stoic human mind."[21] There is much to recommend in Leinsle's interpretation. It is difficult, though, to trace a line from the Stoics to Ramus. Gabriel Nuchelmans succeeded in finding important results with regard to the theory of judgment.[22] However, this is not the subject of this paper. True, the modern concept of subjectivity found its roots in Ramus but it needed Bartholomäus Keckermann to make it explicit.[23] In the following section I will argue against Leinsle that Ramus did not intend to justify the real reference of logic and, therefore, of metaphysics, by referring to the human mind. Rather, Ramus intended to go beyond Aristotelian semantics, and in order to do so, he developed a special operator of humanistic logic, *thema*. Human mind plays a role, according to Ramus, in so far as it is up to the subject to thematize the domain he is referring to and the related operative conditions.[24]

3. Philippists

Among the immediate disciples of Melanchthon, it is David Chytraeus who may be credited with founding the philosophical school of the Philippists. This had been acknowledged by Johann Hermann von Elswich as early as 1720.[25] Chytraeus' *Regulae studiorum* were published twice, in 1563 and in 1595, and had a noteworthy influence. It is interesting to remark that Chytraeus, following Melanchthon's critique, leaves no place for

20 Ibid. p. 24.

21 Ibid. pp. 23-25.

22 Gabriel Nuchelmans: *Judgment and Proposition: From Descartes to Kant,* Amsterdam: North Holland 1983, pp. 9-35.

23 Bartholomäus Keckermann: *Operum omnium quae extant tomus primus,* Geneva: Albertus 1614, I 1, 90.

24 *Scholae in liberales artes* (as note 18) II 7, 49.

25 Johann Hermann von Elswich: *De varia Aristotelis in scholis protestantium fortuna schediasma,* in: Jean de Launoy: *De varia Aristotelis in Academia Parisiensi fortuna,* ed. Johann Hermann von Elswich, Wittenberg: Hannauer 1720, p. 39.

metaphysics,[26] and that, in the pages dedicated to logic, he advocates a grounding in nature and exercise, quite analogous to that required by Ramus: "the perpetual and certain precepts of the dialectical art descend, by the heroic makers, partially from that natural light of human reason, and partially from a long observation and examples".[27]

Another important Philippist was Matthias Flacius Illyricus Jr., the son of the famous theologian and antagonist of Melanchthon. Notwithstanding his father's dislike of Melanchthon – and diametrically opposed to Chytraeus – Flacius Jr. was determined to interpret Melanchthon as an Aristotelian, rather than as Ramus' predecessor. Flacius was among the first in Germany to have read carefully the logical writings of Jacopo Zabarella and he was eager to point out the gulf separating Melanchthon from Ramus. At issue was a passage written by Melanchthon in his 1537 *De Aristotele* with reference to a French scholar, which could be (and in fact was), mistaken for an attack against Ramus.[28] In this passage, Melanchthon evi-

26 David Chytraeus: *Regulae studiorum,* Lipsia: Grosius 1595, fol. 76v.

27 [...] praecepta artis Dialecticae, perpetua et certa, accedunt, ab Heroicis artificibus, partim ex illa naturali luce rationis humanae, partim ex longa observatione et exemplis. Ibid. fol. 60r.

28 Cf. Melanchthon: *De Aristotele* (1537), in: *Opera quae supersunt omnia,* vol. 11, ed. Karl Gottlieb Bretschneider, Halle: Schwetschke 1843, repr. New York, Frankfort: Johnson and Minerva 1963, p. 367: "Quod autem Gallus quidam contendit, ne tabulas quidem ex illo Aristotelis naufragio receptas ullas esse, ac prorsus amissos esse Dialecticos libros: deinde hos, qui hoc titulo circumferuntur, rudera continere non artis integrae extructionem, seu, ut ita dicam, aedificium, illum hoc suo somnio delectari sinamus: Nos haec monumenta, cuiuscumque sint, etsi Aristotelis esse multae sunt et graves conjecturae, amemus, et eis fruamur. Et hinc veram methodi formam in omni disputatione sumendam esse, sciamus." This passage was identified as an attack on Ramus by Jacob Martini's *Vernunfftspiegel,* Wittenberg: Selfisch 1618, II, chap. 14, 848. But, of course, this is impossible, because in 1537 Ramus was still very young and had published nothing. The question of the identity of the *Gallus quidam* referred to by Melanchthon in *De Aristotele* remains open. In fact, Melanchthon once actually spoke about Ramus in his letter #3259 of 11 June 1543 to Cunradus Embecanus, cf. Melanchthon's *Briefwechsel,* Bd. 3: *Regesten: 2336-3420 (1540-1543),* ed. Heinz Scheible, Stuttgart - Bad Cannstatt: Frommann-Holzboog 1979, p. 400. In this letter, which is conserved in Washington, D.C., in the Folger Shakespeare Library (F.c. a, MSA 9, #441), Melanchthon considers mainly the publisher's point of view – Embecanus was at that time in Cologne in the house of the publisher Johannes Gymnicus – and asserts that

dently sided with Aristotle. Moreover, the content of his statement could be compared with the famous thesis maintained by the young Ramus that the transmitted doctrines of Aristotle "are either fictitious or false".[29]

Melanchthon's explicit nod toward the Aristotelian *corpus* encouraged the Philippists. Some of them, notably Flacius Jr. (in the last version of his *Opus logicum*), even incorporated the results of Catholic Aristotelians such as Zabarella, thereby effecting a curious mixture between the content and the division of Melanchthon's *Erotemata dialectices* and the definitions and the methods of Zabarella's *Opera logica.* First and foremost, the following definition, which became standard in Protestant Germany: "Logic is an instrument, not a part of philosophy. In fact, logic does not teach knowledge of things, it rather teaches a certain way and a method to reach the science of difficult and hidden things."[30] The important innovation is, according to Flacius Jr., that logic no longer treats *natura rerum,* as Melanchthon maintained, but only problems of knowledge.

The third important Philippist we need to mention is Owen Günther. Professor at Helmstedt, he was the first in Germany to define logic as a *habitus organicus.* Günther maintains an ideal and natural grounding for logic. He refers back to Plato's *Phaedrus,* which he considers to provide the correct grounding for any method, "no matter how a definition or a way of explaining is found, it seems hardly difficult to include arranged disciplines by methods, which method or dialectic is to be comprehended under the name of wisdom."[31] Günther's logic is evidently based on metaphysics, and one should therefore agree with Leinsle's stress on Günther's Plato-

(a) Ramus' *Aristotelicae Animadversiones* are void of any scientific value, and (b) that at any rate they did not seem to sell well.

29 [...] quot commentitiae, quot falsae sunt. Ramus: *Aristotelicae Animadversiones*, Paris: Wechel 1543, repr. Stuttgart - Bad Cannstatt: Frommann 1964. Quoted in Risse: *Logik der Neuzeit* (as note 2) p. 123.

30 Logica instrumentum est, non pars philosophiae. Non enim logica rerum cognitionem tradit, sed viam quandam et methodum ad adipiscendam rerum difficilium abditarumque scientiam. Matthias Flacius Illyricus Jr.: *Opus logicum in Organum Aristotelis,* Rostock: Miliander 1593, p. 4.

31 [...] quamvis definitionem seu modum explicandi inventas, constitutasque disciplinas Methodis vix difficile annumerari velle videatur, quod Methodus seu Dialectica, sapientiae nomine comprehendat. Owen Günther: *Methodorum tractatus duo,* Helmstedt: Lucius 1586, fols. 11v-12r.

nism. In fact, according to Günther, logic is not only a *habitus organicus,* but a *habitus princeps* as well, and refers to an ideal grounding: "But logic is said to be the habit of inducing the true, which, by means of investigating the proportion of things, after having artificially begun from the same sensible forms, and thereafter progressing to the intellectual ideas, leads the mind to the nature of what is unknown, and therefore Plato writes in the Philebus [16c] that logic was once carried off to humans by Prometheus, together with the most bright fire."[32]

4. Ramists

A not inconsiderable number of German philosophers declared themselves to be Ramists during the two decades between the *Formula of Concord* and the turn of the seventeenth century. Nathan Chytraeus, David's brother, was one such person but, unfortunately, as far as I know Nathan Chytraeus did not express an opinion on the subject of logic and metaphysics. Other Ramists included Friedrich Beurhaus, Hermann Nicephorus, Georg Busing, and Carl Bumann. Ramism was very popular among the students – Heizo Buscher, one of the leading Ramists, was a former student of the *Alma Julia* – but virtually no faculty member taught Ramist philosophy *ex professo.* Such instruction was provided, on the whole, by untenured instructors such as Otto Casmann, who criticized Melanchthon, or Anton Nothold, who taught Ramist logic without even referring to Melanchthon. Caspar Pfaffrad, too, the assistant of the renowned Daniel Hoffman, taught successfully private courses on Ramist logic at Helmstedt from 1588 until 1592. Other Ramists included Günther Adam and Christian Beckmann.

Confronted with such popularity, the Aristotelian professors at Helmstedt were forced to exhort the youth, as Cornelius Martini did in 1597: "I exhort you, young students of logic, to love the opinion of such eminent men [Melanchthon and Zabarella], and to diligently cultivate Aristotelian phi-

32 Is ipsa autem veri inducendi habitus, *logike* dicitur, quae rerum logous inquirendo, artificiose sumpto initio ab ipsis sensibilibus formis, indeque ad intellectuales quae Ideae dicuntur progrediendo mentem ad cuiusque ignoti naturam perducit, ideoque Plato in Phileb. [16c] scribit, eam per Prometheum aliquando de coelo ad homines latam esse, cum lucidissimo igne. Ibid. sig. B3v.

losophy; and to be entirely persuaded that those [Beurhaus, Buscher, Nothold] who try to make Aristotelian philosophy suspect to the youth, mock the age and the inexperience of the youth, and behave like the fox with the cudgeled tail of the fables, which believed to be alone misshapen. Do not think that philosophy, such a divine good, consists of childish trivia. And who does not appeal to children more than Ramus?"[33]

5. Philippo-Ramists

A consistent program aimed at combining together the philosophies of Melanchthon and Ramus was proposed by the group of the so-called Philippo-Ramists. One has to refer especially to Heizo Buscher's textbook, *Harmonia logicae philipporameae* (1597). But the idea originated with Friedrich Beurhaus' collation of Melanchthon's and Ramus' logical positions a decade earlier.

According to Buscher, the harmony he was speaking about found the following systematic placement. On the one hand, the Philippo-Ramists denied that metaphysics had an important role to play in philosophy: "neither is metaphysics a peculiar science – it is rather a certain extraordinary commentary – nor are categories (which are the distillate of certain artificial concepts) or beings as beings proper to the consideration of metaphysics, because nothing remains to be considered in those explained by the arts."[34]

33 Vos autem Logicae studiosi iuvenes hortor, ut de consilio tam praestantium virorum [Melanchthon and Zabarella], ametis, et diligenter colatis Aristotelicam philosophiam: omninoque vobis persuadeatis, eos [Beurhaus, Buscher, Nothold], qui hanc iuventuti suspectam reddere conantur, aetati et imprudentia eius illudere, et hoc agere, quod in fabellis, vulpes cauda mulctata, ne sola deformis esset. Nolite existimare philosophiam tam divinum bonum, pueriles esse ineptias: et cui magis quam pueris placet Ramus? Cornelius Martini: *Programma cum responsione studiosorum Rameae philosophiae,* Helmstedt: Grothenius 1597, sig. B8v.

34 [...] nec Metaphysica est scientia peculiaris, sed extraordinaria quaedam commentatio, nec Categoriae (quae sunt peculiaris noematum artificiosorum digestio) aut entia, qua entia, sunt Metaphysicae considerationis propria: quod in illis per artes explicatis nihil considerandum reliquum maneat. Friedrich Beurhaus: *Cornelii Martini Antvverpii adversus Ramistas disputatio de subjecto & fine logicae,* Lemgo: Holstenius 1596, sig. C2v-C3v.

On the other hand, the Platonic grounding of the logic was implied by the fact that logic is an art, and art "is a true knowledge of things to be done, like architectonic knowledge […]. In fact, in the order of nature, art is prior to the habit of the artist."[35] This definition of *habitus* goes back to Melanchthon's *Erotemata dialectices,*[36] which was rather obvious. The interesting problem was given by the alternative to *habitus,* i.e., *system.* Because, a *system,* according to the definition by Zeno of Citium reported by Lucianus,[37] is like an "art or order of certain propositions, learned by exercise, toward a goal *useful* in life", as Melanchthon used to translate it, implying a reference to the Platonic theory of ideas.[38]

6. Peripatetics

The Peripatetics established themselves in Helmstedt around 1590, i.e., after Johann Caselius had been appointed professor of eloquence and ethics at both the Arts and Law Schools. Caselius urged his followers and pupils to devote themselves to logic and metaphysics. These included Albert Clamp, Duncan Liddel and, especially, Cornelius Martini. Caselius was also among the first in Germany to perceive the necessity of a thoroughgoing controversy against Ramism. He advised his followers to refer to the philosophical positions of Zabarella, whom he heard of in Bologna, in order to find weapons likely to succeed in combating Ramism.

Martini should be credited with recognizing that the real issue with Ramist logic and metaphysics involved its proponents' crypto-Platonism. He makes this clear when commenting on the Ramist interpretation of the definition of *system* given by Zeno of Citium. "If no man were capable of an art", Martini asks, "and if therefore the art were in no man's mind, where would your arts live? Perhaps near Plato's ideas? Or in Utopia to-

35 […] est notitia vera faciendorum, ut Architectonica […]. Ars enim ordine naturae prior est artificis habitu. Heizo Buscher: *Harmoniae logicae philipporameae libri duo,* 2nd ed., Wittenberg: Holstenius 1597, sig. F6v-F7r.

36 Melanchthon: *Erotemata dialectices* (as note 9) p. 537.

37 *De parasito* IV 2 = STV I 21, 6.

38 […] ars seu ordo certarum propositionum, exercitatione preceptarum ad finem in vitam utilem. Melanchthon: *Erotemata dialectices* (as note 9) p. 537.

gether with your philosophy? If there were no tailor in nature, where would be the art of tailoring?"[39]

Martini's strategy was simple in its efficacy. He taught logic by adopting the first book of Melanchthon's *Erotemata dialectices.* But, in order to distance himself from both Ramists and Philippo-Ramists, he added Zabarella's introductory treatise, *De natura logicae* (which was a sort of prolegomenon to the Italian's *Opera logica,* 1577). An impressive example of this strategy can be found in Martini's lectures on logic delivered between February 1599 and March 1601, which are known as the *Rudimentum logicum.*[40] Indeed, Beurhaus made it also clear that Ramus' approach to philosophy has nothing to do with Aristotle, but rather corresponds to Plato.

The issue was quite important as it involved some very practical needs. For if one accepts that the subject of philosophy is both *ens et non-ens,* one will quickly generate inconsistencies. Actually, the Ramists postulated that the subject of philosophy lies *extra mentem,*[41] whereas the Aristotelians maintained, it is *mentis habitus.*[42] Beurhaus often stated that philosophy is "art […] even if learned by no one",[43] and Martini felt obliged to ask, "what monster can nature not suffer? Such as that the accident were outside of its subject, although the being of the accident is to be inherent?"[44] Ramus' error, according to Martini, consists in having neglected to distinguish between first and second intentions. "Hence was born that horrible confusion

39 Si nullus sit homo capax artis, et per consequens in nulla hominis mente ars, ubi tuae artes habitabunt? an apud ideas Platonicas? vel in Utopia simul cum vestra philosophia? Et si nullus sit sutor in rerum natura, ubi erit ars sutoria? Martini: *Programma* (as note 33) sig. B4v.

40 See Cornelius Martini: *De natura logicae.* Prolegomeni ad un corso di logica del 1599, in: *Rivista di storia della filosofia* 44 (1989) pp. 499-527. The manuscript is now in the Herzog August Bibliothek in Wolfenbüttel, Cod. Guelf. 816 Helmst.

41 Cornelius Martini: *Disputationum logicarum adversus Ramistas prima de subjecto & fine logicae,* Helmstedt: Luchtenius 1594, p. 2.

42 Ibid.

43 [...] artem [...] etiam si a nemine discatur. Ibid. p. 3. Cf. Friedrich Beurhaus: *Dialecticae Petri Rami* [...] *quae Paedagogia logica pars prima,* 3rd ed., Dortmund: Colinus 1587.

44 [...] quale monstrum rerum natura non patitur; ita enim accidens extra suum subiectum sit, cum tamen accidentis esse sit inesse? Martini: *Disputationum logicarum adversus Ramistas prima* (as note 41) p. 3.

between primary and secondary notions in Ramus' logic; a miserable mutilation of most splendid doctrines, which gave us the corpse of dialectic rather than dialectic or logic itself; because Ramus' logic neither leads to that goal nor wishes to deal with anything and in any way."[45] Rather, the correct answer according to Martini (closely following Zabarella), is the following: "all things, certainly not insofar as they are things, but insofar as they subsist by means of second notions, i.e., by means of those notions which serve our cognition to attain the truth."[46] Martini concluded his argument by stating that the Ramists' generalization of the subject of philosophy is false both because it renders impossible any meaningful semantic limitation, and because it involves sophisms: "when one reads that being as being were the subject of metaphysics, one would not know the difference between metaphysics and dialectic, and one would slip into madness by stating that being, as being, were the subject of logic and moreover by putting this dullness on Aristotle."[47]

A close reading of Ramus' metaphysics is provided by Martini in a second disputation, published in 1596. In this text, Martini's charge against Ramus is of having illegitimately confused *praedicabilia* and *praedicamenta*. "In fact, the fourth book of the *Scholae* [*in liberales artes*], chapter 5, after having rejected the ten genera according to substance, quantity, quality, etc., maintains that the genera of logical arguments, causes, facts, subjects, adjuncts, contraries, compared, nominal reasons, divisions, definitions, witness, are the first differences of being and the first species. This is something which is both rotten and absurd, something which would merit severe punishment."[48]

45 Ex horum quippe neglectu turpis illa in Logicis Rami primarum et secundarum notionum confusio nata est; tum miserabilis illa praestantissimarum doctrinarum mutilatio, ut cadaver nobis Dialecticae potius, quam Dialecticam seu Logicam reliquerit; Quia neque in quem finem illa traderet, neque circa quae, et quomodo, versaretur attendit. Ibid. p. 7.

46 [...] res omnes, non quidem quatenus sunt res, sed quatenus secundis notionibus substant, iis videlicet, quae cognitioni nostrae ad veritatem adipiscendam inserviunt, Ibid. p. 8.

47 [...] qui cum legisset, Ens qua Ens Metaphysices subiectum esse, neque differentiam inter Metaphysicen et Dialecticen sciret, eo vesaniae prolapsus est, ut Ens, qua Ens, subiectum Logicae esse statueret, et hanc eandem stoliditatem Aristoteli quoque adfingeret. Ibid. p. 11.

48 Nam quod lib. Schol. 4. cap. 5. reiectis decem generibus substantia, quantitate, qualitate etc. asserit Logicorum argumentorum genera, caussas, facta, subiecta, adiuncta,

Martini refers to Agricola and Melanchthon when he remarks that Ramus simply missed the point, when he set up his *topica universalis.* In fact, "if you examine either Ramus' or Agricola's, or Melanchthon's *loci:* definition, genus, species, diverse, disparate, conflicting, cognate, etymology, whole, part, cause, effect, etc., they are proper or appropriate to any specific thing. Such as only the animal would have that, or only the body, or only the physical thing, or the ethical thing, etc., but the affections of notions of this mode are such that they regard all things or almost certain things or the species of being."[49] Thus Ramus cannot rely on a truly apt and particular science, because "their universality fades away, so that the medium term, which they consider to be proper and appropriate, since by way of the same argument it could be used for very different things."[50] The solution to the question about categories and *loci* cannot be dealt with within logic, as Ramus wishes, but needs a properly scientific treatment: "Three such disciplines you will find apt according to our opinion; the first is the one that considers being as being, and thereafter the affections of being with its principles; the second is the one which forms notions, by means of which one gets to truth in all matters, which is logic; the third is grammar, which concedes the use of topic by itself and also because of some needs."[51] Mar-

dissentanea, comparata, nominum rationem, divisionem, definitionem, testimonium, primas Entis differentias et primas species esse tam est putidum et absurdum, uti severam castigationem mereatur. Cornelius Martini: *Disputationum logicarum adversus Ramistas secunda de constitutione logicae & natura locorum in genere,* Helmstedt: Luchtenius 1596, p. 26.

49 [...] si examines locos, sive Rami, sive Agricolae, sive Philippi: definitio, genus, species, diversa, disparata, pugnantia, comparata, coniugata, etymologia, totum, partes, causae, effecta, etc. non uni alicui rei domestica vel propria sunt. Ut vel solum animal haec habeat, vel solum corpus, vel solae rei physicae, vel Ethicae etc. sed tales sunt affectiones istiusmodi notionis, quae vel omnes vel certae plerasque res sive entis species respiciunt. Ibid. p. 82.

50 [...] universalitas eorum obest, ne medius terminus, qui ab eis sumitur domesticus sit sive proprius: quia eodem argumento etiam in alia diversissima re uti potes. Ibid. p. 83.

51 Tres vero tales reperias huic nostro instituto aptas, unam quae Ens qua Ens considerat, et proinde Entis affectiones cum principiis, quae eo conveniunt, qua est Ens. Alteram quae eas notiones format, quibus in omni re ad veritatem pervenitur, quae logica est. Tertia est grammatica, quae et ipsa de suis quoque opibus aliquid in usum concedit topicae. Ibid. p. 116.

tini makes clear that Ramus illegitimately confused the following three ontological realms: "Here, though, a monster is born. If the *loci* are species of being, and however no beings are contained in the *loci*. No things are or are contained in the *loci*, and however the *loci* are species of being. Just like if you would say: brute and rational are true species of animals, but between brute and rational there is no animal."[52] This is the starting point of every sophistry and skepticism. "Why don't we move to the camp of the Pyrrhonists? Why don't we confess that we are not sure about anything, if we cannot show the connection among necessary principles and conclusions of a certain thing?"[53] Martini's conclusion could not be more effective: "We showed therefore that the *loci* of Ramus and the arguments of these *loci* are of no relevance at all, we do not recognize any invention, and no certain part of dialectic, so that this very division by Ramus, following the understanding by the Ramists, is vicious."[54]

Martini's critique of Ramus made it clear that it was high time to sweep away the ambiguity of Renaissance Platonism. Philosophy was too important to be left in the hands of sophists and skeptics. For this reason, Martini advocated the immediate restitution of metaphysics to Protestant universities. He himself was the first to put this claim into practice, commencing as early as 1597 a course of metaphysics at Helmstedt, which closely followed the model of Francisco Suárez' *Disputationes metaphysicae*.[55]

52 Hic autem mirabile nascitur monstrum. Loci sunt species Entis, et tamen Entia in locis nulla continentur: Nullae res sunt vel continentur in locis, et tamen loci sunt rerum species, Quasi tu dicas: brutum et rationale, veras esse animalis species; sed inter bruta et rationalia nulla tamen esse animalia. Ibid. p. 173.

53 Cur igitur non ad Pyrrhoniorum castra transimus? cur non fatemur nihil certi esse, si nexum principiorum necessariorum et conclusionum inter se de re aliqua ostendere non possumus? Ibid. p. 181.

54 Cum itaque loci illi Ramistici et argumenta istorum locorum, tam sint nullius momenti sicuti demonstravimus, Inventionem hic non agnoscimus, neque aliquam Dialecticae partem, ut ita partitio Rami ista, Ramistico modo intellecta, sit vitiosa. Ibid. p. 240.

55 On some results of Martini's positions on logic and metaphysics in sixteenth century philosophy cf. Jean-François Courtine: *Suarez et le système de la métaphysique,* Paris: Presses Universitaires de France 1990, p. 416.

The Usefulness of Ramism

ERLAND SELLBERG

All humanists expressed their loathing for the scholastics' proclivity for philosophical subtleties. Although Petrus Ramus and his many followers cannot be described as typical humanists, in certain respects they all distanced themselves from medieval university culture and sided with the ideals of the humanists. We have several examples of conflicts in the late sixteenth century between university professors, who were nominally Aristotelians (or rather scholastics), and the Ramists – which sometimes could not be separated from other opponents of philosophy. But by the turn of the seventeenth century the conflicts faded and the Ramist kind of condensed philosophical school system became less valued. Or perhaps it would be more correct to say that a certain congruence occurred; the scholastics conformed to the humanist movement and few seventeenth-century professors could be described as either Ramists or Philippists.

Yet there exists an interesting exception to such a trend. In Sweden, as far as we know, there existed no conflict during the sixteenth century over philosophical teaching, and there are only hints at Ramus' influence. In the aftermath of the Reformation Sweden had rather poor schools while the university was closed for many years. In the first half of the seventeenth century, on the other hand, we find a bitter and lengthy conflict between Aristotelians and Ramists, which in some ways differed from earlier debates. The conflict in the growing Kingdom of Sweden involved the Government and the Church and by studying Swedish examples we could learn more about Ramism and its impact.

In November 1639 the chancellor of Uppsala University in Sweden, Johan Skytte, attended the examination of the young students. His visit to Uppsala was an integral part of his duty and normally there would have been nothing extraordinary about the presence of one of the most distinguished politicians of the reign at such an event. But on this occasion everything went wrong. Skytte became furious with several professors – it was rather an examination of the professors than of the students – and he even appears to have threatened to thrash some of them. Skytte and the vice-chancellor, Archbishop Laurentius Paulinus Gothus, were explicit about the reasons for the clash between the professors and the authorities: They were displeased with the professor of logic, Laurentius Stigzelius. Stigzelius was aware of the dangers in teaching logic and had requested to be released from his duties as professor of logic. His request, however, was denied and he now faced the consequences of his determination not to submit to the instructions – formulated primarily by Skytte himself – who demanded that the lecturer in logic follow Ramus.

Stigzelius attempted to avoid direct confrontation and emphasized the fact that he had taught logic according to both Ramus and the Aristotelians and, therefore, had broken no rules: "It couldn't be wrong to explain to the students how the Aristotelians interpret the same parts of logic as Ramus, and it is not wrong to use other authors to elucidate Ramus more clearly and thus make the students understand the works of good authors, of whom several are Aristotelians."[1]

Despite his prudence and his anxiety to placate his critics, the professor of logic failed to satisfy the latter. The archbishop bluntly asked him whether he was an Aristotelian or a Ramist? To which question the professor refused to give a straight answer. But following another direct question, whether he thought Ramist logic would suffice on any level of education, Stigzelius responded: "It may be enough for beginners, but for the more learned and for those who endeavor to get a sound knowledge of logic, it is not satisfactory."[2] According to the professor, then, one could admit that

1 C. Annerstedt, ed.: *Upsala universitets historia. Bihang 1: handlingar 1477-1654,* Uppsala 1877, pp. 348-362. Nihil incommodi habet auditoribus indicare, quibus vocibus et terminis Aristotelici eadem praecepta, quae Ramus habet, exprimant, nec inutile est pauca addere, quibus Ramea uberius declarantur, et auditores ad scripta bonorum auctorum intelligenda expeditiores reddantur, inter quos plurima ex Aristotelica sunt disciplina. p. 355.

2 Ibid. p. 356: Incipientibus sufficere potest, doctoribus autem et iis qui ad exquisitam logicae cognitionem adspirant, non item.

Ramist logic was a useful introductory manual, but insufficient for those who desired to go further in their studies. Thus Stigzelius tried to present himself as a kind of semi-Ramist or Philippo-Ramist, but to no avail as the instructions were explicit. As the archbishop reminded him, the teaching of logic should be according to nothing but Ramist principles. And while the professor claimed that he thought he acted in accordance with the rules, the furious Skytte retorted that Stigzelius expounded Ramus as if he were the devil, further asking him whether he included metaphysical trifles in his teaching. This was a delicate issue because a Ramist teaching of logic had to include some metaphysics – though not according to the Aristotelian-scholastic tradition. The archbishop once again intervened in the discussion and confronted the professor by asking whether metaphysics was part of logic or an independent discipline. Stigzelius realized that he could no longer avoid taking a position. He tried to be elusive by claiming that this issue should be directed to the entire faculty and not just to him, but to no avail, and he was forced to declare that the two disciplines should be kept apart.

The archbishop was not satisfied with the answer and repeated the Ramist position that logic was about being and non-being and therefore the professor was wrong. Furthermore, a metaphysics that included teaching about God and angels was surely the devil's discipline.[3] There was not much Stigzelius could say to that, but he could not refrain from ironically pointing out that if he, as the professor of logic, was to teach about being and non-being then he must include everything else, including theology and jurisprudence. Like many Aristotelian critics of Ramist theory Stigzelius opined that it would have been against the three basic methodological laws to mix ontological issues with logic. This offhand remark on basic Ramist principles aroused Skytte's anger and made him to threaten the professor with reprisals.

Skytte was unambiguous in letting the professor know how little he and the archbishop appreciated this last comment as well as the rest of his defense. He then asked Stigzelius what he thought about Justus Lipsius' famous pronouncement that no one could ever become a great man who held Ramus in high esteem. To this Stigzelius responded that he not only agrees with such a statement, but that he believed Lipsius simply wanted to criticize only

3 Ibid. p. 357: Logica agit de omni Ente et non Ente, ergo illa recte in ea proponuntur. De Deo autem et Angelis in Metaphysica doctrina diabolica est.

those who have held Ramus as the only acceptable authority and who rejected the ancient tradition of philosophers. The chancellor could not conceal his disapproval of the way logic was being taught, but refrained from continuing the exchange.

This episode shows how bitter the conflicts between the Ramists and Aristotelians were. The exchange revealed not only some important aspects of Ramism, but also gives us some hints at the reason why anyone became a Ramist. I shall return to the Swedish scene, but first I should like to draw attention to the manner in which this episode serves as an example of the strong passions and bitter conflicts which Ramus and his philosophical ideas have incited.

During the past few decades much attention has been accorded to Ramus and his philosophy by such scholars as Ong, Risse, Hooykaas and, later, Bruyère and Meerhoff among others.[4] Of course, opinions differ concerning specific aspects of Ramus' philosophy, his aims, and his background. He was sometimes called *Plato Gallicus,* but he should perhaps be viewed rather as an eclectic humanist, who still worked within a scholastic, and hence Aristotelian, frame of reference. Most scholars, however, seem to agree on one point: Ramus was not a prominent philosopher; indeed, he might well have been a poor thinker. Not surprisingly, in most textbooks on the history of philosophy Ramus and his ideas are given little, if any, space.

I cannot address the merit of such a claim here, nor will I discuss whether Ramus should be considered a Platonist, a scholastic, or an Aristotelian – or whether he was a typical humanist. Personally, I think that notwithstanding the traces of Renaissance Platonism in his works, Ramus was basically an offspring of the Aristotelian tradition. Similarly, however different from the familiar Italian Renaissance humanism, Ramus was a typical transalpine humanist, who fully adhered to ancient literature, and who left the culture of the scholastic university seeking support from other employers, such as the French king.

Yet while we may debate how best to define it, it is impossible to deny the importance of Ramism. It is essential, I believe, not to confuse Ramus' philosophy and pedagogical ideas with the different ways in which they

4 Peter Sharratt offered a broad survey demonstrating the vigor of Ramist scholarship: Recent Work on Peter Ramus (1970-1986), in: *Rhetorica* 5 (1987) pp. 7-58.

were interpreted throughout Europe and North America. And there remains a great deal to be said – as I have indicated above – about the reasons for the Ramists' ability to provoke such strong feelings. It is not until we understand the practical uses of Ramism that we can fully understand the nature of Ramist ideas. And in order to do that, we have to study Ramus' interpreters and especially those who were willing to defend him, sometimes at all costs.

Who was a Ramist?

I will begin by sketching out some important aspects of Ramism, especially those that prevailed in Lutheran countries, i.e. the Scandinavian and North German countries, as it is important to separate Lutheran from Calvinist Ramism. Here the Melanchthonian school reform paved the way for Ramism and the prestige of the German reformer was so great that it is difficult to find any learned person in those countries during the period around 1600 who did not pay tribute to him. But Melanchthon's eclectic Lutheranism, as is well known, was synthesized into a kind of Lutheran scholasticism with strong influences from Aristotelians such as Jacopo Zabarella and the Martinis, Jacob and Cornelius, in Wittenberg and Helmstedt. The outcome was a new concentration on metaphysical studies to which neither Ramism nor Philippism could satisfactorily contribute.

Yet the impact of Ramism and Philippism persisted, at least for a while. Many professors in Northern Germany and Scandinavia tried even to reconcile the difference between them. Johannes Bilstenius in Basle and Friedrich Beurhaus, the famous headmaster of the Dortmund *Gymnasium,* were prominent examples of what usually was called semi-Ramism or Philippo-Ramism.[5] They both wrote books on logic where they compared, point by point, Ramus' theories with those of Melanchthon. It is obvious that the two are basically Ramists, though they maintained ardently that Melanchthon would have endorsed their position. Yet, Ong characterizes Bilstenius as semi-Ramist while regarding Beurhaus as a Ramist. Similarly,

5 Ch. G. Jöcher: *Allgemeines Gelehrten-Lexicon,* tome 1, Leipzig 1750, col. 1094, which characterizes Bilstenius as a Ramist although it also indicates *Syntagma Philippo-Rameum* (1623, first published Basle 1588) as his most important work.

while Johannes Rigerus of Frankfort is described by Ong as a typical semi-Ramist, he was recommended as an excellent Ramist by one of the most prominent Ramists of the late sixteenth century, Henning Rennemann (1567-1646) of Erfurt. And Rennemann was correct as Rigerus' logic is definitely a Ramistic one. But one must also remember that he also attempted to prove that Melanchthon had been a good Ramist at heart.[6]

More significant was the case of Severinus Sluterus (1571-1648). He is one of the more striking examples of a Ramist who was rather a syncretist or an eclectic. He was a Ramist insofar as he defended Ramus and his philosophy, while not adhering to strict Ramist positions. Most noteworthy was his defense of the autonomy of metaphysics. There is no need to dwell here on the well known Ramist position on the subject. Ramus did not acknowledge any particular system of metaphysics since logic according to him was a science concerned with being *ens qua ens.* The Ramists did not distinguish between *intentiones primae* and *secundae.* In logic one learned about *ens,* being, but the *objectum* of logic could equally incorporate *non-ens.*

This was a point that was much criticized by the Aristotelians and it is hardly surprising that the logic professor at Uppsala was unable to satisfy Skytte and Archbishop Paulinus. Stigzelius was trapped. According to him it would be impossible to transfer ontological issues to logic. Logic was not an *ars* or a *doctrina,* but a *habitus intellectualis.*

The Ramists emphatically rejected metaphysics as a separate discipline. Logic could provide instruction about being. But they could not define logic plainly as a *doctrina de ente,* for if one accepted that logic concerned itself only with *ens qua ens* it would have been difficult to explain how could logic also comprise *res fictae.* After all, it was possible to determine on problems which never happened or could not have happened. Therefore, Ramus and his followers, like the Swedish archbishop, were of the opinion that logic included *non-ens.* But to the Aristotelians this was an absurd and contradictory theory: The Ramists had to chose: Either logic was about *ens*

6 Henning Rennemann: *Ramus sive Enodatio perspicua totius Philosophiae Rameae,* Francof. 1603, pp. 144-145. Cfr. Johannes Rigerus: *Petri Rami dialectica cum utilissimis observationibus ex Ramo ipso et Philippo Melanchthone desumptis,* Francof. 1588, p. 6: [...] ut cui palmam deferas, non facile diiudices, nisi [...] facilitati et certitudini iudicium committas. Quod si feceris, constabit Ramum si non multis parasangis, haud tamen parum Philippum antecedere.

reale or it was a question of *ens qua argumentum.* The Ramists, according to the Aristotelians, violated elementary logical principles.

Sluterus was fully cognizant of the dilemma and sought to resolve it, by subjecting it to Aristotelian principles. He declared that *non ens* was included in the *objectum* of logic according to Ramus, but, on the other hand, he denied that logic concerned *ens quatenus ens,* relegating that aspect to metaphysics, which he thus accepted as a separate discipline. As a Ramist he was not able to use the scholastic terms, but he maintained that logic was about *ens* but *quatenus disserabile.*[7] I am unable to examine Sluterus' theory further, but we can easily notice how anxious he was to use Ramistic terms while maintaining that metaphysics existed as a separate branch of science, with logic being the *doctrina de ente qua disserabile.*[8]

It should be clear by now that it is not feasible to define Ramism strictly on the basis of essential elements of Ramus' theory. Invariably, one must classify certain authors of Ramist textbooks as semi-Ramists or Philippo-Ramists – or some such thing – simply to save the definition. But the question is whether it really matters if Sluterus or Bilstenius are defined as Ramists or as something else? We know for certain that his textbooks were read and, therefore, the interesting question is why?

The issue of *objectum logicae* was crucial. Sluterus' attempt to equivocate on this issue was of course a bold act. Many Ramists, like Archbishop Paulinus, were intimidated by what was regarded as the scholastic leaven. Among the Lutherans there was a tradition of suspicion of philosophy – perhaps even a tradition of anti-intellectualism. Before long, however, there arose a need of finding a solution to many theological problems. By the late sixteenth century many Protestant theologians turned back to Thomas Aquinas and other scholastics to resolve ontological theological issues. We thus encounter a kind of Lutheran scholasticism which, of course, was quite controversial. Many Lutherans objected to such a wide use of metaphysics. Among them Paulinus maintained the older Luther-Melanchthon position, according to which metaphysics and Aristotelian ethics were philosophical

7 Severinus Sluterus: *Anatomica logicae Aristoteleae sive syncrisis logica,* Francof. 1610, pp. 137-142.

8 About Sluterus see Erland Sellberg: *Filosofin och nyttan I. Petrus Ramus och ramismen,* Göteborg 1979, pp. 84-85. 95-96.

disciplines which one could and should avoid. They were – according to the archbishop – the devil's disciplines.

While Ramus strove to make logic ontological partly because he was suspicious of the theological implications of metaphysics, he rejected metaphysics as an *ars* primarily because it simply did not fit into his pedagogical ambitions and his philosophical curriculum. This should substantiate the conclusion of how Ramists – at least among the Lutheran – followed Ramus for their own, and not his, reasons. Paulinus refused to allow the return of scholastic subject matter to Swedish schools and Uppsala University; it would have jeopardized the Reformation. Ramus thus provided him with a viable alternative to the scholastic curriculum.

Usus

Before discussing Sweden and the way Johan Skytte interpreted Ramus, I would like to emphasize an essential feature of Swedish Ramism – its application. Ramus was severely criticized for having confused the arts in his educational program. He believed that it was meaningless to study the arts separately. One needed to be able to read the same text in both rhetoric and logic. These arts were symbolized in the medieval tradition with an open palm and a closed fist. Only in rhetoric did the student resort to the ancient orators, because rhetoric was directed, like the open hand, to all. Logic, on the other hand, was intended for the learned only. Ramus opposed such an approach. Regardless of the discipline, he wanted the students to use the same text – for example a speech by Cicero – though, of course, concentrating on different aspects. He pointed out that in the scholastic tradition the student had confused certain logical elements when studying rhetoric and vice versa. In rhetoric as in logic one needed to find adequate *loci,* which necessitated resorting to *inventio* in both disciplines. For Ramus this was tantamount to confusing the arts. He thus assigned *inventio* and *iudicium* to logic or dialectic alone, while rhetoric treated the art of embellishment and delivery.

It would be a mistake to view the ensuing debate over this innovation as a storm in a teapot. It discloses different approaches to the arts and to learning. Ramus was convinced that each art correlated to a part of nature. The whole of nature could be divided into parts which represented an *ars* or *doctrina*. Each *doctrina* formed, therefore, an exclusive part of the whole and it would be against nature to confuse the arts. Furthermore, *doctrina* or *ars* must always conform to nature. Every doctrine was an exclusive part of the whole and it would be against nature to confuse the separate arts. Consequently the content of all doctrines could never in any way be changed as nature itself could never change.[9] Hence, it was incumbent on the logician to always find the natural way to think. The syllogistic deductions were hardly natural in this respect as we don't spontaneously think in such an artificial way. Ramus emphasized instead, as more natural, the dialectic way to reach conclusions.

This, however, did not automatically make Ramus sceptical of everything ontological. He emphatically rejected such interpretation of his criticism of Aristotle. Nor did his adherents think that the Ramist ideal of *libertas philosophandi* necessitated such a view. Rather it was equivalent to the freedom of not belonging to any philosophical school. The goal was to be a philosopher, not the slave of the *philosophus,* i.e. Aristotle. This was the main reason why Ramists shunned the label: in most cases they preferred to be known as *socratici.* Like Socrates they wanted to reveal the ignorance and conceit of those who pretended to know more than others. From this followed their conviction that Ramus simply taught that which Aristotle, Cicero and others had already said. The Ramist contribution was restricted to form, not content. For that reason Rodolphus Snellius warned against those pseudo-Ramists who substituted Ramus for Aristotle. The only authority a philosopher should have, according to Snellius, was reason itself.[10]

The link between *natura* and *ars* was thus essential for any level of education. It was important that the student avoided learning the precepts of the different disciplines as the Ramists accused the scholastics of having done. Instead the Ramists concentrated on practice. We can therefore say that the most significant aspect of Ramism was the third element or *exercitatio.* The first textbook of Ramus in dialectic, *Dialecticae institutiones* (1543), was

9 See for example Rodolphus Snellius: *Snellio-Ramaeum philosophiae syntagma,* Francof. 1596, p. 15. According to Snellius one could argue that the content of physics is its *praecepta* as well as *res naturales.*

10 Sellberg: *Filosofin och nyttan* (as note 8) pp. 104-108. See also Ramus: *Physicae praefationes III.* Rennemann declared that *libertas philosophandi* was quite different from scepticism, indeed its opposite: *Ramus sive Enodatio perspicua totius Philosophiae Rameae,* Francof. 1603, pp. 122-123.

accordingly divided into three parts: *natura, doctrina* and *exercitatio.* By immersing himself in the *exercitatio* the student learnt not only theory, but also *applicatio.* The young student should learn logic not only by studying the theories in *doctrina,* but by reading how famous ancient authors had made adequate and correct conclusions in speeches, poems and other texts. Through this analysis he learnt how natural the theories were. Ramus maintained that the teaching of logic should begin by studying famous speakers' and authors' texts and thoughts. This had already been recommended by Aristotle. In analyzing the text the student should scrutinize how the author had composed the text according to the rules of logic, which question was at stake, which arguments had the author chosen and for what reasons?

In 1639, Johan Skytte rebuked the professor of eloquence for not including a logical and a grammatical analysis in his examination of Cicero's *Pro Marcello.* The professor responded by saying this would be too elementary and result in a confusion of the disciplines. The chancellor retorted by stating that those arts were so close that it would be wrong not to include all of them in the practical treatment of the text. The angry archbishop then quipped that perhaps the professor never learnt any logic and therefore was unable to make such an analysis. Though the episode is indicative of the tension between the central government and the university, it also shows how crucial was the practical confusion of arts for the Ramists, and it was obvious that the chancellor thought that the professor's method of teaching was neither efficient nor adequate.[11]

Having studied several ancient texts and learned how to analyze them properly the student was next expected to embark on imitation, composing his own text according to these principles. This method of *exercitatio* was then to be followed in all arts. According to such practical view, Virgil or Cicero were as good logicians as was Aristotle. The latter was just more accomplished as a theoretical logician who, in his *Organon,* had written *praecepta* to the *logica artificialis* as a complement to natural logic. Ramus took over the Lucian definition of *ars* as a *comprehensio praeceptorum in rebus vitae utilibus* and from the start he and his adherents stressed the usefulness of arts. Students should not strive for theoretical happiness. They should rather use their knowledge for the benefit of something.

11 Annerstedt: *Upsala universitets historia* (as note 1) p. 349.

By imitating the natural course which he detected in the texts he read, the student could and should adjust himself to the natural human way of thinking, talking, writing and seeing. It was not, however, a question of imitating a natural physical process, as the alchemists tried to do. In that respect Ramus was as uninterested as the scholastics in the practical use of physics. His curriculum was strictly restricted to the study of texts.

Ramus was sometimes called *usuarius,* a reminder of the most important aspect of his philosophical or pedagogical intentions. This aspect was the lowest common denominator that united Skytte, Sluterus, Paulinus, Rennemann, Snellius, and Freigius – all those who considered themselves, and were considered by others, as Ramists or semi-Ramists. Of course, Ramists were not the only ones to stress the need for useful studies. After all, could anyone claim that his intellectual activities or philosophy was useless? But Ramists were distinguished from others by their excessive references to usefulness. Why was it so? What did the Ramists mean by use? And did they all interpret it the same way? And did their interest in usefulness explain their success?

Whenever one discusses usefulness – and this is as true today as it was 400 years ago – one is at once apologetic and aggressive. One maintains that his program is useful while, at the same time, arguing that others are inadequate, or a waste of time. Over the past few centuries scientists have often described their studies and researches as useful in comparison with the activities of those who are studying humanities. They do so because it is the best way to obtain funding from the politicians. The talk about usefulness is a kind of weapon and means to finance one's activities, as well as means for self-respect. When Ramists thus stressed the usefulness of their offerings they may have criticized scholastic education more than characterized their own program. But there was more. In his stress on *exercitatio* Ramus developed his theory and his pedagogical ideas in contrast to scholasticism. And he was certainly not unaware of the possibility of getting alternative support to that received by the scholastics in the traditional institutions of learning. Stressing the usefulness of his curriculum was an effective way of obtaining support. Before trying to prove my case I would like to say something about the concept "use" or "usefulness". It is easy to point out that Ramists talked about *usus* more often than others. But it is necessary to examine closely the content of this concept in order to understand its meaning. Today we often use this concept to demonstrate that schools and higher education require a great deal of public money. It is therefore important to make the education efficient. Every discipline, every kind of education is expensive and therefore must be useful.

Since the eighteenth century the predominant interpretation of the word "useful" has been that knowledge or science must be transformed into material needs, increased prosperity and economic growth. The knowledge of the learned should be used to better mankind. We may call this kind of usefulness a *materialistic* one, although the concept is a little bit deceptive. The scholastics did not often talk about usefulness, but that does not of course mean that they thought their studies were useless. Thomas Aquinas or Johannes Scotus would emphatically reject such a rude accusation. Their scholarly endeavors were certainly – as we all know – directed by something they thought was the most important and useful, the salvation of the human soul. Neither wealth nor health was as important as this. We might label this kind of "use" as *ideological.* It concerned everyone. Today, however, we seldom refer to such usefulness, but certainly it existed in the curriculum for former socialist states where it was regarded as very useful to spend years studying Marx, Lenin or Mao.

The humanists could have criticized the scholastics for failing to achieve their goal, and some of course did, but the humanist attack on the scholastics went further than mocking at the lack of cleverness of the scholastics. Humanists like Erasmus were disgusted by the logical hairsplitting of the Scotists and they rejected the very notion that it was possible to reach salvation by reason and intellectual sharpness. On the contrary, such endeavors would hinder it. The dialectical approach to the text became significant of the humanists' attitude: they tried to establish probable rather than absolute truth.

An important factor in understanding this development is to recognize the political changes that occurred as medieval Europe with a unified Catholic Church was gradually replaced by separate states and national churches in transalpine Europe. These early modern states witness the rise of a new and large bureaucracy.

Traditionally education had been one of the most important responsibilities of the Church, with most students being trained for the ministry. But in the course of the sixteenth and seventeenth centuries, schools and universities found themselves trapped between State and Church, as the former required learned officials. In many cases schools or *Gymnasia* in the Hanseatic cities of Westphalia were established as free not only from the Prince's authority but from that of the Church as well. Here Ramism frequently won considerable influence, because the essence of the Ramist textbooks satisfied demands for tuition especially for commerce and trade. The Dortmund *Gymnasium* is a good example of such an establishment. And we know that many of those who were called Ramists, semi-Ramists, or Philippo-Ramists, were persons connected to these new schools – like Otto Casmann in Stade, Friedrich Beurhaus in Dortmund and Lazarus Schöner in Lemgo.

Ramus abandoned the traditional medieval culture of learning as well as the Catholic Church, becoming a Huguenot and professor at the recently founded *Collège Royal.* Thereby he became further removed from traditional university culture. This may have had a symbolic importance for the spread of his ideas. When he became one of the victims of the St. Bartholomew's Day Massacre he was made a martyr by his many adherents.

There was a paradox in the 1572 Massacre. The king was a good Catholic and before his accession even a devout one. He could not prevent the Catholic league from proceeding with their plans, but we have reason to believe that he did not encourage it, indeed, did not wish the Duke of Guise to be victorious. The Catholics represented in many ways the viewpoint of the old feudal nobility whose goal it was to conserve or rather recreate medieval order. The king on the other hand strove towards a consolidation of a strong State under a strong and powerful monarch. There was thus more which united the Catholic king and his Huguenot successor, Henry of Navarre, than separated them. And the king's professor being a Huguenot was accordingly a natural goal to attack for the Catholic league.

The situation in Sweden was somewhat different from that of France. During the decades around 1600 it ceased to be a little and unimportant country in Northern Europe, almost beyond civilization, and became a great power. With the annexation of Estonia in 1561 the remarkably quick transformation of Sweden had begun. This aggressive development could not happen without problems. One of the most distinguished Swedish historians of today describes seventeenth-century Sweden as a *military* state.[12]

12 See Sven A. Nilsson*: De stora krigens tid. Om Sverige som militärstat och bondesamhälle,* with an English summary: *The Era of the Great Wars. Sweden as a*

Sweden was at war almost continuously until 1718 when it lost almost all of its conquered territories and was again reduced to a modest size. Almost everything in Sweden during these years of power and ongoing wars was directed and influenced by the needs of warfare. But although the development in Sweden may have been faster and more radical than in other countries we can see the similar tendencies in several states in Northern Europe and France and England. All early modern states faced similar problems.

One of these problems involved a shortage of educated government officials. Charles IX and his son Gustavus Adolphus often complained about the difficulties in finding educated and suitable officials and diplomats. The Reformation deprived the Church and the educational institutions of the means with which to provide education for young boys and Gustavus I imported the officials he needed from Germany. But it is easy to understand that this was only a temporary solution. The Church was short of money and could not afford to keep schools open, while the Government was unwilling to invest in them without tangible returns. Thus, from the middle of the sixteenth century we witness a continuous struggle between the kings and their powerful bishops about what role the Church would have in a Lutheran State. Sometimes the most formidable bishops were attacked by the Government attempting to restore Catholicism when they actually fought for their privileges, though the terms of the conflict were for the most part more subtle.

The University of Uppsala was reopened in 1593 after a political compromise, but this did not put an end to the struggle to win power over the educational system. Charles IX was elected king of Sweden in 1604 and his goal was to transform the university from a seminary for priests to a training ground for future government officials as well. Yet the professors were adamantly opposed to such change. The number of noblemen in Sweden was small and most of them were little more than farmers with large holdings. The noblemen were hardly desirous of a scholarly education in order to assume administrative duties, as they believed that their real

Military State and its Agrarian Society, Stockholm 1990, especially pp. 150-177. 226-270. For a more perspicuous survey see for instance Michael Roberts: *Gustavus Adolphus and the rise of Sweden,* London 1973; Idem: *The early Vasas, 1523-1611,* Cambridge 1968.

calling was to wage war and exercise knightly virtues. When Axel Oxenstierna in 1612 became chancellor of the state, he was a remarkable exception to this tradition. He had a good university education and he was a skilled politician rather than a warrior. Oxenstierna attempted to encourage other noblemen to follow his example by telling them that the life of a politician was as *magna ars* as the military life.

For the most part, however, Charles IX did not turn to the nobility for help in these matters. He had good reasons to question their loyalty to him and to his family, as he had deposed the legitimate king and his nephew, Sigismund Vasa, who also was king of Poland and devoted to the Counter Reformation. Also, as part of his *coup d'État* Charles beheaded several of the most prominent and most learned noblemen. To replace them the king had to trust young students of lower social rank, whom he encouraged to study and paid for their education in German, Dutch, or Swiss universities. When they returned home Charles employed them to conduct his correspondence with foreign countries and to defend him as a legitimate king. His critics thought this was a kind of secretary regime, but it nonetheless enabled the king to avoid dependence on the small class of noblemen.

Johan Skytte was one of the most distinguished of these secretaries, and whose skills in rhetoric and Latin soon became well-known. He turned out to be an exceptional advocate of the new king's policy. On his missions to the foreign courts he delivered superb Latin speeches in which he explained and defended the king's proceedings and legitimacy. We have already noted that Skytte was also a Ramist. He had studied at different universities in Germany during the last decade of the sixteenth century and he had also visited Switzerland, England, and France during his peregrinations. One of his friends relates the turmoil Skytte aroused among the professors of the Sorbonne when he paid tribute to Ramus and accused them of the death of the philosopher. Although this report is probably an exaggeration Skytte's academic dissertations certainly demonstrate his Ramism.

Skytte could well have embarked on an academic career had he not previously been selected by Charles for service in his Government. He was later appointed to supervise the education of the young prince, Gustavus Adolphus, and he held several important offices during Charles' reign, including his being commissioned to write a new code of law. His endeavors, however, invariably failed because his ideas and his suggestions provoked much resistance. Churchmen in particular became agitated by his project for placing the Church under full Government control. But it is nevertheless worth pointing out that Skytte's deeds reveal his basic attitude towards politics, through which we can explain his Ramism. Furthermore his designs were ultimately carried out at the end of the century, when the Church, as a consequence of the 1686 ecclesiastical laws, became a closer partner of the monarchy, in a way which characterizes Sweden to this day.

Under Gustavus Adolphus, Skytte embarked on another career. He was elected a member of the council and made baron. For some time he was responsible for the treasury, and later he was appointed governor of the Baltic countries. In 1622 he also became the first chancellor of the university. Johan Skytte, then, was an excellent example of the model advocated by the monarchs: intelligent, well educated and dependent on the king for his career and who, to the bitter end, saw his goal as strengthening the new state and crushing any separatist tendencies. Consequently, he was as detested by the old nobility as he was appreciated by the king.

The other Swedish Ramist, Archbishop Paulinus, was quite an ill-matched mate to Skytte. He was an old man in 1639 (nearly seventy years old), and had studied in Germany – especially in Rostock where he was influenced by the famous David Chytraeus. He backed the winning side in the political struggle between Sigismund and Charles during the 1590s and was elected professor at Uppsala University. There he soon became a prominent representative of the professors who wanted to defend and preserve the right of the Church to control schools and the university. Paulinus was among those clergymen who emphatically defended the ecclesiastical privileges and resisted the king's and – for that matter – Skytte's endeavors to control the university.

There were two occasions in early seventeenth-century Sweden when Ramism appeared to confront Aristotelianism or scholasticism. The first occurred in 1615 when Paulinus demanded that the teaching of Aristotelian ethics should be forbidden in schools and at university. In his polemic he invoked Ramus and was incensed at anyone who disagreed with him, particularly the professor of ethics who combined learning and shrewdness in defending his discipline. Paulinus, then bishop of Strengnäs, tried to get the support of the Church, but the king asked him to desist. It is difficult now to reconstruct the various elements of this conflict which lasted for more than a year. But it is clear that it was the Government which forced the professorship of ethics on the university in order to facilitate education in politics, and that the king had no intention of giving up his plans to take control of the university. The conflict, then, was not about ethics. The bishop had attempted several years earlier to prevent another interference from the Government and now he could see how, by enlarging the number of professorships, the Government slowly turn the curriculum away from the old ecclesiastical tradition of the academy – and that it did so in order to get control of higher education.

The professor of ethics, Jonas Magni, was younger. He had studied in Wittenberg and was influenced by Jacob Martini's Aristotelianism. It seems as if the conflict was between Ramism and Aristotelianism, but this was not the case. Ramus was hostile towards ethics and the bishop had to seek theoretical support from several anti-philosophical Lutheran groups. It is easy to see that for him Ramism represented a bulwark against Lutheran scholasticism which began to infiltrate German universities. Paulinus appreciated Ramism, with its efficient educational program that offered assistance in the battle against the dreaded regenerated metaphysics.

In the conflict over ethics Paulinus certainly could not count on the support of the other Ramist, Johan Skytte, who defended a policy aimed at subjecting the university to the needs of the State. As the first chancellor of the university, Skytte endowed a professorship of politics and Jonas Magni served as professor *pro tempore* until Skytte appointed a permanent one. It may appear peculiar that Skytte would let a prominent Aristotelian hold his own professorship; yet for him it was far more important that Jonas Magni taught the students politics in a way that could prepare them for service of the State. Magni also supported a less aggressive policy towards the Church than Paulinus did. He maintained that it was important to restrict the power of the Church to spiritual matters, and it was the king who was to be considered as the foremost member of the Church.

The second conflict occurred when Skytte and Paulinus inspected the university in 1639. The archbishop was anxious to crush any tendencies towards a renewal of scholasticism, while Skytte was primarily incensed because he found that the professors did not obey his intentions. For him Ramism was primarily a pedagogical project which would best produce an efficient curriculum when no time was wasted on hairsplitting questions. Skytte thus stressed the fact that not only the Church was in need of learned officials but the State too. One of his most severe criticisms against the professors' teaching was the absence of young noblemen in the examination. It seemed obvious to Skytte that the professors had not encouraged them enough to attend the lectures and that they stuck to the old scholastic curriculum.

We can better understand Skytte's opinion if we study more closely his last speech as chancellor in 1640. He was then over sixty and nearing death. In his speech Skytte reflected on his own student days and expressed his concern over the spread of new philosophical "novelties" – Aristotelianism and Lutheran scholasticism. His message to the students was to stick to Ramism and always privilege usefulness. In his own political endeavors Skytte certainly placed the needs and interests of the State before anything else. Such thinking was actually "modern" and the students also heard about the more private benefits which his own professors reaped from their studies and learning. They had all, of course, been Ramists. Several had become bishops or prominent clergymen. We must not separate their importance from those of the State. In both cases it was the usefulness of efficient education which mattered. Without neglecting the religious needs of the people, Skytte would have liked to reduce the worldly power of the Church, and later on in his speech Skytte elaborated on what he meant by *usus*.[13] He insisted that what the students studied could not be considered a private concern. The State had a right and an obligation to control and examine the teachings at schools and universities. He then offered a rather surprising example to underline his message by referring to the situation in China, where no one – according to Skytte – had a right to spend his life in a town without giving account of his whereabouts and how he earned his living.

Skytte thus wanted the State to supervise schools and universities and he demanded that usefulness to the State be the foremost consideration. Having mentioned earlier materialistic usefulness and ideological usefulness, we can now add *administrative usefulness,* where the needs and interests of the State and the political authorities inform the educational process. This was

13 Skytte's speech is printed in C. Nettelbladt: *Schwedishe Bibliotec IV,* Stockholm and Leipzig 1730, pp. 122-135.

the main reason why Skytte defended Ramism with such rigor. According to him *methodus socratica* was the best way to learn and the *philosophia socratica* was the most efficient and useful curriculum – without metaphysical and other useless nonsense. In Skytte's vocabulary the word *usus* is very frequent. When he visited the university in 1627 he even suggested that all arts which could not prove their usefulness be removed from the curriculum. He did not point at any specific discipline but told the professors that some disciplines, such as astronomy, could be exempted from those demands because of their prestige. In the rules governing the university, Skytte and Chancellor Axel Oxenstierna expressed the same demand that "the philosophers should be more the men of action than of words; that everyone should direct his study in such a way that rhetoric should be practiced to prevent anyone from leaving school after having studied eloquence, though still being unable to make a speech, or after having studied mathematics still being unable to put his knowledge into practice. The same should apply to the other disciplines."[14]

In what way was Ramism useful? The humanists often talked about usefulness. It also became almost an attribute to Ramus and all his adherents and a weapon in their censure of scholasticism. Yet in many ways Ramus himself belonged to the same scholastic curriculum which he so strenuously opposed. He could hardly free himself from the medieval Aristotle how much he yet tried to. The impact of Ramism is not a theoretical one. In vain we often look for his name in the textbooks of the history of philosophy. The importance of Ramism is instead directly connected to some significant facts of early modern Europe. The development from a traditional, medieval and mainly ecclesiastical directed curriculum to a modern and political directed one was slow and not without conflicts and backlashes but yet inevitable.

Ramism was a significant step on this course and gained support from those who endeavored to fulfill new demands of the State. We thus can observe how the two Swedish Ramists in the same aspects were unanimous

14 Annerstedt: *Upsala universitets historia* (as note 1) p. 292. About the laws of the university see ibid. p. 280: Corondisis loco tales erunt philosophi inque sedulo incumbent, ut rerum sint magis actores, quam verborum fabricatores, utque quisque cum philosophiae studio eloquentiam in usu coniunat [...].

but in others very far away from each other. They were designated anti-scholastics but on grounds of different reasons, the bishop of theological and extremely Lutheran reasons, Skytte because he was a humanist but mostly because of his urgency of constructing a modern state. But no one wanted the students or the pupils to spoil their studies by learning what could be unuseful. And certainly they were both convincted that Ramus' pedagogical ideas would be the best to answer to their demands.

English Ramism: A Reinterpretation

MORDECHAI FEINGOLD*

More than four decades ago Walter Ong characterized "the pattern of Ramism" in the English-speaking world as "juvenile". It manifested itself in "a good deal of shouting for or against Ramism by sophisters or other youthful university disputants" and in students' reading of Ramist texts. Beyond such "ebullient youthful" display, however, there existed "practically no serious and mature scholarship oriented by Ramism." Ong was hardly perturbed by his discovery since the profundity of Ramism was irrelevant to the more momentous revolution he claimed to have unraveled:

> This does not make the Ramist influence in England any the less telling. Quite the contrary. Insofar as it shows that rhetoric and, more significantly, logic were being assimilated independently of a well-worked-out philosophical foundation, the condition of Ramism in the British Isles shows how much Renaissance logic there and elsewhere derived not from an insight into reality but from a state of mind [...]. The pervasiveness of the Ramist experience can be seen in the circulation of Ramist terms.

It was precisely the derivation of such a "state of mind [...] from the study of logic as a fairly elementary discipline" that accorded the experience greater permanence. Ong further speculated that "the greater age of the English universities gave Aristotle a more secure hold in England and prevented Ramism from producing in the British Isles any comprehensive approach to the higher branches of knowledge comparable to what it produced in Germany."[1] The verity of Ong's broader thesis, which encountered

* I wish to thank William Newman, Richard Serjeantson, and Peter Sharratt for comments on an earlier version of this paper.

1 Walter J. Ong: *Ramus, Method, and the Decay of Dialogue,* Cambridge, Mass. 1958, pp. 303-304. As Ong stated elsewhere, "despite its intrinsic inconsequentiality, Ramist logic is exceedingly important in the psychological and cultural development of

considerable criticism, is beyond the scope of this article. But his portrayal of the English situation is, I believe, quite accurate. Ong erred only in ascribing the failure of a more thorough diffusion of Ramism into English institutions of higher learning to an alleged pervasiveness of Aristotelianism already there.

Unfortunately, subsequent scholars tended to ignore Ong's insight concerning the English case, opting instead to celebrate Ramism as a major intellectual movement that both revolutionized and democratized education – by making it accessible to the lower echelons of society – while at the same time, ushering in "modernity". The latter assumed varied forms. Ramus, it was argued, contributed to a "profound change in the intellectual process, and in the discovery and the dissemination of knowledge" while the movement that he gave birth to "contributed significantly to the growth of the critical spirit." Others credited Ramism with providing an alternative to the elite learning of the universities, thereby contributing to the emergent philosophy of merchants and artisans which, in turn, gave rise to early modern English religious, social, and political radicalism. Analogously, the purported single-minded Ramist emphasis on utility sanctioned scholars to credit the movement with the efflorescence of the mathematical sciences, experimentalism, and utilitarianism.[2]

Curiously, given its alleged role as the harbinger of capitalism, reform, and the new science, it is odd that historians have been reticent to define Ramism. Whereas the *Oxford English Dictionary* renders the suffix "ism" as denoting "a form of doctrine, theory, or practice having, or claiming to have, a distinctive character or relation", to date no one has been willing (or able) to document the existence of such a coherent Ramist movement. Instead, we encounter a rather promiscuous use of the term and an equally indiscriminate attribution of adherence to various individuals.

mankind." John Milton: *Complete Prose Works,* ed. D. M. Wolfe, 8 vols., New Haven 1953-1980, vol. 8, p. 148.

2 Louis A. Knafla: Ramism and the English Renaissance, in: *Science, Technology, and Culture in Historical Perspective,* ed. Louis A. Knafla et al., Calgary 1976, pp. 26-50, on p. 34; Hugh Kearney: *Scholars and Gentlemen: Universities and Society in Pre-Industrial Britain 1500-1700,* Ithaca 1970, pp. 46-70; Christopher Hill: *Intellectual Origins of the English Revolution,* Oxford 1965; Reijer Hooykaas: *Humanisme, science et réforme: Pierre de la Ramée,* Leiden 1958.

Undoubtedly, the vagueness in specifying those credentials warranting identification as a Ramist is attributable to the anecdotal nature of the evidence adduced to support sweeping claims regarding membership in the brotherhood, or the centrality of the movement more generally. Recitation of contemporary, or near contemporary, labeling of individuals as Ramists, the ownership of a Ramist text or two, or an author's utilization of charts and dichotomies in his writings seem to suffice. Guilt by association is also a viable method. Since certain Cambridge colleges are "known" to have been hotbeds of Ramism, mere attendance seems proof enough for membership in the fraternity. And, of course, a Puritan could always be presumed to have embraced Ramism. To my mind, however, such recourse to random facts and rash presuppositions is incapable of ascertaining full-fledged Ramism on the part of any individual – let alone demonstrating the existence of a coherent and purposeful movement.

Most remarkable of all, claims concerning the broad revolutionary impact of Ramism were espoused at the same time that the derivativeness and shallowness of Ramist ideas were acknowledged. Such dissonance could be circumvented by elevating other issues to prominence, as in the case of Ong when he argued for the impact of the typographical revolution on the modern mind. More recently, Grafton and Jardine have found Ramist shallowness reassuring in substantiating their thesis regarding the breakdown of humanism. Ramism, they argued, brought about "the final *secularisation* of humanist teaching – the transition from 'humanism' to 'the humanities'."[3] Irrespective of the merit of such retrospective theses concerning the long-term trends effected by Ramism, the explicit recognition of the triteness of Ramus' teaching must caution us against criticizing contemporaries for viewing the movement as banal, if not outright inimical to learning. And it is the purpose of the second part of this article to offer a more discerning evaluation of the nature of the opposition to Ramism.

Concentrating on the reaction to, and reasons for, rejecting Ramism prevents me from discussing certain recent attempts to endow Ramus' thought with greater profundity than has hitherto been common. However, I should express my concern lest such claims – even if justified – become the start-

[3] Anthony Grafton, Lisa Jardine: *From Humanism to the Humanities,* Cambridge, Mass. 1986, p. 168.

ing point for scholarship on the diffusion of Ramism. For example, Nelly Bruyère's careful study of the evolution of Ramus' thinking about logic and method, based on a meticulous textual analysis of Ramus' manuscripts and every edition of his textbooks, purports to have identified an unsuspected coherence and acumen in his work. Yet she aims even higher, devoting one fourth of her book to suggest that many noted philosophers for a century after Ramus' death read Ramus the way she did and were profoundly influenced by him.[4] But even assuming that her enthusiasm is not misplaced, I doubt whether any reader of Ramus ever engaged in anything remotely resembling Bruyère's enterprise. Early modern readers were neither interested in detecting the manner in which Ramus' views evolved in the course of his determined effort to counter each and every objection raised against him, nor were they seeking to unravel a deep coherency that allegedly pervaded his dialectic. When applied to the domain of dissemination of ideas, Bruyère-Robinet's approach is liable to create false genealogies based on artificial similitudes, further implying that either creative individuals were incapable of independent thought or that they needed to avail themselves of Ramus for novel ideas.[5]

As if bolstering were needed for the penchant to detect the ubiquitous presence of Ramus! Scholars have frequently resorted to detecting "traces", "tendencies", and "shared outlooks" in the writings of consequential figures to compensate for the absence of concrete evidence substantiating Ramus' influence. An anti-Aristotelian stance, for example, has habitually elicited attribution of indebtedness to Ramus, a case in point being Vasoli's claim for Ramus' influence on Francesco Patrizi. In so doing, scholars have often mistaken semblance for derivation. In their eagerness to privilege Ramus they failed to consider other sources that were as capable of providing inspiration. Indeed, even explicit admission of debt is insufficient to brand one a Ramist. Pierre Gassendi, for example, recalled in 1624 that it was the reading of Vives, Mirandola, Charron and "especially" Ramus that

4 Nelly Bruyère: *Méthode et dialectique dans l'œuvre de La Ramée,* Paris 1984; Kees Meerhoff*: Rhétorique et poétique au XVIe siècle en France: Du Bellay, Ramus et les autres,* Leiden 1986.

5 André Robinet: *Aux sources de l'esprit cartésien: L'axe La Ramée-Descartes: De la Dialectique de 1555 aux Regulae,* Paris 1996.

had emboldened him to write against Aristotle. Yet, Ramus' chief contribution to the *Exercitationum paradoxicarum adversus Aristoteleos libri septem* seems to consist primarily in the uncharacteristic vehemence with which Gassendi couched his work.[6]

Sharratt was correct, therefore, to reiterate Ong's apt assessment that "Ramus's primary concern was with teaching and not with thinking" and though he was certainly "a great educator", Ramus was not "a great philosopher".[7] That Ramist texts served as the backbone of many Protestant *school* curricula is a well-documented phenomenon, and no mean achievement. But the situation in the universities was altogether different, and in this context, the English universities provide a cautionary tale against over zealous generalizations.

Since the production of English Ramist textbooks was negligible compared to the Continent, the most important source for analyzing the penetration of Ramist logic and rhetoric into the universities are students' library catalogues and commonplace books. Unfortunately, the few historians who have availed themselves of such resources have culled only those titles that interested them without paying much attention to the general nature of the library catalogue or the course of study, thereby presenting a distorted impression of what the source actually teaches us. As I have pointed out elsewhere, not only was the early modern Oxbridge curriculum predominantly literary in character – and hence the relegation of logic, and philosophy more generally, to a subordinate position in it – but undergraduate instruction was done through the utilization of several introductory manuals, not a single one, with the expectation that the students would further improve themselves by their own industry later.

To illustrate the point, we may scrutinize the library catalogues of Cambridge scholars who died while at university and whose probate inventories included at least one Ramist text. Such a survey quickly reveals that logic and philosophy textbooks were far outnumbered by classical texts – in both

6 Cesare Vasoli: De Pierre de la Ramée à François Patrizi: Thèmes et raison de la polémique autour d'Aristote, in: *Revue des sciences philosophiques et théologiques* 70 (1986) pp. 87-98; Pierre Gassendi: *Opera omnia,* Lyon 1658, vol. 3, p. 99, cited in: Barry Brundell: *Pierre Gassendi: From Aristotelianism to a New Natural Philosophy,* Dordrecht 1987, p. 15.

7 Peter Sharratt: Recent Work on Peter Ramus (1970-1986), in: *Rhetorica* 5 (1987) p. 17.

Greek and Latin – or theological works in the case of more senior scholars. More significantly, Ramist texts rarely, if ever, dominated the small section devoted to logic and rhetoric. Quite the opposite. Thomas Layton (MA Peterhouse 1562, died 1565), for example, owned Ramus' *Dialectica,* but also Aristotle's *Organon,* Sturm's *Partitionum dialecticarum libri quatuor,* Périon's *De dialectica libri III,* and even Boethius' commentary on Porphyry. William Lyffe (MA Trinity 1566, died 1569) owned, in addition to Ramus' *Dialectica,* Matthisius' commentary on Aristotle's *Organon* and Niphus' commentary on the *Topica,* as well as Seton, Périon, Clicthoveus, and Agricola's *De inventione dialectica* – the last being a very popular text in sixteenth-century Cambridge. His rhetorical texts included Hermogenes – another Cambridge bestseller – Trapezuntius, and Melanchthon. For his part, Robert Smith (MA St. John's 1571, died 1571) owned the dialectical works of Seton, P. Varius, and Agricola, in addition to Aristotle's works, as well as two of Ramus' mathematical works and Talaeus' *Rhetorica.*[8]

The pattern continued uninterrupted in subsequent years. Edmund Rudd (MA Clare 1576, died 1577) owned a predominantly literary collection that included rhetorical works by Valla, Sturm, Hermogenes and Talaeus. Likewise, his copy of Ramus' *Dialectica* co-existed with several commentaries on Aristotle, and the dialectical works of Gorscius and Willichius. The modest collection of Edmund Robert (MA St John's 1576, died 1580) included Ramus' *Dialectica* as well as Seton, Agricola, and Aristotle in Greek. Ambrose Barker (MA Christ's 1575, BD 1582, died 1583), who appears to have been a Puritan, owned Ramus' expositions of Virgil's *Bucolica* and *Georgica,* his *Dialectica* and Talaeus' *Rhetorica,* but also several of Aristotle's works, including the *Organon,* as well as Sturm's dialectic. A shorter (and anonymous) strongly Calvinist catalogue included Ramus' *Dialectica,* Aristotle's *Organon,* Seton's logic, and Wecker's *Organum logicum.* The scholar also owned Peter Lombard and Thomas Aquinas. Abraham Tillman (MA Corpus Christi 1587, died 1590) left behind a very small collection with Ramus' *Dialectica* and Piscator's commentary on Ramus, balanced by the Louvain *Commentarii in Isagogen Porphyrii et in omnes libros Aristotelis de dialectica.* A much larger collection belonged to Richard

8 E. S. Leedham-Green: *Books in Cambridge Inventories,* 2 vols., Cambridge 1986, vol. 1, pp. 282-284. 302-305. 313-314.

Motte (MA St John's 1579, BD 1586, died 1592). In addition to a great deal of Calvinist literature we also find Ramus' *Dialectica,* Niphus' commentary on Aristotle's *Topica,* Aristotle's *Problemata,* Valerius' logic, Lever's *The Arte of Reason,* Simonius' *Quaestionum dialecticarum fragmentum,* and Agricola. Three more libraries should be mentioned. Stephen Thompson (MA St. John's 1591, died 1599) owned Ramus' *Scholae in liberales artes* as well as *G. Tempelii epistola de dialectica P. Rami ad J. Piscatorem.* He also owned Aristotle's *Organon* (in Greek), Seton, Melanchthon, and Rivius' *Compendium in universam dialecticam.* Thompson's rhetorical texts included Hermogenes, Valla, Sturm, Agricola and Melanchthon. The Calvinist John Shaxton (MA Trinity 1591, BD 1598, died 1601) owned, in addition to Ramus' *Commentariorum de Religione Christiana libri IV,* rhetorical works by Talaeus and Melanchthon as well as Ramus' *Dialectica,* Temple's *Epistola,* Agricola, and Caspar Olevianus' *De inventione dialecticae liber.* Finally, Godwin Walsall (MA 1605 Pembroke, died 1608) owned Ramus' *Dialectica* and Talaeus' *Rhetorica,* several works by Aristotle, including the *Organon* in Greek, Agricola's dialectic and Niphus' commentary on the *Topica.*[9]

I felt it necessary to offer such a tedious recitation of inventories in order to show that while Ramist textbooks in logic and rhetoric were quite common in Cambridge, they were used in conjunction with other texts, which makes it necessary to reevaluate the significance of their incidence in scholars' libraries. Conversely, even the architects of the most Puritan of all Cambridge colleges, Emmanuel, would not elevate Ramus to a position of prominence. Mildmay's statutes prescribed that the head lecturer expound on the traditional authorities, Plato, Aristotle, or Cicero, and in 1588 Chaderton and the fellows further enjoined that Aristotle and Ramus should be used in tandem in teaching logic, while the Stagirite's writings were assigned as the required texts in the teaching of ethics, politics, and physics. Equally intriguing is the absence of any copies of Ramus' works from the College library during its first half century of existence, while other logic texts – such as Aristotle, Agricola, Schegk – were represented. And though it is correct to observe that this may reflect an expectation that students own

9 Ibid. vol. 1, pp. 322-325. 344-346. 351-354. 385-387. 481-483. 508-522. 536-540. 540-543. 557-559. The inventories of booksellers offer a similar expansive occurrence of textbooks.

their own copies, nevertheless Ramist texts were acquired by other College libraries and their absence from that of Emmanuel provides yet another indication of the need to avoid sweeping generalizations concerning the primacy of Ramus in the teaching of logic or the banishment of Aristotle.[10]

On the basis of this survey, therefore, it may be safely concluded that Howell was wrong to assume that Ramism was the "prevailing logical system in England.[11] In fact, within the overwhelmingly humanistic literary program, no single text was allowed to dominate the teaching of logic. As I have argued elsewhere, by the late sixteenth century, instruction in the entire arts and sciences curriculum was compressed within the undergraduate course. In addition, there occurred a privatization of studies, which meant that the supervision a student received from his tutor in the study of logic (or any other subject) was confined to joint "reading" of an introductory text, which outlined the domain of study and acquainted the student with the primary terminology. The young undergraduate was then left to proceed on his own, and the function of the tutor became more supervisory, an "oracle" available for further information or a watchman to monitor progress. During this stage of independent study the student acquired (or borrowed) an assortment of other textbooks. Obviously, within a period of less than a year one could not expect to attain any profundity, but it was expected that the student continue to improve himself later. In the absence of any authoritative text, personal taste determined the choice of introductory textbooks (often they were manuscript manuals prepared by tutors, which later found their way into print). And though certain authors were more popular than others, none was considered canonical; hence, the primary objective of tutors was to compile a list of approved authors from which students could select their texts.[12]

The same trend can be detected across the Atlantic, as Harvard tutors who had been educated in Cambridge and Oxford proceeded to implement that

10 *The Statutes of Sir Walter Mildmay* [...] *for the Government of Emmanuel College*, trans. Frank Stubbings, Cambridge 1983, pp. 70. 101; Sargent Bush, Carl J. Rasmussen: *The Library of Emmanuel College, Cambridge, 1584-1637*, Cambridge 1986, pp. 27-28. 134.

11 Wilbur S. Howell: *Eighteenth-Century British Logic and Rhetoric*, Princeton 1971, p. 16.

12 Mordechai Feingold: The Humanities, in: *The History of the University of Oxford*, ed. Nicholas Tyacke, vol. 4: *Seventeenth-Century Oxford*, Oxford 1977, pp. 211-357.

with which they were familiar. Perry Miller was, of course, aware of the commixing of a variety of logic texts at Harvard, yet he chose to equivocate: "The Peripatetic system was indeed read at Harvard," he thundered, "but the Ramist was believed." Such unshaken assurance in the ubiquity of the Ramist presence at Harvard was perpetuated by subsequent historians who were just as willing to slight the evidence for the persistence of the traditional approach in teaching logic and other subjects in the American colonies. Yet George Starkey (BA 1646) recalled that he was "altogether educated" in "the vulgar" logic and philosophy, "though never satisfied with it; at length *Aristotles* Logick I exchanged for that of *Ramus,* and found myself as empty as before." A decade later President Chauncy, "upon a Dislike of the *Ramaean* Strains in which our Young Disputant [Increase Mather] was carrying on his Thesis", rose to interrupt the harangue, and only the pleas by the youth's tutor dissuaded him from silencing Mather.[13] The temptation to claim too much for Ramism is carried into the portrayal of teaching at other institutions. The modern biographer of Samuel Johnson of Connecticut, though cognizant of Johnson's exposure to Ramist and non-Ramists texts alike – the latter included Keckermann, Burgersdijk, and Heereboord – still claimed that Ramus was "more important than any of the others". Johnson himself, though, attested that his academic diet consisted of "nothing but the scholastic cobwebs of a few little English and Dutch systems that would hardly now be taken up in the street, some of Ramus and Alstad's works was considered as the highest attainments", and also recalled his skill in making "synopses or abridgements" of old scholastic books.[14]

Consequently, we should conceive of English Ramism less in terms of embracing a particular attitude toward logic – though this existed, too – and more in terms of a broader attitude toward learning. Judicious educators refused to condone a narrow approach based on a single modern author and esteemed a

13 Perry Miller: *The New England Mind: The Seventeenth Century,* Cambridge, Mass. 1961, pp. 116. 121; William R. Newman: *Gehennical Fire: The Lives of George Starkey, an American Alchemist in the Scientific Revolution,* Cambridge, Mass. 1994, p. 33.

14 Joseph J. Ellis: *The New England Mind in Transition: Samuel Johnson of Connecticut, 1696-1772,* New Haven 1973, pp. 19-20; Herbert and Carol Schneider: *Samuel Johnson President of King's College His Career and Writings,* 4 vols., New York 1929, vol. 1, pp. 6-7.

Ramist compendium in a manner not inconsistent with William Hamilton's judgment of Henry Aldrich's *Artis logicae compendium:* "As a slender introduction to the after-study of logic [...] it is not to be despised; as a full course of instruction – as an independent system of the science, it is utterly contemptible." The veritable Ramists, on the other hand, and by this I mean the more zealous Puritans, were those who believed – to paraphrase Hamilton again – that the minimum of Ramus is the maximum of the schools.[15]

For this reason I consider the argument concerning a progressivist unity of purpose between Puritans and Ramists to be unfounded. The Puritans were not interested in a thorough secular educational reform and their critique of the ancients was not motivated by any desire to promote novel philosophical or literary programs. They embraced a Ramist cast of mind because, on the one hand, it offered them a dogmatic approach that particularly suited the Puritan mentality and, on the other hand, assured them that a lengthy and arduous application to profane studies was unnecessary. Not surprisingly, as we shall see below, it was precisely Puritan dogmatism and its perceived devaluation of erudition and literature that evoked much of the hostility Ramism encountered. However, insofar as the specifics of Ramist offerings are concerned, what was adopted were the charts, dichotomies, and disjunctive syllogisms that better facilitated the propagandist needs of Puritans – from the inculcation of doctrine and dogmatic habits of thought, to the expeditious preparation of youth for the ministry and effective sermonizing.

Of course, the diagrammatic tools could be (and were) used by non-Puritans as well, as both textbooks writers and doctrinaires of all sorts appreciated their facilitating presentation.[16] An interesting example of the lat-

15 William Hamilton: *Discussions on Philosophy and Literature,* New York 1855, pp. 126-127.

16 The cavalier assumption that those availing themselves of charts and tables must be Ramists dies hard. For the proliferation of such aids before Ramus, and on their utilization by non-Ramists, see K. J. Höltgen: Synoptische Tabellen in der medizinischen Literatur und die Logik Agricolas und Ramus', in: *Sudhoffs Archiv* 49 (1965) pp. 371-390; Charles B. Schmitt: *Aristotle and the Renaissance,* Cambridge, Mass. 1983, pp. 56-57; Idem: *John Case and Aristotelianism in Renaissance England,* Kingston, Montreal: McGill-Queen's University Press 1983, pp. 145-146; Stephen Ferguson: System and Schema: Tabulae of the fifteenth to eighteenth centuries, in: *Princeton University Library Chronicle* 49 (1987) pp. 9-30. See also Maclean's article in this volume.

ter type is Hobbes who was taken by Ong to be "a Ramist at heart." As Fisch put it,

> Hobbes had learned from the Puritans the trick of the disjunctive syllogism. He could dichotomize as well as any Puritan logic-chopper – hence the forensic quality of his style – and dismiss as mere allegory all that he could not treat baldly and simply at the literal level. He employs the simplicity of the steam-roller; he irons out (in the Puritan-Ramist fashion) all the imaginative overtones of the text before him, all its spiritual suggestions, all its symbolism; and he is left with a bald, matter-of-fact statement which becomes at once amenable to the type of construction he wishes to impose upon it.[17]

For good measure we may also point to the nice "Ramist" chart that Hobbes appended to chapter nine of the *Leviathan* to encapsulate the "Register of Sciences". Among textbook writers, mention can be made of the Oxford Aristotelian philosopher John Case, who made extensive use of *tabulae,* dichotomies and divisions in his pedagogical works in order to alleviate the complexity of the subject matter. Similarly, Peter Heylin borrowed from Freigius the idea of presenting the "General Praecognita of Historie" in his 1621 *Microcosmus* in the form of a Ramist chart.[18]

Since English Ramism manifested itself almost exclusively in terms of format, I believe it would not be inappropriate to define this wide-spread diagrammatic predilection as "low-grade Ramism". And contemporaries found little to object to in this instrumentalist approach. Nevertheless, any effort to claim more for Ramism runs the risk of imputing to it a greater significance than it actually had and more seriously, of branding those who failed to demonstrate allegiance to Ramism as conservatives or worse. To reiterate, though I shall argue for a close correlation between Puritanism

17 Walter J. Ong: Hobbes and Talon's Ramist Rhetoric in English, in: *Transactions of the Cambridge Bibliographical Society* 3 (1951) pp. 260-269; Harold Fisch: *Jerusalem and Albion: The Hebraic Factor in Seventeenth-Century Literature,* London 1964, pp. 244-245.

18 Schmitt: *John Case* (as note 16) pp. 144-146; D. R. Woolf: *The Idea of History in Early Stuart England,* Toronto 1990, pp. 182-183. For his part, Robert Burton mischievously "used the tools of Ramism for utterly different ends from those for which they had been forged." Everything "that Ramus had sought to banish – digressiveness, inconsistency, copia, the fusion of rhetoric with dialectic – ran riot over and through the meticulous pattern of the 'method'." David Renaker: Robert Burton and Ramist Method, in: *Renaissance Quarterly* 24 (1971) pp. 210-220.

and Ramism, I consider the repeated claims concerning the intellectually revolutionary character of both to be unfounded. The Puritans exhibited little interest in secular learning for its own sake, found reason generally suspect, and showed themselves inimical to the new philosophies. Ramism provided them with the desired indoctrinarian mentality as well as with a semblance of universal learning – an important armor against the persistent (and effective) charges that Puritans lacked learning.

There were perfectly good reasons, therefore, to oppose certain broader features that many deemed anti-intellectual both for Ramism's attitude toward learning and its bolstering of dogmatic Puritanism. Hardly surprising, then, that scholars and literary figures were at best politely on the fence. But before turning to such opposition, it is necessary to describe a factor that contributed to making Ramus the Puritan philosopher of choice.

The circumstances surrounding Ramus' death, it is generally agreed, greatly facilitated the diffusion of his writings. Regardless of whether Kelley's claim that "Ramus's attack on Aristotelian authority concealed an attack on Roman Catholic orthodoxy" is wholly accurate – Ramus, after all, launched his attack long before his conversion to Protestantism – it is undeniable that by the late 1560s Ramus had made explicit that argument his followers would later propagate.[19] Shortly before his death, Ramus prepared a new preface to his *Dialectica* suggesting that since God had blessed his work, his logic necessarily advanced the cause of true religion. His brutal murder promptly enshrined his memory within Protestant martyrology, as the St. Bartholomew's Day Massacre sent shock waves throughout the Reformed Churches in Europe. At Oxford, for example, on the anniversary of the Massacre "all public buildings in Oxford remained closed in commemoration."[20] The event also elicited an outpouring of sentiment that elevated Ramus to an iconic stature. One of the earliest poetic responses was Gabriel Harvey's nativity ode on St. Stephen's day, 1574, which drew clear parallels between Ramus' fate and that of the protomartyr, as well as

19 Donald R. Kelley: Horizons of Intellectual History: Retrospect, Circumspect, Prospect, in: *Journal of the History of Ideas* 48 (1987) p. 158.

20 M. R. A. Bullard: Talking Heads: The Bodleian Frieze, its inspiration, sources, designer and significance, in: *Bodleian Library Record* 14 (1994) pp. 461-500, on p. 481.

lavished praise on Ramus' method and its applicability to all learning. Several years later another Cambridge scholar (designated only by the initials J. C.) included "An Epitaph vpon the death of the famous Phylosopher, P. Ramus, who was murdered in the persecution of Fraunce, 26. Augusti. Anno Domini. 1572. Etatis circiter. 57" in a miscellaneous volume that he published in 1579.[21] However, in view of the enormity of the horror that the Massacre evoked among Protestants, we should be careful not to over hastily conclude that such eulogies indicated strict adherence to Ramism. Daniel Rogers, for example, who composed more than one such panegyric, was hardly a Ramist.[22]

Several seventeenth-century commentators noted the symbiotic relationship between Ramus' radical Protestantism and the espousal of his pedagogic ideas by fellow-travelers in England. Herbert Thorndike suggested that the Puritans' predilection for the "democratization" of the Church may well have had its roots in Ramus' endorsement of such a position, since his "credit in our Universities was the first means to bring this conceit in religion among us: for about the time that he was most cried up in them, Brown and Barrow published it."[23] Richard Baxter was even more blunt. When attacking the Independents in 1650, he warned them to desist from advocating Church democracy lest they bring on England a fate similar to that experienced by the French Huguenots: "Is it not strange that so learned a man as Pet. Ramus should be advocate for the multitude's authority in

21 Gabriel Harvey: *Ode Natalitia, vel opus eius feriae, quae S. Stephani protomartyris nomine celebrata est anno 1574. In memoriam P. Rami, optimi, et clarissimi viri,* London 1575; Warren B. Austin: Gabriel Harvey's 'Lost' Ode on Ramus, in: *Modern Language Notes* 61 (1946) pp. 242-247; J. C., gent.: *A poore Knight his Pallace of private pleasures. Gallantly garnished, with goodly Galleries of strang inventions and prudently polished, with sundry pleasant Posies, and other fine fancies of dainty devices, and rare delightes. Written by a student in Cambridge. And published by I. C. Gent,* London 1579.

22 Huntington Library MS HM 31188. Rogers' "In Petri Rami fatum" will be published in a forthcoming paper by Roger Kuin and Anne Lake Prescott: Some Unpublished Manuscript Poems Concerning the Sidney Family. I'm grateful to Dr. Kuin for drawing my attention to this eulogy.

23 *The theological works of Herbert Thorndike,* 6 vols., Oxford 1844-1856, vol. 1/2, pp. 445-446. Thorndike offered an alternative explanation as well: "Unless it be more probable to fetch it from the troubles of Frankfort."

church-government? But that God must use so sharp a cure for those contentions, as that bloody French massacre, methinks should make England tremble to consider it, lest the same disease here must have the like cure."[24]

For his part, Strype opined that the popularity of Ramism in the late sixteenth century "prevailed much; and perhaps brought into the college to be read" on account of "Ramus being a protestant, as well as a learned man." He further speculated that "Digby seemed to oppose Ramus's philosophy chiefly out of a prejudice against him upon the account of religion."[25]

Be that as it may, for those Protestants who, in the course of the late sixteenth and early seventeenth centuries, became increasingly apprehensive of utilizing Catholic textbooks – given their conviction that the Papists, and especially the Jesuits, adulterated their philosophy with their superstitious theology – the manuals of the Protestant saint offered a natural solution.[26] Indeed, the joining of martyrdom and "safe" philosophy was made explicit by the very titles of vernacular Ramist works. Roland MacIlmaine's translation into English carried the simple and powerful title *The Logike of the Moste Excellent Philosopher P. Ramus Martyr,* while Anthony Wotton's 1626 *The Art of Logick* advertised in the subtitle that it was "Gathered out of Aristotle, and set in due forme, according to his instructions, by Peter

24 Richard Baxter: *The Saints Everlasting Rest,* in: *The Practical Works,* ed. William Orme, London 1830, vol. 23, p. 135, cited in Robert M. Kingdon: *Geneva and the Consolidation of the French Protestant Movement 1564-1572,* Geneva 1967, p. 134.

25 John Strype: *Annals of the Reformation and establishment of religion and other various occurrences in the Church of England during Queen Elizabeth's happy reign,* 4 vols. in 7, repr. New York 1968, vol. 2/2, pp. 405-406. Richard Montague's hostility toward both Puritanism and Ramism incited him to denounce dichotomies "which being the darling of the Father of Novellists in Grammar, Philosophie, and unless hee had died opportunely, in Theologie, must needs be the dotage of all such as he." Richard Montague: *Diatribae upon the first part of the Late History of Tithes,* London 1621, p. 341.

26 A sober expression of such an anxiety was made as late as the 1650s by Thomas Barlow. He warned prospective theology students that "Because the most of these Books of Logique wee see, were written by Popish Authors, who that they may defend their wild, and senselesse position of Transubstantiation in Divinitie, are necessitated to maintaine many irrationall and inconsistent Assertions in Logick, especially about the nature of Proprium and Accidens, and of Quantitie; It will concerne the Younger Students to consider before they consent to what they say in those particulars." *A Library for Younger Schollers,* ed. A. Dejordy, H. F. Fletcher, Urbana 1961, p. 2.

Ramus, Professor of Philosophy and Rhetoric in Paris, and there Martyred for the Gospell of the Lord Jesus". Other writers expounded the theme in their prefaces, as did Nathaniel Baxter in his strongly anti-Catholic *In P. Rami Dialecticam quaestiones et responsiones* (London 1585), wherein he all but deified that "martyrem celeberrimum" who had vanquished the verbiage of both the scholastics and modern papists.[27]

Surely, then, Perry Miller hardly exaggerated when he concluded that for the Puritans Ramus "died equally for the cause of logic and of Christ."[28] For a long time they continued to revere and lionize the French martyr. William Ames – who, his biographer tells us, believed that "a true understanding of Ramism was the beginning of good religion" – thought Ramus to have been not only the "greatest master of the arts", but one who "was no less pious than prudent."[29] During the early 1670s, the dissenter Theophilus Gale, who made a formidable effort to construct a comprehensive Christian philosophy, accorded Ramus a place of honor among his intellectual ancestors. "None deserves a more illustrious character among the first reformers of Philosophie," Gale wrote, "than *Peter Ramus,* who followed the steps of *Faber Stapulensis,* for the Reforming of Logic and Philosophie, as also of other ingenuous Arts; which lay buried under the Barbarisme of Scholastic Theologie." And it was Ramus' opposition to Aristotle that "gained him the hatred of *Jacobus Carpentarius* [...] whose scholars imbrued their hands in *Ramus's* bloud, in the general massacre at *Paris,* 1572." Moreover, Gale continued, Ramus "was a person of a very generose and flagrant zele for the Reformation of Ecclesiastic Discipline, as well as Philosophie", and served as a perfect example of "how much our first zelose Reformers thought themselves concerned for the Reformation of Philosophie, in order to a more pure Theologie. And indeed, no wonder, since the main corruptions in Theologie have their origine from some corruptions in Philosophie." Several years later another dissenter, Thomas

27 Nathaniel Baxter: *In P. Rami Dialecticam quaestiones et responsiones,* London 1585, cited in Schmitt: *John Case* (as note 16) p. 51. William Temple made use of the theme as well.

28 Miller: *New England Mind* (as note 13) p. 117.

29 Keith L. Sprunger: *The Learned Doctor William Ames: Dutch Backgrounds of English and American Puritanism,* Urbana 1972, p. 15; William Ames: *The Marrow of Theology,* trans. John D. Eusden, Boston 1968, p. 226.

Lawson, hailed Ramus for "his Love to Protestants, and Testimony against Aristotle", while for Increase Mather he remained "that Great Scholar and Blessed martyr".[30]

Not all Puritans, of course, and certainly not most Anglicans, followed the logic of such adulation into a wholesale adoption of Ramist textbooks. Yet the fact that Ramus died a martyr forced, at least for a few years, a marked deference toward him, including a certain immunization from attacks. Even the leaders of the French Protestant Church, who were set to expel Ramus for his subversive advocacy of Congregationalism, felt obliged following the Massacre to shift the blame for his insurgency to others.[31] English Protestants faced a similar dilemma and the virtual duty, in the aftermath of the event, to pay homage to Ramus makes it even more difficult to evaluate the exact nature of his influence. John Chamber, university lecturer of astronomy at Oxford, offers one such instance. When in 1574 he delivered his inaugural address, Chamber felt compelled to solemnize Ramus "quem omnes amare meritissimo pro eius eximia sanctitate debemus."[32] Whether such an obligatory statement makes Chamber a Ramist, though, is another matter.

Even more telling is the case of John Rainolds. Having embarked on his Rhetoric lectures shortly after the St. Bartholomew's Day Massacre, Rainolds' grudging commendations of Ramus are illustrative of his willingness to exculpate the rhetorical transgressions of the French martyr on confessional grounds, even though as a scholar he was ill at ease with Ramus' innovations. Rainolds found it possible to partially condone Ramus' criticism of Aristotle by arguing that much of that criticism was actually directed against the Sorbonnists and not against Aristotle himself. More importantly, a special dispensation should be allowed for Ramus: "above all, as a holy man, and a martyr in the cause of Christ, he is deserving of love." Rainolds

30 Theophilus Gale: *The Court of the Gentiles,* London 1672-1677, Part four, Preface; Thomas Lawson: *A Mite into the Treasury,* London 1680, p. 16. Another nonconformist divine, John Quick, extolled that "most Godly and Learned Scholar" who "received the Crowne of Martyrdome." Cited in Keith L. Sprunger: John Yates of Norfolk: The Radical Puritan Preacher as Ramist Philosopher, in: *Journal of the History of Ideas* 37 (1976) p. 698.

31 Kingdon: *Geneva* (as note 24) p. 113. See also pp. 96-111 and *passim* for Ramus' seditious, if short-lived, activity as Protestant provocateur.

32 John Chamber: *Astronomiae Encomium,* 1601, p. 6.

impressed upon his auditors that "Ramus was harrassed by the Scotists, driven from the Schools, heaped up with calumny and placed in peril of his life." Waxing eloquent on one occasion, Rainolds wrote: "These reasons moved Peter Ramus, an excellent man of piety and erudition, to assail Aristotle perhaps a little too sharply. He certainly was not moved by a desire for calumny, so far as I can surmise from conjecture, but by a righteous grief of the soul that the Sorbonnists still believed in Aristotle even when Aristotle was wrong, while other writers who spoke the truth where suspected of heresy."

Yet, the specific contribution that Ramus was credited with in the lecture – setting up "only two forms of adducing proofs", the syllogism and the enthymeme – is a dubious compliment, for Rainolds immediately proceeded to add that "Rudolf Agricola seemed almost to have thought the very same notion." Clearly, then, while ideologically aligning himself with Ramus against the hated Sorbonnists, the denominational charity that shielded the Frenchman from specific criticism was not extended to making him an intellectual authority. Rainolds was unwilling to endorse more than certain minor contributions made by Ramus – hardly a significant recommendation given his prior commitment to Agricola and Vives, the sources of many of Ramus' own reforms. As Green pointed out, "nowhere does Rainolds go so far as to urge Ramus's approach to the practice of rhetoric and dialectic, and Ramus finds no place in Rainolds's crucial summary in the final lectures." Ramus ends up as a marginal figure in Rainolds' lectures and this should warn us against turning everyone who mentions Ramus into a disciple. Certainly, on several occasions, as we shall see below, Rainolds cautioned his auditors not to emulate Ramus' conduct.[33]

By the 1590s, though, the memory of the St. Bartholomew's Day Massacre had sufficiently faded to allow for more pronounced censures. But before chronicling the debate over Ramism and learning it is important to elucidate that other factor critical in shaping the reaction to Ramism. Earlier I cited Ong's observation that "the pattern of Ramism" manifested itself

33 *John Rainolds's Oxford Lectures on Aristotle's Rhetoric,* ed. Lawrence D. Green, Newark 1986, pp. 181. 87; James K. McConica: Humanism and Aristotle in Tudor Oxford, in: *English Historical Review* 94 (1979) pp. 306-307. In his subsequent dialectic lectures Rainolds "faults his co-religionists for being somewhat too sharp in disagreeing with Aristotelians [...] and even too acid with Aristotle himself." *John Rainolds's Oxford Lectures,* p. 55.

rather in "a good deal of shouting for or against Ramism by sophisters or other youthful university disputants". Unfortunately, historians have failed to appreciate the significance of the context of strife to the fortunes of the movement in England.

The gateway to this issue should be an evaluation of what early modern Englishmen perceived to be Ramus' contribution to logic. Most modern commentators share Ong's position that Ramus was a veritable humanist who situated logic firmly within the domain of language and the arts of discourse. Ramus' definition of logic as *ars bene disserendi,* Ong insisted, should be understood strictly in terms of "the art of discoursing well" – "expatiate" being the nearest English equivalent – though other meanings of *disserere,* such as to examine and to argue, were also implied.[34] I'm not so sure. Not only did Ramus himself equate in the 1543 *Dialecticae institutiones* the verb *disserere* with *disputare,* but the 1555 French edition of the *Dialectique* also defines logic as the "art de bien disputer"; it was the editor of the posthumous 1576 edition who substituted "l'art de bien raisonner".

More important for our purpose here, though, is that English readers of Ramus understood him to offer a facile tool for argumentation. Indeed, it was precisely that feature that made Ramist logic attractive to Puritans – as it did to other reformers on the Continent[35] – in the final decades of the sixteenth century. It offered them a compact and practical manual that promised success in engaging members of the established Church in disputes, without any considerable investment of time and labor. Appropriately, Roland MacIlmaine, Ramus' first English translator, defined logic as "an arte which teacheth to dispute well", as did the Puritans Robert Fage and Henry Ainsworth several decades later. Pierre de la Primaudaye's *The French Academy* also advertised that logic taught "the truth of althings by disputation." Indeed, it was precisely such an advertised objective of logic that prompted the humanist man of letters Abraham Fraunce to reject Ramus' definition. Fraunce repudiated Ramus (and Beurhaus) for using the verb *disputare,* and even *disserere,* to denote the overall aim of logic for, he

34 Ong: *Ramus* (as note 1) pp. 178-180. I shall present below my doubts concerning the appropriateness of regarding Ramus as a true humanist.

35 As Elsmann points out in his article, Johannes Molanus advocated the embracing of Ramism in no small part for its being better suited for confessional confrontations.

argued, both fall short of encompassing "the generall use of this art", limiting as they do the aim "to those two functions belonging to methode" alone. Only the "art of reasoning", Fraunce was convinced, was capable of denoting the subject in its entirety. Fraunce's less than wholehearted allegiance to Ramism is further evident in his purging from the Ramist definition of logic the adverb "well" which, he muttered, "might as well have been left out.[36] An equally committed humanist, John Milton, concurred. He was as explicit as Fraunce in rejecting Ramus' conceiving of logic as "ars bene disserendi" because, Milton argued, *disserere* "generally has no wider sense" than *disputare;* therefore he, too, opted for "ars bene ratiocinandi".[37] Further to substantiate my claim that contemporaries understood Ramist logic to promise success in disputations can be inferred from Christopher Marlowe's repudiation in *Doctor Faustus* (written ca. 1588) of what the Cambridge Ramists most esteemed in logic:

> Sweet Analytics, 'tis thou hast ravished me.
> *Bene disserere est finis logices.*
> Is 'to dispute well logic's chiefest end'?
> Affords this art no greater miracle?[38]

I should hasten to add that I do not mean to argue that Ramus and at least some of his followers had no interest in discourse more broadly.

36 Abraham Fraunce: *The Lawiers Logicke,* London 1588, pp. 2-4; R[obert] F[age]: *Peter Ramus* [...] *his Dialectica,* London 1632; Zachary Cook: *The Art of Logick: or the Entire Body of Logick in English unfolding to the meanest Capacity the way to dispute well, and to Refute all Fallacies whatsoever,* London 1654. J. S. Measell: The Authorship of the Art of Logick (1654), in: *Journal of the History of Philosophy* 15 (1977) pp. 321-324 made the case for Ainsworth being the author. However, as Serjeantson demonstrated, the work is, in fact, an abbreviated (and unacknowledged) translation of Bartholomäus Keckermann's *Systema Logicae*. See R. W. Serjeantson: Testimony and Proof in Seventeenth-Century England, in: *Studies in History and Philosophy of Science* 30 (1999) p. 207, n. 63. Pierre de la Primaudaye: *The French Academy,* London 1602, p. 72. An exception to the rule may be found in the Puritan Dudley Fenner who also defines logic as "an Art of reasoning." *The Artes of Logike and Rethorike, plainelie set foorth in the English tounge, easie to be learned and practised,* 1584.

37 Milton: *Complete Prose Works* (as note 1) vol. 8, pp. 217-218.

38 Christopher Marlowe: *The Complete Plays,* ed. J. B. Steane, Harmondsworth 1969, p. 266.

Rather, the point I wish to make is that in addition to proving deleterious to formal logic and erudition more generally – an issue to which I shall return – Ramism actually turned into a wanton vehicle for fomenting disputes. For by dispensing with large portions of formal logic and its terminology – generating instead an axiomatic art requiring little learning – and by introducing the hydra of disjunctive syllogism, the Ramists facilitated the acquisition of argumentative proficiency, thereby multiplying disputes in and out of the schools. Ramism, in other words, could be charged with the same vice that the humanists found to pervade scholasticism: namely, that the sole purpose of the art was to triumph in debates rather than search for truth![39]

The promise of such facile dexterity endeared Ramism to Puritans and contributed to their becoming virtually indistinguishable. In 1574, for instance, John Barebone of Magdalen College, Oxford, was denied his MA because in his exercises for the degree the young Ramist vehemently attacked Aristotle. The moral of the tale, however, can be found not in the requirement that Barebone defend "the Opinions of Aristotle in three questions against the minds of all other Philosophers" before he be allowed to repeat his disputations, but in the injunction that he begin his exercises by apologizing to the other Regent Masters for the offense he had formerly given in disputing against them with excessive acrimony. Surely Howell was correct to conclude that rather than reflecting the authorities' hostility toward Ramus, the episode demonstrates a "lack of enthusiasm for a contentious young colleague who had already spent three years among them as fellow." The verity of such a judgment is borne out by Barebone's persistence in his contentious behavior in his own college, where he appears to have been the ring-leader of a group of young Puritan fellows who waged a rearguard campaign against President Humphrey. Such activity ultimately

[39] Brucker commented on this feature of Ramist logic: It had "the merit of turning the art of reasoning from the futile speculations of the schools to forensic and common use; but his plan is defective, in confining the whole dialectic art to a single object of disputation." William Enfield: *The History of Philosophy* [...] *drawn up from Brucker's Historia Critica Philosophiae,* 2 vols., London 1819, vol. 2, p. 530. Howell, too, noted that Ramist logic "was conceived, organized, and taught less as a master guide to philosophical enquiry than as a master guide to controversy". *Eighteenth-Century British Logic* (as note 11) p. 16.

resulted in Barebone's expulsion from the college in 1578.[40] That Ramism served as an extension of religious militancy can further be surmised from John Penry's career at Peterhouse, Cambridge. Thomas Nashe's "biographical" account of Penry is undeniably biased and abusive, yet his description of the manner in which Penry joined an attack on Aristotle, in the name of Ramism, with an attack on President Andrew Perne in the name of Puritanism, must nonetheless be taken as essentially factual. While still a first-year undergraduate who "had scarce tasted, as we say, of *Setons modalibus*", Nashe wrote, Penry

> began to affect factions in art, and shew himselfe openly a studious disgracer of antiquitie. Who then such an unnatural enemie to *Aristotle,* or such a new-fangled friend unto *Ramus*? This one thing I am sure of, hee never went for other then an asse amongst his companions and equalles, yet such a mutinous block-head was he alwaies accounted that through town and Colledge he was commonly called the seditious dunce. For one while he wold be libelling against *Arist.* And all his followers he knew, another while hee would all to be-rime Doctour *Perne,* for his new statutes, and make a by-word of his bald pate, yea, had the Dean, President, or any other officer never so little angered him, they were sure ere the weeke went about to have hard of it, in some libell or other.[41]

Small wonder, then, that just as Barebone was expelled from Oxford for his intransigence, so did Penry's bellicosity result in his forced migration to Oxford, shortly after graduating BA in 1585, in order to complete his studies.

Another example of the stirring up of disputes in the schools can found in the well known anecdote involving William Gough, who entered King's College, Cambridge, in 1595 and graduated MA in 1602. According to his biographer

> at the first entrance of his studies, he applied himself to *Peter Ramus* his Logick, and grew so expert therein, that in the publick schools he maintained and defended him, insomuch as when on a time divers *Sophisters* set themselves to vilifie *Ramus,* for which end the Respondent had given this question, *nunquam erit magnus, cui Ramus est magnus,* which some of the *Sophisters* hearing, and knowing the said *William*

40 Anthony Wood: *History and Antiquities of the University of Oxford,* ed. John Gutch, 2 vols. in 3, Oxford 1792-1796, vol. 2, p. 176; Howell: *Eighteenth-Century British Logic* (as note 11) pp. 189-190; C. M. Dent: *Protestant Reformers in Elizabethan Oxford,* Oxford 1983, pp. 55-58.

41 Thomas Nashe: *An Almond for a Parrat,* in: *The Works of Thomas Nashe,* ed. Ronald B. McKerrow, 5 vols., London 1903-1910, vol. 3, pp. 367-368.

> *Gough* to be an acute disputant, and a stiff defender of *Ramus,* they went to the *Divinity* Schools, where he was then hearing an *Act,* and told him how in the other Schools they were abusing *Ramus,* he thereupon went into the *Sophisters* Schooles, and upon the Moderators calling for another Opponent, he stepped up, and brought such an argument as stumbled the Respondent, whereupon the Moderator took upon him to answer it, but could not satisfie the doubt: This occasioned a *Sophister* that stood by to say with a loud voice, *Do you come to vilifie Ramus, and cannot answer the Argument of a Ramist?* Whereupon the Moderator rose up, and gave him a box on the ear, then the School was all an uproar; but the said *William Gough* was safely conveyed out from amongst them.[42]

I shall dwell on the significance of the topic for disputation below but it is important to comment first on contemporary perceptions of controversies as generating attitudes toward Ramism. The fracas contrived by Gough – a seasoned master of arts who had already embarked on his theological studies – during the exercises of second or third year undergraduates, as well as other incidents, are indicative of the insatiable belligerence that was endemic to Puritan Ramism.[43] Such quarrelsomeness, as is well known,[44] was a European-wide phenomenon, and it was the main reason for the official efforts to curb Ramism. In fact, even had there existed no affinity between Ramism and Puritanism, Oxbridge authorities would have become apprehensive at the sight of controversies in the schools. After all, the preservation of law and order, almost as much as religion, was uppermost in the minds of university and state officials, and injunctions aimed at ensuring it were issued regularly. Thus, the Cambridge Chancellor, Sir William Cecil, exhorted in 1566 the Vice-Chancellor to tighten up discipline and repress

42 Samuel Clark: *Lives of Thirty-Two Divines,* London 1677, p. 235.

43 I recognize, of course, that the topic of disputation involved anti-Ramist provocation, for which there exists more evidence. For example, the ruling of the Vice-Chancellor's Court that "Albeyt wee of the universitye for Philosophye matters have preferred Aristotle, yet Ramus ys not to be abused with reproachfull words." Yet I believe that it was the Ramists' provocations that were more unruly and which set a reaction in motion. Not long after the incident involving Gough, two young scholars of Clare pasted a picture of Ramus on a rope and "rayle uppon Christs College, Emmanuel and Kynges College". And it did not take long before "from knocking they came to blows." The quotes are drawn from Sarah Bendall, Christopher Brooke, Patrick Collinson: *A History of Emmanuel College Cambridge,* Woodbridge 1999, p. 76.

44 "Rámus" is still a Czech locution for "noise".

"the lightness and disordered behaviour" of the youth.[45] The flare up of Puritan and then Ramist debates in the following decade merely aggravated matters and prompted officials to take peremptory measures. Furthermore, it was the provocation of the Puritans and the Ramists that unsettled even potential sympathizers, who both shunned the notoriety sought by their more radical colleagues and were genuinely apprehensive of the effects of reckless disputes on adolescents. Accordingly, Thomas Cartwright's notorious lack of moderation and his utilizing the Lady Margaret Professorship at Cambridge as the platform from which to attack the establishment, was a major cause of alarm to even such sympathizers as Archbishop Grindal, who complained in 1570 that the "youth of the universitie [...] frequente his lecture in a greate number and therfore in daunger to be poysoned by hym with love of contention and lykynge of novelties and to become hereafter not only unprofitable, but also hurtful to the Church."[46] Coming on the heels of both the debates over Puritan teachings and the related debates over the 1570 statutes – both involving predominantly younger members of the university – the Ramist agitators could not but be seen as fanning the fire.[47]

Understandably, in England as well as on the Continent, many came to doubt whether what was valuable in Ramism was worth the strife and dissension that invariably followed. It is instructive in this respect to recall that when Ramus informed Cardinal de Lorraine that he was invited to teach in Bologna, his patron responded: "Go [...] and you will free France from great worry, trouble and fear; you will also utterly perturb the whole of Italy by going there and teaching, and perhaps bring a great danger on yourself."

45 Charles Henry Cooper: *Annals of Cambridge,* 5 vols., 1842-1908, vol. 2, p. 230.

46 J. B. Mullinger: *The University of Cambridge,* 3 vols., Cambridge 1873-1911, vol. 2, p. 216.

47 It is noteworthy that William Temple's fierce advocacy of Ramus elicited several attempts at refutations – apart from those by Everard Digby's – though none were published. In 1581, for example, a year after graduating BA from Christ's College, Cambridge, Andrew Willet "undertook the defence of his master Aristotle, in a Latin tractate against Temple, a lover of Petrus Ramus." Two years later, an Oxford student informed a correspondent that Thomas Savile of Merton College and John Williams of All Souls were also writing against Temple. Thomas Fuller: *Abel Redevivus,* ed. William Nichols, 2 vols., London 1867, vol. 2, p. 317; R. W. Jeffrey: History of the College: 1547-1603, in: *Brasenose College Quatercentenary Monographs,* ed. F. Madan, 2 vols., Oxford 1909, vol. 2, monograph 10, p. 57.

Hardly surprising, then, that Joseph Scaliger opposed the introduction of Ramism into Leiden on the grounds "that the inevitable quarrels between Ramists and Aristotelians would destroy the institution" or that other influential statesmen on the Continent – Beza in Geneva and Ursinus in Heidelberg – thought the same. Even those who were well disposed toward Ramus were made ill at ease by his and his followers' contentiousness. Though revering Ramus' adherence to the cause of Protestantism, John Rainolds was bothered by the vehemence of Ramus' attack and, as he loved moderation and feared for Oxford's tranquility, he warned the auditors of his lectures on dialectics not to follow such an example, going so far as to cite Beza's decree at Geneva which mandated that "in interpreting authors, violent abuse is to be avoided in favour of a faithful explanation of their meaning."[48]

As I hinted above, the strife and dissension generated by the Ramists was exacerbated by the fact that undergraduates, who were in possession of much passion but little judgment – and who were visibly delighted to witness any attack on authority, justified or not – were most often involved in such debates. Gabriel Harvey, who knew more about student sub-culture than most people, evoked his own experience when defending himself against the appeal that Nashe's attack on him held for such a constituency: "Come young Sophisters, you that affecte raylinge in your disputations, and with a clamorous howte would set the Philosophy schooles *non plus.*" The pubescent predilection for contention and fashion was also wryly commented upon by John Case:

> I cannot but acknowledge that the youthful ardor of mind in both universities has of late been fighting it out to determine whether in the mastering of the arts the great acuteness of Aristotle is of more worth than the flowing genius of Ramus. But, as I expect, the young exalt such apostasy from the true experience of age and from the wise old custom of philosophizing, because beardless youth often does what white hair denies to have been rightly done [...] I do not blame Ramus in this, for he was

48 Peter Sharratt: Nicolaus Nancelius, *Petri Rami Vita*. Edited with an English Translation, in: *Humanistica Lovaniensia* 24 (1975) pp. 253-254; Anthony Grafton: *Joseph Scaliger: A Study in the History of Classical Scholarship,* 2 vols., Oxford 1983-1993, vol. 1, p. 336, n. 160; E. Winkelmann: *Urkundenbuch der Universität Heidelberg,* 2 vols., Heidelberg 1886, vol 1, pp. 311-312; McConica: Humanism and Aristotle (as note 33) p. 307. See also Rother's article in this volume.

learned; I rather exalt Aristotle, for he stands out above all. But perhaps the young will hold my work the poorer because I name in it the old interpreters of Aristotle.[49]

It is against the background of official anxiety, then, that we need to read the 1586 Oxford statute: "it was ordered for the future that all Bachelaurs and Undergraduates in their Disputations should lay aside their various Authors, such that caused many dissensions and strifes in the Schools, and only follow Aristotle and those that defend him, and take their Questions from him, and that they exclude from the Schools all steril and inane Questions, disagreeing from the antient and true Philosophy."[50] McConica's gloss on the decree is undoubtedly accurate: while prescribing Aristotle in an effort to rein controversies, the statute also made it explicit that it discouraged "the pursuit of scholastic subtleties."[51]

Closely related to youthful zeal in engendering disputes was the desire for glory. At least ever since Erasmus, contemporaries commented on the propitiousness of notoriety to the advancement of one's fortunes. He who craves "immortality of Fame by setting out Books", Erasmus quipped, should desist from incessant labor, to the ruin of his health and wealth, in the hope of producing a learned tome. The only author to be read and admired is he "who never studies for anything but puts in writing whatever he pleases or what comes first to his head, though it be but his dreams; and all this with small waste of paper, as well knowing that the vainer those trifles are, the higher esteem they will have with the greater number, that is to say all the fools and unlearned."[52] Taking his cue from Erasmus, Robert Burton's youthful play *Philosophaster* introduces as the quintessential shallow philosopher the character of Polupragmaticus who, I believe, targeted not

49 *The Works of Gabriel Harvey,* ed. Alexander B. Grosart, 3 vols., repr. New York 1966, vol. 2, p. 42; John Case: *Speculum moralium quaestionum,* Oxford 1585, fol. 5, quoted in Wilbur S. Howell: *Logic and Rhetoric in England 1500-1700,* Princeton 1956, pp. 190-191. See also McConica: Humanism and Aristotle (as note 33) p. 300.

50 Wood: *History University of Oxford* (as note 40) vol. 2, p. 226. For the original see Strickland Gibson: *Statuta Antiqua Universitatis Oxoniensis,* Oxford 1931, p. 437. In 1589 Chancellor Christopher Hatton sent a letter reiterating earlier injunctions. Wood, vol. 2, p. 239.

51 McConica: Humanism and Aristotle (as note 33) p. 301.

52 Desiderius Erasmus: *The Praise of Folly,* trans. John Wilson, London 1668, repr. Ann Arbor 1967, pp. 88-89.

only the Jesuits but Ramus as well. At one point, Polupragmaticus – the very name is telling – advises Simon Actius that "the shortest road to fame" would require the following: to "declaim frequently and with great force in the public schools", to persist in making public appearances, and to "publish the work of someone else in your name with certain notes and additions". Furthermore, "if you know of some famous writer, oppose him. Whatever he says, deny it. Tear him to pieces, reproach him, rebuke him, scorn, whistle, or hiss him out of the place." Better still, "found a new sect. You will immediately have many followers." It would help to advance paradoxes and then "disregard or ridicule anyone who attacks" you. In brief, "if you want to be distinguished and considered a first-rate philosopher, you must be sharp, fierce, bold, impudent, a pimp, a parasite, vain, a fox, an ape, a haughty defender, one who tosses about all knowledge, a scoffer, a speaker of lofty things, a man for all seasons. Thus will you grow rich and become famous most quickly."[53]

Gabriel Harvey acknowledged in 1592 his own craving for reputation a decade and a half earlier: "I was then yong in yeares, fresh in courage, greene in experience, and as the manner is, somewhat overweening in conceit: and for varietie of study, and some deeper intelligence in the affayres of the world, otherwhiles reading invectives, and Satyres, artificially amplifyed in the most exaggerated and hyperbolicall kinde." Elsewhere, the onetime Ramist admitted that "there was a time, when I floted in a sea of encountring waves; and devoured many famous confutations, with an eager, and insatiable appetite [...] in that hott contradictory course of Logique, and Philosophy."[54] Now, Harvey was neither a Puritan nor, as we shall see below, could his deep-seated humanism easily coexist with unadulterated Ramism. In his case, it was an uncontainable desire to distinguish himself that prompted Harvey to jump in the mid-1570s on the Ramist band wagon, after he had convinced himself that not only did Ramus (and Talaeus) offer innovative reforms in the domain of rhetoric, but that tying his star to the

53 Robert Burton: *Philosophaster,* trans. Connie McQuillen, Binghamton: Medieval and Renaissance Texts and Studies 1993, pp. 137-141.

54 *Works of Harvey* (as note 49) vol. 1, p. 178. vol. 2, p. 45. Collinson aptly described Harvey as "impetuous if not factious". See Patrick Collinson: Andrew Perne and his Times, in: Patrick Collinson, David McKitterick, Elisabeth Leedham-Green: *Andrew Perne: Quatercentenary Studies,* Cambridge 1991, p. 23.

new fashion would assure his academic promotion. Whether he was aware of it or not, Harvey even adopted some of Ramus' mannerisms, both in his apparel and disputations. Unfortunately for him, Harvey miscued in estimating the fortunes of the new fashion and his career never took off. Still, it is certainly justified to generalize from his experience on the correlation between Ramism and controversies. As late as 1640 one of Samuel Hartlib's correspondents commented that it was precisely because Jacobus Acontius would not "dispute [and] oppugne some famous Adversarie and because hee was not a Professor making many disciples therfore hee could make no sect but remaines obscure which is also the reason why Verulam Herbert Cartes are no more taken notice of. The contrary is in Ramo who quarrels etc. and hence more celebris in a short time."[55]

In recognizing that quarrels bestowed renown (or infamy) on their protagonists, we should also contemplate whether they also created an optical illusion relative to the significance of Ramism. After all, Puritan radicalism was often masqueraded as debates over Ramism and, consequently, should not be considered as necessarily indicative of a deep concern for Ramist ideas, especially as much of the evidence for the identity of Ramists is derived from anecdotes in the lives of various Puritans. And the fact that the corroborative evidence comes primarily in the form of Puritans' utilizing Ramist charts or engaging with a logical analysis of a scriptural text should further impress upon us the need to avoid hasty conclusions concerning the nature and scope of the movement.

While obstreperousness accounted for some of the notoriety gained by Ramism – as well as helps explain the official reaction to it – boisterousness alone is insufficient to elucidate the deeper causes for disapprobation of the movement and what it represented. Such censure was grounded on the perceived pernicious effects that Ramism held for erudition. This was articulated by numerous scholars and it was this fear which ensured that, at least in England, Ramism never transcended the propaedeutic level. Grafton and Jardine somewhat obscure this when they argue that the *succès fou* enjoyed by Ramism derived from the fact that Ramus "deliberately discarded the difficulty and rigour of high scholastic schooling and thereby attracted those who regarded education as a means to social position rather

55 Sheffield University Library, Hartlib Papers, Ephemerides (1640) 30/4/39B.

than as a preparation for a life of scholarship (or of theological debate)."[56] Ramus did indeed dispense with erudition, but the general reception of this attitude was not to occur for another century. Late sixteenth- and early seventeenth-century educated men still held learning in very high regard and it was such wide esteem, not alleged Aristotelianism, that was inhospitable to Ramism.

Part of the problem in recognizing that it was genuine concern for learning that animated the opponents of Ramism has to do, of course, with the iconic image of Ramus as the slayer of Aristotle and scholasticism. Necessarily, any defense of tradition could easily be construed as evidence for the persistence of obscurantism. An English case in point is Godfrey Goodman's vitriol against Ramus who, he thundered, was

> a man of meane knowledge, without any ordinary parts either of nature or learning; yet though his owne impudency and pride, (desiring an innovation of all Arts, to cast all things in a new mould, to build his own fame in the ruins of others), hath opposed himself, not against any particular opinion, but against the whole streame and current of all ancient learning; overthrowing the grounds, principles, and rules of that most illustrious and thrice renowned *Aristotle,* whom all ages doe reverence and acknowledge to have bin natures chiefe secretary, or best counseller of estate, the father of all humane knowledge: and that which is yet more strange, this *Ramus* hath left followers and Disciples behind him, who seeme to be like affected to the ancient learning, reducing all things, (even the very scope and foundation of arts), to their own fond invention, and barbarous innovation. A shame it is, that Colledges and ancient foundations, should give harbour to such a wretched brood; who as they are first nourished up in obstinacy and wilfulnes, so they will prove to bee firebrands in the Church; who neglecting the fathers, and all ancient orders and decency, shall with their owne furious and witles conceits, set the whole world in combustion.[57]

Goodman was an inordinately conservative clergyman, noted as much for his impassioned defense of the idea of the Decay of Nature as for his insistence that the "moderns" could contribute virtually nothing new to learning. Consequently, it is tempting to relegate his judgment of Ramus, like his belittling of other neoterics, to oblivion.

Nevertheless, at least as regards Ramus, Goodman was neither unjustified in his criticism nor unrepresentative of the opinion expressed by far greater scholars. The notable Continental *érudits,* for example, were never

56 Grafton, Jardine: *Humanism* (as note 3) p. 168.

57 Godfrey Goodman: *The Fall of Man,* London 1616, p. 268.

more than civil when writing about Ramus. Thus, it was undoubtedly deference for the ultimate sacrifice endured by Ramus in the cause of Protestantism that prevented Scaliger – who loathed Catholicism with palpable passion – from becoming more caustic against his compatriot in print. He did abuse Ramus in one of his poems and so little esteemed his work that the very fact that several publications by Ramus were included in Franciscus Gomarus' library fortified Scaliger in his opinion that Gomarus was not truly learned. More generally, Scaliger held Ramus and his disciples responsible for producing pernicious effects on scholarship as well as for effecting perilous disputes in the schools.[58] Isaac Casaubon was even more explicit in his censure, partly because he produced in 1590 an edition of Aristotle. There he deplored the impoverished, albeit fashionable, "philosophy" of Ramus, and derided such logicians who had forgotten that logic is merely an instrument, "not the be-all and the end-all." As late as 1613 Casaubon refused to intercede with King James I on behalf of Alexander Hume, who had sent Casaubon a copy of his Ramist grammar to further his application, since – Casaubon bluntly told the supplicant – he disapproved of each and every one of Ramus' innovations. However, as pointed as his derogatory remarks concerning Ramus' "contribution" to philosophy might have been, Casaubon's contempt for Ramus' (and his followers') ignorance of *bonae literae* was far more profound. As Glucker pointed out, Casaubon was impressed neither by Ramus' pretensions to be a great humanist nor by his claim to have returned "ad fontes" when he attempted to purify Aristotle and other ancients. For Casaubon, the fashionableness of one who claimed to have "reformed" Aristotle though ignorant of Greek and who mutilated classical texts was a sad reflection on the impoverishment of genuine humanist criticism. In this respect, it is telling that when he carped in 1605 at Jacobus Carpentarius' diminutive philosophical and linguistic skills, Casaubon referred to the one who had been Ramus' most bitter adversary as "ille Petri Rami aemulus."[59] The third literary giant of the late sixteenth century, Justus Lipsius, concurred. He endorsed Casaubon's ver-

58 *Scaligerana,* Cologne 1695, p. 358; Carl Bangs: *Arminius: A Study in the Dutch Reformation,* Grand Rapids 1985, p. 247. See p. 150 above and pp. 173-174 below.

59 J. Glucker: Casaubon's Aristotle, in: *Classica & Mediaevalia* 25 (1964) pp. 274-296.

dict on Ramus in one of his letters and told a correspondent that "you will never be a great man if you think that Ramus was a great man."[60]

Such a crushing dismissal of Ramus brings us back to the Cambridge disputation that was interrupted by William Gough. Though often cited, scholars have failed to recognize that the proposition defended by the respondent – "nunquam erit magnus, cui Ramus est magnus" – was actually lifted out of the above-mentioned Lipsius letter, and this should further help to illuminate the nature of the English dismissive attitude toward Ramism. In other words, rather than being a debate over specific logical propositions or method, by the 1590s the context within which Ramism was discussed was more often than not that of erudition (and Puritanism). As I've shown elsewhere, it was precisely during the last two decades of the sixteenth century that a true renaissance of classical learning and philosophical freedom began establishing itself in the English universities and this made both the disparagement of learning and the dogmatism inherent in Ramism the more glaring and, hence, the more reprehensible.[61]

Before proceeding to further substantiate an essentially humanistic interpretation of the English response to Ramism, I should like to defend the more controversial claim that Ramus can not be truly regarded as a humanist according to the expansive discernment of the ideal by his contemporaries. As a teacher of certain humanist disciplines he was certainly phenomenally successful. Yet, not only did Ramus exclude such key humanist fields as history, poetry, and moral philosophy from his system, but he was insensible (not to say hostile) to the most characteristic features of late Renaissance humanism: the passion for books and literature. There is no need to rely on the accusations of his many enemies in order to substantiate

60 *Iusti Lipsi Epistolae,* ed. A. Gerlo, M. A. Nauwelaerts, H. D. L. Vervliet, Brussels, 1978ff., vol. 1, pp. 439-440. vol. 3, p. 296. In a similar manner, Cardano had probably Ramus in mind when inveighing against "those who impudently condemn great men and found new sects without having first digested the ancients, to the great detriment of the human race and of good letters."

61 Mordechai Feingold: Reversal of Fortunes: The Displacement of Cultural Hegemony from the Netherlands to England in the Seventeenth and Early Eighteenth Centuries, in: *The World of William and Mary,* ed. Mordechai Feingold, Dale Hoak, Stanford 1995, pp. 234-261; Feingold: Schooling in Erudition: The Classics in Early Modern English Culture (forthcoming).

such a lacuna. It is Ramus' close friend and biographer, Nicolas de Nancel, who provides us with ample testimony for Ramus' dull imagination as well as his indifference to poetry and books.

Nor was it just that Ramus was ignorant of (and uninterested in) Greek, a condition that Nancel attempted to gloss over by saying that Ramus "applied himself to Greek late, and dabbled in it rather than got to know it thoroughly [...] scarcely ever read or studied any other Greek author" than Plato, Aristotle, and Demosthenes. Rather, the creative energy itself was adjudged missing by Nancel: "He was perhaps lacking in imagination, since he rarely did anything but make use of other people's ideas, polishing them up and embellishing them" – undoubtedly because of his "hankering after novelty." More deficient still was Ramus in poetry: "just as he excelled in eloquence, so he was very much a novice in poetry and little versed in it" wrote Nancel, adding that Ramus hardly ever composed verse. Significantly, the biographer also informs us that Ramus "rarely associated with poets, as though they did not have common interests." Once he did attempt "like Apollo" to preside over a group of famous poets that he had invited for dinner, but the result was a fiasco and never repeated. Certainly, beyond the little service rendered for him by the Pléiade poets, the relations between Ramus and the literati were negligible. Indeed, said Nancel, Du Bellay and Rabelais were among the poets who attacked him. And yet, having completed his rhetoric, the pugnacious Ramus

> decided to turn his attention to poetics [...] (though he had less natural inclination for it) and just as he had originally intended to run through and comment on all Cicero's speeches in order, he now proposed to explain in the same way the whole of Virgil [...]. Starting from the *Bucolics,* he went on the *Georgics,* in spite of the laughter and mockery of some envious detractors, and made a distinguished attempt at unravelling and explaining each of these works.

Equally damning is Nancel's testimony that Ramus lacked the most characteristic trait of a humanist: "from the time I first knew the man I scarcely ever saw him merely reading books in order to cultivate his mind, but always writing away."[62]

62 Sharratt: Nancelius, *Rami Vita* (as note 48) pp. 249. 207. 254. 193. 231. The attack by the authors of the *Port Royal Logic* is noteworthy in this respect. After dismissing Ramus' method as injurious to sound judgment they jeer at his claim that poets com-

One additional element requires consideration. Ramus has often been accused – and not without reason – of failing to acknowledge his deep indebtedness to the very ancients whom he had so savagely attacked as well as of pillaging material from more recent authors and passing it off as his own. If we further remember that Ramus was engaged in a massive collaborative enterprise – and modern scholars have yet to acknowledge the extent to which assistants, students, and colleagues were, if not the authors, at least vital co-workers on significant portions of his rhetorical and mathematical works[63] – then we are in a better position to understand the attribution to Ramus, by Turnèbe and other contemporaries, of the nickname *usuarius.* Again, Nancel attempts to make a virtue of necessity by rendering a positive gloss on what he himself admitted to have been a calumny: Ramus, he wrote, was styled "'usufructuary' by his opponents, who wanted to brand him with infamy, because he showed that all the liberal arts were to be directed to the benefit of human life." Undoubtedly, the intention of his opponents was quite different. Ramus was rather seen as "the usufructuary, the man living off the income of property belonging to someone else".[64]

If Nancel recognized the limitations of Ramus' humanism, one might imagine what non-admirers thought. Being a fluent, animated and effective Latin speaker and teacher were certainly necessary qualifications, but they were not sufficient. It has been said that "readiness of speech and grace of

posed according to rules: "Had Virgil stopped to run through the Topics in search of an argument the lines would never have been written. To write such noble and spirited verse Virgil had to forget not only these rules (had he ever known them) but, in a sense, even himself that he might capture the passion he would represent." Antoine Arnauld: *The Art of Thinking: Port-Royal Logic,* trans. James Dickoff, Patricia James, Indianapolis 1964, pp. 238-239. I thank Richard Serjeantson for reminding me of this reference.

63 Ramus' mathematics alone is of value, Scaliger opined, yet he is not its author. *Scaligerana* (as note 58) p. 333.

64 Sharratt: Nancelius, *Rami Vita* (as note 48) p. 213; Walter J. Ong: Ramus: Rhetoric and the Pre-Newtonian Mind, in: *English Institute Essays,* ed. Alan S. Downes, New York 1954, p. 152. Such was certainly Turnèbe's intention. Convinced, as he was, that Ramus' ignorance precluded him from understanding classical authors and culture, Turnèbe gave Ramus the *sobriquet* '*usuarius*' in order to denote someone who had appropriated "to his own self-indulgent ends" Cicero and other ancients. John Lewis: *Adrien Turnèbe (1512-1565): A Humanist Observed,* Geneva 1998, pp. 224-225.

delivery have often been known to co-exist with but a slender amount of deep and solid learning", and this holds true for Ramus. In this context, Robert Burton seems again to have had Ramus in mind when he mocked his unlearned countrymen who could not distinguish between a scholar and "him that by reason of a voluble tongue, a strong voice, a pleasing tone, and some trivantly polyanthean helps, steals and gleans a few notes from other men's harvests, and so makes a fairer show than he that is truly learned indeed: that thinks it no more to preach than to speak, 'or to run away with an empty cart,' as a grave man said: and thereupon vilify us and our pains; scorn us and all learning."[65]

Nowhere was the correlation of Ramist ideas and intellectual impoverishment more pronounced than in the realm of petty compendia that he and his followers produced. Ramus, of course, did not invent the genre but in his (and his disciples') hands, such manuals became the scourge of learning. As Gilbert observed, by the time Ramus published his "even more elementary and perfunctory set of textbooks on the liberal arts, some realization of the Pandora's box that had been opened began to dawn upon startled scholars. Soon Europe was deluged with small, convenient school manuals in which the life of traditional subjects was systematically and methodically eradicated and reduced to rules so simple that any child, literally, could learn them."[66]

The issue was widely discussed in England, one of the earliest exponents being Roger Ascham. As early as 1550 Ascham wrote Johannes Sturm that Joachim Périon's efforts were regarded superior to those of "the big mouth of Cephas Chlononius, who by impudently attacking both the prince of Greek learning and the schoolmaster of Rome has insolently wronged their good work and misused his own intellect." A couple of years later, having learnt that Ramus was informed of his quip, Ascham wrote Sturm at greater length, professing (too much) his respect for the Frenchman's learning and even his approbation of Ramus' "intentions, since I think he was flailing inept and dull Aristotelians rather than refuting Aristotle himself." Ascham

65 Robert Burton: *The Anatomy of Melancholy,* ed. Holbrook Jackson, London 1978, vol. 1, p. 316; *The Correspondence of John Cosin,* ed. G. Ornsby, Durham 1868-1872 [The Publications of the Surtees Society, vols. 52. 55] vol. 1, p. xi.

66 N. W. Gilbert: *Renaissance Concepts of Method,* New York 1960, p. 73.

added that what he had objected to was Ramus' "loose tongue" and determination "to follow a strange and churlish way of thinking rather than not to fight with Aristotle about everything." Ramus' problem is his reluctance to follow anybody lest he appear as an imitator. Then Ascham proceeded to absolve Ramus on account of "his diseased mind. For just as the sick always reject the best remedies, so also when they never lie quiet, they have the severest relapses."[67]

In *The Schoolmaster,* Ascham thought about Ramus when reflecting on how epitomes have always proved the bane of learning. Their durability facilitated, if not directly caused, the loss of the more substantial works that they abbreviated: "For by it have we lost whole *Trogus,* the best part of *T. Livius,* the goodlie Dictionarie of *Pompeius festus,* a great deale of the Civill lawe, and other many notable bookes, for the which cause, I do the more mislike this exercise, both in old and yong." More specifically, Ascham thought that while epitomes might be serviceable, in private, to some students, they were "ill commonlie for all other that use other mens labor therein: a silie poore kinde of studie, not unlike to the doing of those poore folke, which neyther till, nor sowe, nor reape themselves, but gleane by stelth, upon other mens growndes. Soch, have emptie barnes, for deare yeares."[68]

Several years later Richard Hooker made famous the association between "Ramystry" and the absence of erudition:

> In the povertie of that other new devised ayde two things there are notwithstanding singular. Of marveilous quicke dispatch it is, and doth shewe them that have it as much almost in three dayes, as if it dwell threescore yeares with them. Againe because the curiositie of mans wit, doth many times with perill wade farther in the search of things, then were convenient: the same is thereby restrayned unto such

[67] *Letters of Roger Ascham,* trans. Maurice Hatch, Alvin Vos, New York 1989, pp. 161. 198-199.

[68] Roger Ascham: *The English Works,* ed. William A. Wright, Cambridge 1904, repr. 1970, p. 259. In the same work Ascham denounced Ramus and Talaeus for their following Quintilian (on the issue of translation) who "having a lust, to dissent, from *Tullie* [...] doth greatlie commend *Paraphrasis,* crossing spitefullie *Tullies* judgement in refusing the same: and so do *Ramus* and *Talaeus* even at this day in *France* to. But such singularitie in dissenting from the best mens judgements, in liking onelie their owne opinions, is moch misliked of all them, that joyne with learning discretion and wisedome." Ibid. p. 243.

generalities as every where offering themselves, are apparent unto men of the weakest conceipt that neede be. So as following the rules and precepts thereof, we may define it to be, an Art which teacheth the way of speedie discourse, and restrayneth the minde of man that it may not waxe over wise.[69]

Like his teacher Rainolds, Hooker was convinced that "the search of knowledg is a thing painful"[70] and not the outcome of quick and effortless perusal of epitomes. And such an outlook sharply contrasted with the extravagant promises of ease made by the Ramists. MacIlmaine made the issue of brevity and expediency explicit: "thou shalt understand that there is nothing appartayning to dialectike eyther in Aristotles xvij. booke of logike, in his eight bookes of Physike, or in his xiiij. bookes of Philosophie, in Cicero his bookes of Oratorie, or in Quintilian [...] but thou shalt fynde it shortlie and after a perfecte methode in this booke declared," he announced in the introduction to his one-hundred page manual. Later he added that his book, "beyng well perused is able to bring more profytt to thee [...] then all thy fower yeares studie in Plato or Aristotle as they are now extant." Dudley Fenner's even shorter manual avowed to be a logic that was "easie to be learned and practised." Half a century later, Anthony Wotton still promised in his subtitle to provide "a short exposition of the praecepts, by which any

69 *The Folger Library Edition of the Works of Richard Hooker,* W. Speed Hill, general editor, 7 vols., Cambridge, Mass., Tempe, 1977-1998, vol. 1, p. 76. Curiously, Hakewill attempted to salvage something of Ramist reputation by perverting Hooker's meaning. In the preface to his *Apologie* Hakewill praised Ramus' studiousness, and later, after denouncing the scholastics' excessive preoccupation with logic, he added: "*Latter ages* finding this intolerable inconvenience, have well compacted the body of this Art into a lesser compasse (yet so as *Aristotles* Text is not to be neglected) and to this body have they not improperly added the doctrine of *Methods* as a necessary limbe thereof [...]. Even *Hooker* himselfe (though otherwise no friend to *Ramystry*) acknowledgeth that it is of marveilous quicke dispatch, shewing them that have it as much almost in three dayes, as if it dwelt three score yeares with them." Telling, however, is the caveat concerning the blessing to religion by the abandonment of "serious" study: "and againe, that the minde of man is thereby restrained, which through curiosity, doth many times with perill wade farther in the search of truth then were convenient." George Hakewill: *An Apologie or Declaration of the Power and Providence of God in the Government of the World,* 3rd ed., Oxford 1635, sig. B4r-b4v. pp. 298-299.

70 "and the painfulnes of knowledge", continued Hooker, "is that which maketh the will so hardly inclinable thereunto." *Works of Hooker* (as note 69) vol. 1, p. 81.

one of indifferent capacitie, may with a little paines, attaine to some competent knowledge and use of that noble and necessary science." Robert Fage announced that he constructed his translation in the form of a dialogue "that so my booke may even learne to stoupe to the weakest capacities." Running between 25 and 100 pages each, these and other authors made good on their promise.[71]

Let me hasten to reiterate a point I made earlier. It was not that contemporaries denied the serviceableness of manuals. These have been an integral part of undergraduate education.[72] What was rejected was the premise – associated with Ramists and Puritans – that epitomes furnished all the necessary knowledge in the respective disciplines. And on these grounds we find numerous critics denouncing the pernicious effects of epitomes, invariably condemning Ramus as the chief culprit in their proliferation. Sir Henry Wotton devoted one of his pedagogical aphorisms to the topic:

> Epitomies are helpful to the memory, and of good private use, but set forth for publick Monuments, accuse the industrious Writers of delivering much impertinency, and divert many to close and shallow cisterns, whose leasure might well be acquainted with more deep and open springs. In brief, what I heard sometimes spoken of *Ramus,* I believe of those thrifty Compendiums; They shew a short course to those who are contented to know a little, and a sure way to such whose care is not to understand much.[73]

71 *The Logike of the Moste Excellent Philosopher P. Ramus Martyr,* trans. Roll. Makylmaine, London 1574, pp. 7. 14; Dudley Fenner: *The Artes of Logike and Rethorike, plainelie set foorth in the English tounge, easie to be learned and practised,* Middelburg 1584; Anthony Wotton: *The Art of Logick,* London 1626; *Peter Ramus* [...] *his Dialectica,* London 1632, sig. A3. Ramus, of course, was not alone in promising shortcuts to learning. Francisco Sánchez "claimed that he could teach Latin in eight months, Greek in twenty days, astronomy in eight or ten days, philosophy and music in a month or less." Cited in Hans Aarsleff: *From Locke to Saussure: Essays on the Study of Language and Intellectual History,* Minneapolis 1982, p. 106. Understandably, many university professors were incensed. Frey lashed out at "the enemies of the academies and professors who accuse us of teaching in a very long time what can be learned equally well in a few days or months". Janus Caecilius Frey: *Opera,* Paris 1645, pp. 454-455. I owe this reference to Ann Blair.

72 Feingold: The Humanities (as note 12) *passim.*

73 *A Philosophical Survey of Education,* ed. H. S. Kermode, Liverpool 1938, pp. 22-23. 70. On the eve of the Civil War, Adam Martindale, unable to attend a university, studied logic on his own. As he found Keckermann too copious – and Aris-

Wotton summarized a mental attitude that had become commonplace by the first decades of the seventeenth century. Anticipating the outcry of his *Historie of Tithes,* John Selden referred those who doubted the verity of his argument to the sundry sources that he utilized: "He may be there further satisfied, and shall then see also that it is not of the Pitch of the Doctrine of the Breviarie, or within the compasse of Pocket-learning. Nor will it, I think, looke like what were patcht up out of *Postils,* Polyantheos, common place books or any of the rest of such excellent Instruments for the advancement of Ignorance and Laziness."[74] Richard Montague, who wrote a whole book refuting much of what Selden had to say about tithes, nonetheless concurred on the matter of epitomes:

> The Abridgements that have beene made long since, and of late, are held to be one of the chiefe plagues of Learning, and learned men. It maketh men idle, and yet opiniative, and well conceited of themselves. He that can carry an Epitome in his pocket [...] imagineth mightily, that he knoweth much, and yet indeed is but an *ignaro.* In a day he is taught, but to little purpose, as much as others can learne in a whole yeere. Lately the World went a madding this way, for *Systemaes, Syntagmas, Synopseis* and I know not what, both for the Handmaids and Mistresses of Arts. *Lipsius* and *Ramus* swayed all: but soone they, perceiving their owne folly, left them, *turbidos rivulos, & foeculentos,* and retryed, with some losse of time and travell, unto the Fountaines.[75]

Nearly half a century later we find Méric Casaubon to have been every bit as censorious of Ramus and his followers as his father had been. Learning has decayed, he complained in a manuscript work of 1668, because it had reached such a height "that for a man to make himselfe considerable, (besydes competent witt and judgment, without which nothing can

totle too difficult – for a beginner, he pursued Ramus and the compendiums of Smith and Brerewood instead, but "found them over-curt and maimed, leaving out many things that I thought very good in my old Keckermann, the first designedly, the other two by affected brevity." A young Cambridge scholar obtained for him a copy of Burgersdicius "of which I was very fond." *Life of Adam Martindale,* ed. Richard Parkinson, Manchester 1845, pp. 49-50.

74 John Selden: *Historie of Tithes,* London 1618, pp. i-ii. Not irrelevant to his attitude was Selden's belief that "the Jesuits and the lawyers of France, and the Low Countrymen, have engrossed all learning. The rest of the world make nothing but homilies." *The Table Talk of John Selden,* ed. Samuel H. Reynolds, Oxford 1892, p. 102.

75 Richard Montague: *Diatribae upon the first part of the Late History of Tithes,* London 1621, pp. 415-416.

be done, and are not, to that degree, every mans happines) soe much labour, soe much industrie was required, as is enough to fright any, whom God hath not endowed with extraordinarie courage, and strength of bodie withall." Consequently, any "new project, promising new discoveries of a shorter way, must needs be very acceptable unto most. Ramus attempted somewhat, and men began to run after him: but the age was too learned, for him, or the creditt of his attempts, to hold long." In another work, written (and published) in the same year, Casaubon hammered home his message again: "it must be acknowledged, that many conceited, whimsical men, as *Lullus, Ramus,* and the like, have projected wayes of improvement, which if generally received, would have proved very destructive to learning."[76]

Lest one assume that pedants alone were cognizant of (and carping at) such a phenomenon, there existed a profusion of contemporary satires that mocked the indistinguishableness of Ramism from Puritanism and excoriated the banality of Puritans/Ramists' learning. One author cleverly coupled the Ramists' boast of learning without pain – and the concomitant calumny against Aristotle – with the near conspiratorial process that Puritans employed in luring young undergraduates to their camp. No sooner had the tyro arrived at the university than he encountered "some fantasticke *Ramist*" who "welcomes" the "yonger brother" with the following speech:

Be wist, and warie of that prating sect
Which strives 'gainst *Ramus;* lest it thee infect.
For tidy *Peter* like a pritty primmer,
May well be learned ere thou go to dinner.
Hee's pithie, deep, succinct, methodicall,
A *Carnucope,* a volume all in all.
But *Aristotle* is a ridling *Sphinx,*
A river poysonous to him that drinks.
Hee's blunt, unpolisht, tedious, harsh, obscure,
Fraught with vile stuffe, and sentences impure:
The childe is tourn'd, and claps him on the backe,

76 Méric Casaubon: *Generall Learning: A Seventeenth-Century Treatise on the Formation of the General Scholar,* ed. Richard Serjeantson, Cambridge 1999, p. 146; *A Letter of Meric Casaubon* [...] *to Peter du Moulin* [...] *Concerning Natural Experimental Philosophy, and some books lately set out about it,* Cambridge 1669, p. 14. Ramus, of course, produced many progenies including, Casaubon said, Comenius, the framer of a "project of making all men wise." *A Letter,* p. 13.

And swears, that *Ramus* foes shall go to racke:
Making (forsooth) a sad and solemne vow,
That he will reverence the *golden Bough.*
When *Boyes* in *age,* or *wit* have said their fill,
Old *Organon* must be best *Logike* still.[77]

Similarly, in the celebrated Cambridge dramatic production of *Parnassus,* the character of the Puritan "Stupido" is given the following memorable lines:

> I have (I pray God prosper my labours) analised a peece of an hommelie according to Ramus, and surelie in my minde and simple opinion Mr Peter maketh all things verie plaine and easie. As for Setons Logique, trulie I never looke on it but it makes my head ache.

To compound the mockery, Stupido proceeds to admit that he spent ten years under the "map" of Ramus but to little effect – echoing, perhaps, Thomas Nashe's quip that it had taken Ramus sixteen years to compose his "pettie Logique." The author of *Parnassus* proceeded to rebuke the corollary rejection of learning by the Puritans, by making Stupido further advise the travelers:

> studie not these vaine artes of Rhetorique, Poetrie, and Philosophie: there is noe sounde edifying knowledge in them [...]. If you will be good men indeede, goe no further in this way, follow noe longer these profane artes, that are the raggs and parings of learning, sell all these bookes, and buye a good Martin, and twoo or three hundreth of catechismes of Ienevas printe, and I warrant you will have learning enoughe. Mr Martin and other good men tooke this course.[78]

Against such a background we can better appreciate Christopher Marlowe's gruesome thrashing of Ramus' philistinism and inane innovations, for which he was condemned to pay with his life – at the conclusion of a disputation! "O good my Lord," cries the scholar, "Wherein hath Ramus been so offencious?" Guise's answer is as scholarly crushing as it is lethal:

Marry, sir, in having a smack in all,
And yet didst never sound anything to the depth.
Was it not thou that scoff'st the *Organon,*
And said it was a heap of vanities?
He that will be a flat dichotomist,
And seen in nothing but epitomes,

77 *Tyros Roring Megge,* London 1598, sig. A4v.

78 *The Three Parnassus Plays,* ed. J. B. Leishman, London 1949, p. 112; *Works of Nashe* (as note 41) vol. 3, p. 313.

Is in your judgment thought a learned man;
And he, forsooth, must go and preach in Germany,
Excepting against Doctors' axioms,
And *ipse dixi* with this quiddity,
Argumentum testimonii est inartificiale.
To contradict which, I say, Ramus shall die:
How answer you that? your *nego argumentum*
Cannot serve, sirrah. – kill him.[79]

The satirical vein persisted throughout the period in which Ramism was equated with Puritanism, and both movements were accused of philistinism. In his *Wits Miserie* (1596) Thomas Lodge mocked that vainglorious who in "his hood and habit [...] will proove *Ramus,* to be a déeper Philosopher then Aristotle, and presume to read the *Mathematiques* to the studious, when he knowes not what either *Axis*, *Equator,* or *Circulus* is." In sum he is "both in soule and body a méere adversary to all Science." Thomas Randolph made merry with the Puritan fanatic who is also the "university Ramist" through the character of "Wild-man" (in *Aristippus, or the Ioviall Philosopher*), who had "grown mad" seeing the popularity of Aristippus who infects all with his Arminianism as well as with his philosophy.[80] His Oxford contemporary, John Earle, made a more up-to-date composite in his character of the superficial "Self-Conceited Man", whose "discourse is all positions and definitive decrees, with *thus it must be* and *thus it is,* and he

79 Marlowe: *Complete Plays* (as note 38) pp. 552-553. My interpretation of Marlowe is at variance with traditional renditions of the scene. I hope to demonstrate elsewhere my conviction that just as *The Massacre of Paris* was not intended to serve as Protestant propaganda, as Julia Briggs demonstrated in a recent article, so too Marlowe was not trying to pay tribute to Ramus. Marlowe, I believe, was deliberate when he made the Guise denounce the poverty of Ramus' logic and his corrupting influence on learning. At the same time, Marlowe also targeted the Puritans who stand accused of similar crimes. Julia Briggs: Marlowe's Massacre at Paris: A Reconsideration, in: *Review of English Studies* n.s. 34 (1983) pp. 257-278. For the claim that Marlowe canonised Ramus, see John R. Glenn: The Martyrdom of Ramus in Marlowe's The Massacre of Paris, in: *Papers on Literature and Language* 9 (1973) 365-379. For another interpretation of certain aspects of this scene, see Richard W. Serjeantson: *Testimony, Authority and Proof in Seventeenth-Century England,* Ph. D. thesis, Cambridge 1998, pp. 72-73.

80 *The Complete Works of Thomas Lodge,* 4 vols., repr. New York 1963, vol. 4, p. 5; *Poetical and Dramatic Works of Thomas Randolph,* ed. W. Carew Hazlitt, 2 vols., repr. New York, London 1968, vol. 1, pp. 9-10.

will not humble his authority to prove it [...]. He has an excellent humour for an heretick, and in these days made the first Arminian. He prefers Ramus before Aristotle, and Paracelsus before Galen, and Lipsius his hopping stile before either Tully or Quintilian."[81]

At stake was not only the Puritan belittling of the importance of secular learning, but also the dire consequence for theology of such an attitude. Not long after he had left Cambridge, Thomas Nashe published a defense of his deceased friend Robert Greene against Gabriel Harvey, in the course of which he addressed both topics and defended erudition against the "abusive enormities" that were inflicted on true learning by the "ravenous rable" at the universities:

> There is no such discredit of Arte as an ignoraunt Artificer, – men of meaner judgement measuring oft times the excellencie of the one by the ignoraunce of the other. But as hee that censureth the dignitie of Poetry by *Cherillus* paultry paines, the majestie of Rhetoric by the rudenesse of a stutting *Hortensius,* the subtiltie of Logique by the rayling of *Ramus,* might judge the one a foole in writing he knewe not what, the other tipsie by his stammering, the thirde the sonne of Zantippe by his scolding: so he that estimats Artes by the insolence of Idiots, who professe that wherein they are Infants, may deeme the Universitie nought but the nurse of follie, and the knowledge of Artes nought but the imitation of the Stage. This I speake to shew what an obloquie these impudent incipients in Arts are unto Art.

"How is it then such bungling practitioners in principles should ever profit the Common-wealth by their negligent paines," asked Nashe, "who have no more cunning in Logicke or dialogue Latine then appertaines to the literall construction of either? Neverthelesse, it is daily apparent to our domestical eyes that there is none so forward to publish their imperfections, either in the trade of glose or translations, as those that are more unlearned then ignorance, and less conceiving than infants." The cause was easy enough to discern: "The efficient whereof I imagine to issue from the upstart discipline of our reformatorie Churchmen, who account Wit vanitie, and poetry impiety [...] so I deeme him farre unworthy the name of a scholer, and so, consequently, to sacrifice his endeavours to Art, that is not a Poet, either in whole or in part."[82]

81 John Earle: *Microcosmography,* Bristol 1897, pp. 29-30. In later editions the final sentence was substituted with "and whosoever with most paradox is commended."

82 Thomas Nashe: *The Anatomie of Absurditie* (1589), in: *The Works of Thomas Nashe,* ed. Ronald B. McKerrow, Oxford 1958, pp. 334. 318-319. 321. In the late 1590s, John Donne exhorted his young friend Samuel Brooke, who had recently

It is not that the Puritan proponents of Ramism were necessarily of an indifferent intellect. Rather, Nashe's criticism reflects the fear that in the name of true religion a proper study of the arts and sciences would be rendered unnecessary. And Ramism provided ammunition to those who sought to undermine the necessity of erudition. Thus, long before Bacon pronounced that "a little *Philosophie* inclineth mans minde to *Atheisme,*"[83] Nashe expounded his version of the danger facing both learning and theology by those who applied themselves too early to the ministry:

> I will not deny but in scholler-like matters of controversie a quicker stile may passe as commendable, and that a quip to an Asse is as good as a goad to an Oxe: but when the irregular Ideot, that was up to the eares in divinity before ever he met with *probabile* in the Universitie, shall leave *pro & contra* before hee can scarcely pronounce it, and come to correct common-weales, that never heard of the name of Magistrate before hee came to *Cambridge,* it is to marvaile if every Alehouse vaunt the table of the world turned upside downe, since the child beateth his father, and the Asse whippeth his Master.

Sir John Cheke and his followers, he pointed out, laid down a solid scholarly tradition at Cambridge.

> But how ill their precepts have prospered with our idle age, that leave the fountaines of Sciences, to follow the rivers of Knowledge, their over-fraught studies with trifling compendiaries may testifie: for I know not how it commeth to passe, by the doting practise of our Divinitie Dunces, that strive to make their pupills pulpit-men before they are reconciled to *Priscian;* but those yeares which should bee imployed in *Aristotle* are expired in Epitomies, and well too, they may have so much Catechisme vacation, to rake up a little refuse philosophy. And heere I could enter into a large fielde of invective against our abject abbreviations of Arts, were it not growne to a new fashion among our Nation, to vaunt the pride of contraction in every manuarie action.[84]

Opponents could effectively charge Puritans with an essentially anti-intellectualist stance that allied itself with Ramist charlatanry. Oliver Ormerod did not mince words: "Because our Puritanes doe thinke, that they

graduated BA, not to desist from composing poetry – having "Lately launch'd into the vast Sea of Arts" – further warning him against "those Scismatiques with you, / Which draw all wits of good hope to their crew". Samuel Brooke: *Melanthe,* ed. Joseph S. G. Bolton, repr. Hamden, Conn. 1972, p. 4.

83 *The Works of Francis Bacon,* ed. James Spedding, Robert L. Ellis, Douglas D. Heath, 15 vols., Boston n.d., vol. 12, p. 337.

84 *Works of Nashe* (as note 41) vol. 3, pp. 315-317.

have Logicke enough, when they have read and conned over Ramus his Logicke: and so consequently, they often use fallacies, before ever they be aware. They cannot (forsooth) endure to read Aristotles golden Booke de sophisticis Elanchis." Ormerod further suggested the efficacy of Ramism in the process of indoctrinating youth into Puritanism: Just "as it was the wisdome of the King of Babylon, to take young Children of Israel, whom hee might teach the learning and tongues of Chaldea rather then their olde men: so it is the wisdome of some Ramisticall Tutors, to season our greene vessels with this liquor of Puritanisme, that they may keepe the taste thereof while life remaineth."[85] Two decades later Richard Montague, too, derided Cambridge Puritans who believed that the mastery of a Ramist manual made them skilled logicians.[86]

The implications of the twin evils for the acquisition of the learning necessary for the Church reverberated in numerous works. In the course of a 1617 sermon at Oxford, the "Ever Memorable" John Hales took issue with Puritanism as he attempted to impress on his auditors the importance of sound learning for the ministry. "So great a thing as the skill of exposition of the word and gospel is, so fraught with multiplicity of authors, so full of variety of opinion, must needs be confest to be a matter of great learning, and that cannot, especially in our days, in short time, with a mediocrity of industry, be attained." Quipping at the Ramists by invoking Quintilian's admonition that one can scarcely consider himself an accomplished orator "merely by being masters of some short and common manual of rhetoric", Hales proceeded to extrapolate such a moral to the domain of theology:

> If he thought fit thus to do in an art of so inferior and narrow a sphere, much more is it behoveful, that young students, in so high, so spacious, so large a profession, be advised not to think themselves sufficiently provided, upon their acquaintance with some *notitia,* or system of some technical divine. Look upon those sons of Anak, those giant-like voluminous writers of Rome, in regard of whom, our little tractates, and pocket-volumes in this kind, what are they but as grashoppers? I speak not this like some seditious or factious spy, to bring weakness of hands, or melting of heart, upon any of God's people: but to stir up and kindle in you the spirit of industry, to inlarge your conceits, and not to suffer your labours to be copst and mued up within the poverty of some pretended method.[87]

85 Oliver Ormerod: *The Picture of a Puritane,* London 1605, p. 36.

86 Richard Montague: *Appello Caesarem,* London 1625, p. 14.

87 John Hales: *A Sermon preached at St. Maries in Oxford. Concerning the abuses of obscure places of scripture* (Oxford 1617), in: *The Works of the Ever Memorable*

A decade later, John Milton, a third-year Cambridge undergraduate, produced, in a letter to his old teacher Alexander Gill, an impassioned indictment of the impoverished intellectual horizons of his fellow students at Christ's College who, in their haste to become divines, had lost sight of the study of good letters. Evidently provoked by a fellow of the house who had commissioned Milton for some verses that the unnamed fellow could pass as his own – "he himself being long past light-minded nonsense of that kind and more intent on serious things" – the episode prompted Milton to reflect on the sad state of learning in the College: "There is really hardly anyone among us, as far as I know, who, almost completely unskilled and unlearned in philology and philosophy alike, does not flutter off to theology unfledged, quite content to touch that also most lightly, learning barely enough for sticking together a short harangue by any method whatever and patching it with worn-out pieces from various sources – a practice carried far enough to make one fear that the priestly Ignorance of a former age may gradually attack our Clergy."[88]

Character writers drove home the same message. The grave divine, wrote John Earle, is he who "knows the burthen of his calling, and hath studied to make his shoulders sufficient; for which he hath not been hasty to launch forth of his port, the university, but expected the ballast of learning [...] Divinity is not the beginning but the end of his studies; to which he takes the ordinary stair, and makes the arts his way. He counts it not prophaneness to be polished with human reading, or to smooth his way by Aristotle to school-divinity, He has sounded both religions, and anchored in the best, and is a protestant out of judgment, not faction."[89]

The critical attitude was not confined to scholars and divines. Mathematicians and natural philosophers shared the same antagonism toward the implications of the Ramist frame of mind for serious study in their respective domains. Francis Bacon was willing to acknowledge Ramus' good intentions, but nevertheless believed his actual program proved harmful.

Mr. John Hales, 3 vols., Glasgow 1765, vol. 2, pp. 20-22. In a sermon he delivered in February 1618, John Donne thundered: "We steale our Learning, if we forsake the Fountaines, and the Fathers, and the Schooles, and deale upon Rhapsoders, and Common placers, and Method-mongers." *The Sermons of John Donne,* ed. George R. Potter, Evelyn M. Simpson, 10 vols., Berkeley, Los Angeles 1953, vol. 1, p. 256.

88 Milton: *Complete Prose Works* (as note 1) vol. 1, p. 314.

89 Earle: *Microcosmography* (as note 81) pp. 8-9.

There is no mistaking the passion that pervaded Bacon's allusions to Ramus. It bespeaks deep-seated concerns that cannot be dismissed as an effort to disparage a rival. Indeed, it was precisely the likelihood that contemporaries might mistake his reforms as an extension of Ramus' – simply because both men assaulted Aristotle – which prompted Bacon to assail Ramus. In an early work, *The Masculine Birth of Time,* composed shortly after he had boasted in a letter to Lord Burghley that he had taken all knowledge as his province, Bacon vehemently disavowed any correlation between his ideas and those of Ramus. The reader

> must not suppose me to be in league with that recent rebel against him, Peter Ramus. I have nothing in common with that hide-out of ignorance, that pestilent book-worm, that begetter of handy manuals. Any facts he gets hold of and begins to squeeze in the rack of his summary method soon lose their truth, which oozes or skips away, leaving him to garner only dry and barren trifles. Aquinas, Scotus, and their followers out of their unrealities created a varied world; Ramus out of the real world made a desert [...]. I rate him below the sophists.[90]

In the *Advancement of Learning* Bacon reiterated his charges with vengeance. He abused Ramist manuals on several occasions, as when he railed against "the corruptions and moths of history, which are Epitomes", the use of which "deserveth to be banished, as all men of sound judgement have confessed, as those that have fretted and corroded the sound bodies of many excellent histories, and wrought them into base and unprofitable dregs."[91] Such an indictment foretold more caustic charges. At one point Bacon lambasted Ramus and his followers, without naming them, by saying that

> there has been also laboured and put in practice a method, which is not a lawful method, but a method of imposture; which is to deliver knowledges in such manner, as men may speedily come to make a shew of learning who have it not; such was the travail of Raymundus Lullius, in making that art which bears his name; not unlike to some books of Typocosmy, which have been made since; being nothing but a mass of words of all arts, to give men countenance that those which use the terms might be thought to understand the art; which collections are much like a fripper's or broker's shop, that hath ends of every thing, but nothing of worth.[92]

90 Benjamin Farrington: *The Philosophy of Francis Bacon,* Chicago 1966, pp. 63-64.

91 *Works of Bacon* (as note 83) vol. 6, p. 189.

92 Ibid. p. 296. Camden maintained a similar position: "To reduce surnames to a Methode, is a matter for a *Ramist,* who should haply finde it to be a *Typocosmie:* I will plainly set down from whence the most have been deduced, as far as I can

Such an attitude is necessary for properly appreciating Bacon's back-handed mixture of praise and condemnation that he directed at Ramus a couple of pages earlier when he stated that

> Ramus merited better a great deal in reviving the good rules of Propositions [...] than he did in introducing the canker of Epitomes;[93] and yet (as it is the condition of human things that, according to the ancient fables, '*The most precious things have the most pernicious keepers*') it was so, that the attempt of the one made him fall upon the other.[94] For he had need be well conducted that should design to make axioms *convertible,* if he make them not withal *circular,* and *non-promovent* or *incurring into themselves:* but yet the intention was excellent.[95]

Bacon did not soften his position over the nearly two decades that separated *The Advancement of Learning* and *De Augmentis scientiarum,* as the corresponding translations cited in the previous notes indicate. Elsewhere he also dismissed Ramus' "one and only method," and its "distribution of everything into two members":

> it is needless to speak of it; for it was a kind of cloud that overshadowed knowledge for awhile and blew over: a thing no doubt both very weak in itself and very injurious to the sciences. For while these men press matters by the laws of their method, and when a thing does not aptly fall into those dichotomies, either pass it by or force it out of its natural shape, the effect of their proceeding is this, – the kernels and grains of the sciences leap out, and they are left with nothing in their grasp but the dry and barren husks. And therefore this kind of method produces empty abridgments, and destroys the solid substance of knowledge.[96]

conceive, hoping to incurre no offence herein with any person, when I protest in all sinceritie, that I purpose nothing lesse than to wrong any whosoever." William Camden: *Remaines Concerning Britain,* London 1657, p. 112.

93 In *De Augmentis* the Latin translation reads "his peculiar method and dichotomies." *Works of Bacon* (as note 83) vol. 9, p. 128.

94 The Latin translation substitutes "the attempt of Ramus to amend propositions drove him upon those epitomes and shallows of knowledge." Ibid.

95 Ibid. vol. 6, pp. 294-295.

96 Ibid. vol. 9, p. 122. Bacon also stressed the inhibiting effects of the Ramist controversies on the advancement of learning: "For the Method of Tradition, I see it hath moved a controversy in our time. But as in civil business, if there be a meeting and men fall at words there is commonly an end of the matter for that time and no proceeding at all; so in learning, where there is much controversy there is many times little inquiry. For this part of knowledge of Method seemeth to me so weakly enquired as I shall report it deficient." Ibid. vol. 6, p. 288.

Bacon's first editor, William Rawley, expressed therefore Bacon's own sentiments when in his introduction to *The Advancement of Learning* he wrote, apropos of the history of method, that as "for the Methods of the Moderns, Ramus and others by the improvement of German Writers, impaired; they knit the limbs of knowledge too soon; have bedwarfed Sciences, and are become an Art (as Learned Hooker expresses it) which teaches the way of speedy discourse, and restrains the mind of man, that it may not wax over-wise."[97]

The extent to which Bacon's negative perception of Ramism was further abetted by his hostility to the Puritans is a topic I cannot discuss here – though his attitude toward controversies is suggestive in this respect. Nevertheless, given his explicit pronouncements against Ramus it would be a mistake to maintain (as Walton and Rossi do) Bacon's considerable indebtedness to the Frenchman. Lisa Jardine was closer to the mark when she pointed out that "Bacon's only direct borrowing appears to be his use of the term 'axiom' atypically as Ramus uses it, for any proposition used as a premise for argument."[98]

The attitude of mathematicians conformed to similar contours. As Nicholas Jardine's and Alain Segonds' paper in this volume makes clear, the Protestant astronomers who attempted to refute the Ramist "challenge" in order to produce an astronomy without hypothesis – which basically amounted to "a denial of the legitimacy of theoretical speculation in astronomy" – were polite but unambiguous in dispensing with such nonsense. Rothmann, Ursus, Tycho and Kepler all dismissed Ramus as uninformed, if not ignorant, with Kepler gibing that "to preserve the honour of both of us, I prefer to call myself Regius Professor than Ramus a fool." The relatively mild handling of Ramus should be understood, I believe, in deference to the Protestant martyr. In this they followed Scaliger who respected Ramus' "devotion to Protestantism" more than he did his learning and, consequently, when preparing his edition of Manilius during the 1570s, he sought to refute Ramus' extravagant notions

97 Francis Bacon: *Of the Advancement and Proficincie of Learning,* ed. Gilbert Watts, London 1674, sig. B3.

98 Craig Walton: Ramus and Bacon on Method, in: *Journal of the History of Philosophy* 10 (1971) pp. 289-302. – Bacon's "mature work contains crucial elements similarly handled by Ramus", p. 299; Paolo Rossi: *Francis Bacon: From Magic to Science,* London 1968, *passim;* Lisa Jardine: *Francis Bacon: Discovery and the Art of Discourse,* Cambridge 1974, p. 8.

concerning ancient astronomy by shifting the blame to "the *cénacle* of Ramist mathematicians" who did not understand the ancient texts they cited.[99]

Prior to, and years after the St. Bartholomew's Day Massacre, such circumspection was less necessary. A couple of years before Ramus was murdered, Henry Savile delivered a series of lectures at Oxford that incorporated large chunks of Ramus' 1567 *Prooemium mathematicum*. Notwithstanding such borrowings, Savile was inimical to Ramus' insistence that practice alone should concern mathematicians. This demeaned and subverted the discipline, Savile insisted, and proceeded to cite Proclus who argued that mathematics

> has the same relation to the knowledge of being or first philosophy as childhood education has to virtue [...] I wish you made *this* the end of the most liberal arts, Ramus, agreeing with Plato, Ptolemy, Proclus, and other philosophers more learned than yourself; instead you made the end of these arts mechanical and illiberal. No one from the common people can be a craftsman who refuses to be seen with a level and plumbline when building walls; and no more can you, who refuse to be seen with Euclid and Pappus, be a master of mathematics, although you laboured for many years in this task. If only you had extolled that ascent to separate substances with your usual rhetorical gifts and had completely proven your eloquence to us – now instead we despise it, immersed as it is in such base things. Immortal gods, that mathematics, which until now has been held inviolable from worldly interests, should now be reduced to mere mechanical skill, as though it had been thrust into some vulgar mill![100]

Almost a century later Isaac Barrow's Cambridge lectures drove home a similar message but in far sharper language. Barrow castigated "that great Censurer (or rather Carper)", for "sharply reprov[ing] *Euclid,* because he undertook to demonstrate many Propositions, which in his [Ramus'] Opinion were sufficiently evident from their own Light, and therefore should have been passed by undemonstrated, and been reckoned among the Number of Axioms." In a subsequent lecture Barrow assaulted even more vehemently the "Censurer", who "seems to catch at all Handles, and every where hunt for Occasions to abuse those ancient Masters of the Sciences,

99 Nicholas Jardine: *The Birth of History and Philosophy of Science,* Cambridge 1984, p. 268; Grafton: *Scaliger* (as note 48) vol. 1, p. 217.

100 Bodleian Library, MS Savile fol. 5r-v, cited in Robert Goulding: Henry Savile and the Tychonic World-System, in: *Journal of the Warburg and Courtauld Institutes* 58 (1995) pp. 153-154.

whom yet it were just not to reprehend without great Cause, and only where they manifestly are in the wrong; it would become us that are indebted to them for such great Inventions to act more mildly with them." Ramus was further charged with being both ignorant and malicious:

> How comes he for such poor Causes and false Arguments to brand these venerable Persons, with Notes of Fallacy, Sophistry, and Absurdity, and defame those Builders, and first Propagators of the Sciences with such uncivil Contumelies; without whom perhaps this mighty Artist at Trifles had had nothing but Fables to apply his wonderful Logic in the Resolution of?[101]

Historians often mistook the emphasis on theory as reflective of elitist criticism aimed at non-university practitioners of mathematics. Such an assumption, as I intend to show elsewhere, is erroneous. What informed Savile and Barrow was not hostility toward practice or mathematical instruments, but a respect for learning and the belief that the principles of art must be inculcated prior to practice. And their view was widely shared. That many practitioners published only practical treatises was a function of the book trade and the persistence of scribal publication, not an indication of disavowal of theory. The point may be illustrated by Thomas Bedwell, who translated Ramus into English. Like Milton in the domain of logic, Bedwell maintained that even on the elementary level, Ramus alone was insufficient.[102] Hence, he amplified the text with additional material. More importantly, Bedwell's adaptation of Ramus was not extended to either a devaluing of theory or an endorsing of Ramist method: "Our Autour, in the declaration of the Elements, hath many passages, which in our judgment doe not make so much for the understanding of the matter in hand, as for the defence of the method here used, against Aristotle, Euclide, Proclus, and others, which we have therefore wholly omitted."[103]

English Ramism, I've argued, failed to make significant inroads into English culture because it represented a frame of mind that was an anathema

101 Isaac Barrow: *The Usefulness of Mathematical Learning,* trans. J. Kirby, London 1734, pp. 59. 339-340. 401-404. See also Leibniz's estimation: "But the good Ramus who had changed Euclid's method has not only lost rigor but also truth and accuracy." *Leibniz: Selections,* ed. Philip P. Wiener, New York 1951, p. 56.

102 "For what is the use of achieving brevity, if this means we must go elsewhere for clarification?" Milton: *Complete Prose Works* (as note 1) vol. 8, pp. 209-210.

103 William Bedwell: *Via Regia ad Geometriam,* London 1636, "To the Reader".

to the heightened scholarly and literary sensibilities of the educated public. Even where it did manage to gain some permanence, namely among the Puritans, it did so in a form I labeled "low-grade Ramism". Ultimately, the changing fortunes of Puritanism in the course of the early seventeenth century would bring about the demise of both movements. John Beale was a witness to the simultaneous demise of Ramism and Puritanism. In a 1666 letter to Robert Boyle he recollected of events four decades earlier:

> When I first entered into *Cambridge,* the *Ramists* were in such great esteem, that we gave very high rates for such of them as began to be out of print; but within three or four years they fell so low in credit, that we might buy them at the rate of tainted fish, or *Irish* beef; shops full in *Cambridge,* and cart-loads in *Sturbidge* fair, at one penny *per* pound. This fell out without the solicitation of any part or faction, or other concernment, merely by the prevalence of solid truth, and reasonable discourses [...]. Soon after, the same fate befel *Calvinism,* in both our universities.

Having described the phenomenon, he turned to analyze its causes:

> even in those my younger days, these *Ramistical* dichotomies seemed to me like a taylor's shears, cutting all our best cloth into shreads, or rags; and the *Calvinian* doctors, dancing in the circles of their nimble distinctions, were as ridiculous, as the maid-morrian and hobby-horse in a morrice dance. And, though those latter did put a countenance of gravity on it, for a while, and masked their riddles under a vizor of devotion, yet when they found, that the younger students (such as were scarcely graduated) did by the aid of stronger arguments, and by the advantage of true and authentic antiquities, so far excel the whiffling doctors in their lectures and disputations, the elder men grew pale at it, and the older truth prevailed.[104]

[104] *The Works of the Honourable Robert Boyle,* ed. Thomas Birch, 6 vols., London 1772, vol. 6, pp. 411-412.

Ramism in Trinity College, Dublin in the Early Seventeenth Century

ELIZABETHANNE BORAN

An investigation of the thorny subject of the influence of Ramism in Trinity College, Dublin in the first decades of that College's existence, while fraught with difficulties, can tell us much about the nature of Ramism, and for that matter Puritanism in early seventeenth century Ireland.[1] The sources are fragmentary at best, but by a combination of examinations of the statutes, staff and student notebooks and, last but not least, the collections of the library of T.C.D., one can attempt to plot the progression and decline of Ramism within the newly founded University of Dublin.

Statutes can be easily misinterpreted and due caution must be maintained in their decoding. Too often, stipulations in the official documentation of early modern universities have erroneously been taken as evidence of what was actually going on at the time, yet it is clear that statutes at best represent the official perception of what *ideally* should happen within a university.[2] That being said, we must be equally wary of disregarding them completely – for sometimes official perceptions are not so very far off from what is actually happening! In the case of T.C.D. there are three sets of statutes for the period in question, 1592-1641: The first was drawn up by the fourth provost, William Temple, ca. 1615.[3] Unfortunately, it only par-

1 For a detailed discussion of this see Elizabethanne Boran: *Libraries and Learning: The Early History of Trinity College, Dublin from 1592 to 1641,* T.C.D. Ph. D. thesis 1996.

2 On the perils of an over-dependence on statutory codes see Mordechai Feingold: The Humanities, in: Nicholas Tyacke, ed.: *The History of the University of Oxford,* vol. 4: *Seventeenth Century Oxford,* Oxford 1997, pp. 213ff.

3 The Temple statutes may be found scattered among the following manuscripts: T.C.D. MUN/P/1/201. MUN/P/1/105. MUN/P/1/168. MUN/P/1/171. MUN/P/1/173.

tially survives and contains little relevant information on the impact of Ramism in the College. The second code, drawn up in 1628 by Provost William Bedell, is far more illuminating, particularly when contrasted with the Laudian statutes of 1637, the third set, which one might call a reaction to the two previous codes.[4]

The relevant chapter in the Bedell code is *De Classium Scholasticis Exercitiis* which gives us the following information about the second year subjects of the undergraduate curriculum:

> Praelector secundae Classis controversa Logicae disciplinae capita explicato, et disceptato. Quae veritati consentanea reperientur, ea Auditoribus suis commendabit: Quae vero falsa fuerint, ea argumentorum viribus convicta repudiabit. Hujus Classis Discipuli aliquam Inventionis et Judicii Analysin per singulas Septimanas instituant.[5]

The reference to invention and judgement in the teaching of 'logic' immediately calls to mind the Ramist arrangement.[6] Evidently the same thought occurred to William Laud, Archbishop of Canterbury, for in his revision of the statutes in 1637 he explicitly stipulated that the *Isagoge* of Porphyry, a well-known introduction to Aristotelian logic, should be the set text – a re-affirmation of Aristotle that pervades the entire 1637 code. That we are not merely dealing with an extension of the Oxford statutes, with little thought to the local setting, is clear from the language that Laud uses when talking about the changes in the T.C.D. curriculum. Though he never mentions Ramus' name he leaves us with no doubt about his chief target:

> And because our Will is, that what relates to Logic shall be handled Logically, that is, Syllogistically, not with the flourishes of Rhetoric, we forbid all previous Positions or Suppositions, as well on the Part of the Respondent as Opponent, but they shall enter on the Dispute immediately.[7]

4 A Latin version of the Bedell statutes may be found in the appendix to J. P. Mahaffy: *An Epoch in Irish History,* London 1903, pp. 327-375. An English translation of the Laudian statutes was made by Robert Bolton: *A translation of the Charter and Statutes of Trinity College, Dublin,* Dublin 1749.

5 Mahaffy: *Epoch* (as note 4) p. 352.

6 For a detailed discussion of the intricacies of Ramism, which has yet to be superseded, see Walter J. Ong: *Ramus, Method and the Decay of Dialogue,* Cambridge, Mass. 1983.

7 Bolton: *Statutes* (as note 4) p. 72.

In this case, then, the Laudian code is very much a reaction to earlier teaching practices within T.C.D., practices which Laud felt were prejudical to the smooth running of the College.

The second source tells us much more about the *actual* state of teaching in the arts faculty. These notebooks were primarily staff notebooks belonging to two members of College, Luke Challoner and James Ussher, who were the two principal book buyers of the College library, and therefore the contents of their notebooks are of immense interest.[8] Challoner as one of the first fellows and Ussher as one of the first scholars, played an extremely important role in the development of the early T.C.D. By this time Challoner was most definitely on the teaching staff, and the same is true of James Ussher for the early seventeenth century, which augments the value of their notebooks, as they give us an enhanced view of the teaching preoccupations in early T.C.D. There are definite problems with dating the several notebooks, but I have tried as far as possible to identify those which appear by internal evidence to date from the first two decades of the seventeenth century, when James Ussher was directly involved in College affairs. Luckily, with very few exceptions the most easily identifiable of the notebooks in this regard are also the most fruitful for the present study – as Ussher demonstrated a marked tendency to outline which were the favoured works in different subjects.[9] Other notebooks include those of Ussher's brother Ambrose, which are primarily concerned with theological matters, while one other notebook, belonging to John Travers, has also proved useful in connection with approaches to Aristotelian politics.[10]

Finally, the early library collections of the College provide us with an invaluable insight into intellectual interests within T.C.D. in the period 1592-1641. The role of the library as the central institution of the early College cannot be overestimated. Unlike the vast majority of College libraries elsewhere, which were the result of benefactions, the library of

8 See for example, the following notebooks of James Ussher: T.C.D. MS 773. MS 774. MS 778. MS 782. MS 785. MS 786. MS 788. MS 790. MS 793. Challoner's notebooks are: T.C.D. MS 357. MS 358. MS 785. For secondary material on Challoner and Ussher see N. J. D. White: *Luke Challoner D. D.,* Dublin 1909, and R. B. Knox: *James Ussher, Archbishop of Armagh,* Wales 1967.

9 See, e.g., T.C.D. MS 782 fols. 2v-9r. MS 788 fols. 61r-66v. MS 790 fols. 35v-38v.

10 For Ambrose Ussher see T.C.D. MSS 285-288; Travers' notebook is T.C.D. MS 391.

T.C.D. was the result of a donation of money that came to the College in 1601.[11] The fact that we are dealing with a *purchased* collection meant that far more control was possible over the style of the library, with a directed policy of acquisition possible from the outset, rather than the weeding process familiar to benefaction libraries in Cambridge and elsewhere.[12] The outlay of so much capital at a time when the College was going through a serious financial crisis indicates that the members of the College regarded the library as the most important element of the institution to develop. As such it served to define the new College-University.

The earliest catalogue of the main library is useful in a number of ways, since it is an accessions catalogue and therefore includes the records of the various book-buying expeditions in a somewhat haphazard way.[13] Once put into shelf-mark order though, a number of interesting trends became evident and it is one of those trends, the dominance of Ramism, that I wish to examine. Needless to say, such a study necessarily demands a concurrent look at both catalogues and notebooks – in order to deduce whether the books in question were actually being used, and this is the approach I have taken.

Looking at the library collections for the subjects of the trivium – and keeping in mind that the main part of the collection was pre 1604 – we see that grammar, the most basic of the disciplines, was the least developed area. However, in the College library and the personal libraries of Ussher and Challoner which supplemented some of the official library's deficiencies, Ramist grammars are the most common. There was relatively little emphasis

11 H. L. Murphy: *A History of Trinity College, Dublin from its foundation to 1702,* Dublin 1951, p. 45. See also T.C.D. MIC 130 MS Add. C. 297 fol. 116r.

12 Naturally one also needs to take into account contributing factors such as the book trade and influences on publishing choices. The literature on this topic is immense but for the book trade in Ireland see Mary Pollard: *Dublin's Trade in Books 1550-1800,* Oxford 1989, and Gerard Long, ed.: *Books Beyond the Pale. Aspects of the book trade in Ireland before 1850,* Dublin 1996. For a discussion of the impact of publishers' decision on the growth of Ramism see Ian Maclean: Philosophical books in European markets, 1570-1630: the case of Ramus, in: John Henry, Sarah Hutton, eds.: *New Perspectives on Renaissance Thought. Essays in the history of science, education and philosophy in memory of Charles B. Schmitt,* London 1990, pp. 253-263 – though perhaps not all of his conclusions are applicable in the Irish context.

13 T.C.D. MS 2.

on Latin grammar though the fact that some was included is surely connected to the youthful age of the T.C.D. students. Challoner in his own collection had Ramus' work on Latin grammar, the *Grammatica,* which had been published in 1559 and translated into English by 1585.[14] Like so many of his works, it proved to be very popular in Puritan circles. But this was not the only work by Ramus in Challoner's library – in November 1595 he also had, for instance, Ramus' *Scholae in liberales artes.*[15] That Ramus' Latin grammar was regarded as an important text may be deduced from a list of authorities drawn up by Ussher, which singles it out as a key text, while Ramus' Greek grammar also found its way into Ussher's personal collection.[16]

In the area of rhetoric the presence in the College library of Rudolph Agricola's *De Inventione Dialectica,* and a host of other humanists texts, indicates that T.C.D. by the end of the sixteenth century was firmly in the humanist camp as regards the understanding of the relationship between rhetoric and logic/dialectic. In borrowing lists we find that the Ramist edition of Cicero was the book in most demand, while in the library collections a Ramist ascendancy is evident from Ussher's catalogue, which includes Rodolphus Snellius' commentary on Talaeus' *Rhetorica* (1596), as well as a 1577 edition by Ramus and Talaeus, and indeed a first edition of the *Rhetorica.*[17] While Ussher has no copy of Fenner's *Rhetorike* he does not entirely ignore English Ramist rhetoricians as his inclusion of Charles Butler's *Rameae Rhetoricae Libri Duo* (1597) testifies.

English Ramists were the most popular authors in the area of dialectic also. The biblicism of English Ramists such as William Perkins, Dudley Fenner and George Downame – all authors whose works were purchased by Challoner – may have rendered the novelty of the approach somewhat more palatable to divines. Dudley Fenner's work, *The Artes of Logicke and Rethorike* contains much unacknowledged translation of the *Dialecticae Libri Duo.*[18] It proved a popular book primarily because of its juxtaposition

14 Challoner's library catalogues are in T.C.D. MS 357 fols. 1r-15v. They date from 1595 to 1608.

15 T.C.D. MS 357 fol. 3r.

16 T.C.D. MS 790 fol. 36r. MS 782 fol. 3v. MS 793 fol. 179v.

17 T.C.D. MS 357 fol. 8v. MS 790 fol. 49r. For Ussher's 1608 catalogue see T.C.D. MS 793 fols. 170r-187r.

18 W. S. Howell: *Logic and Rhetoric in England, 1500-1700,* New York 1961, p. 216.

of Ramist works on dialectic and rhetoric, a combination which proved very popular with preachers who accepted logic and rhetoric as essential elements in evangelization.

English Ramists were not the only authors to be collected – treatises by continental authors also played a dominant role. Particular favourites were works by Friedrich Beurhaus, who managed to get three textbooks published at London – all of which are present here.[19] Other treatises by continental authors which may also be found in the early library of T.C.D. are works by Wilhelm Adolf Scribonius who published two works in England. Indeed, Scribonius is recommended by Ussher as an important author to be consulted, in one of his notebooks.[20] Ussher's own catalogue was slightly less slavishly Ramist but he also had works by Ramus and the *Quaestiones Logicae et Ethicae* of Freigius.[21]

A Ramist dominance is clearly identifiable in the early library catalogue of T.C.D. but this is not the only evidence that points to T.C.D. being a 'Ramist' foundation. An examination of Ussher's notebooks indicates the supremacy of Ramism at T.C.D. during the early decades of its foundation. Not only is interest taken in Ramus' own life, but also in the publishing history of his and Talaeus' treatises, with brief notes on various Ramist authors interspersed in his commonplace-books – not to mention dichotomization gone mad.[22] Perhaps the most startling evidence comes in the form of a study plan, for the months from December to November, which outlines the consistent study of Ramus' works on grammar, dialectic and rhetoric.[23] The inclusion of grammar here, indicates that this course plan must have been for the first year of the B.A. This curriculum is followed by a list of authors of a decidedly Ramist tone as complementary material. This trend is duplicated in other notebooks which list some of the best and, indeed, lesser known commentators on Ramus' output. If any doubts prevail, one only has to examine the list Ussher gives in his demarcation of which authors should be read in certain areas to recognise the dominant

19 T.C.D. MS 2.

20 T.C.D. MS 782 fol. 5r.

21 T.C.D. MS 793 fol. 179v.

22 See, for example, T.C.D. MS 782 fol. 11r. MS 786 fol. 5v.

23 T.C.D. MS 782 fol. 16r.

position of Ramism in the trivium curriculum at T.C.D.[24] Logic has been superseded by dialectic, and the list of recommended authors of 'recent' origin reads like a Ramist's dream: Ramus, Beurhusius, Martinus, Piscator, Snellius, Talaeus, Tempellus, Melanchthon, Alberius, Titelman, Valerius, Agricola, Sturmius, Caseus, Bursa, Trapezuntius, Willichius, Tolletus, Caesarius, and Valla – in that order. Ramus heads the fray in the grammar and rhetoric sections also. In yet another notebook, one finds notes on Ramists and authors whose works complement Ramus' output: Bernardus Copius, Henricus Schorus, Franciscus Fabricius Marcoduranus, Freigius, and a list of commentators on Ramist works published after his death – by such men as MacKilwein and Fenner, though antagonists of Ramus were not neglected:

> Jacobus Carpentarius Ramo hostis insensissimus scripsit contra Dialecticae ca. Rameam lit. ipse in annotationib. in organon Aristis.[25]

The borrowing lists substantiate an interpretation of Ramist dominance, though the trivium subjects were not among the most popular of choices. However, it is clear that if any dialectic system was popular, it was the Ramist one. Of the few references to grammatical, rhetorical, or 'logical' works, we find one all-purpose Ramist *Scholae,* two works by Ramus on Cicero, and Fenner's logic.[26]

But Ramism at T.C.D. was not confined to the sections of the trivium. In his notes Ussher also included a number of comments concerning authors involved in the Cambridge Ramist debate, among whom pride of place was given to Piscator – no doubt because of his involvement in the Digby-Temple debate:

> Joannes Piscator Scripsit animadversiones ad Dialecticam Rami, anno 1580. Huic epistola Tempellus respondit anno 1581. Cui Piscator iterum rescripsit eodemque anno P. Rami scholae cum Pistoris emendationibus. Francofurti typis evulgatae sunt. Et anno 1582 iterum Edidit animadversiones ad Tempell. Huic denuo respondit Tempellus ultimo animadversiones edidit no. 1586 item Exercitationes Dialecticas scripsit anno 1585.[27]

24 T.C.D. MS 790 fols. 35v-38r.

25 T.C.D. MS 790 fol. 84v.

26 T.C.D. MS 357 fol. 8r-v. MS 790 fol. 49r-v. MS 790 fol. 169r-v.

27 T.C.D. MS 790 fol. 84v.

The Cambridge debate centred around the works of a later provost of T.C.D., William Temple. Temple's defence of Ramus was not limited to a discussion of Ramist works on the trivium, however, for the welter of publications of Ramist works in England in the 1580s included a number of works on Ramus' mathematical preoccupations. The subjects of the trivium were by far the most important to Ramus, but he was unwilling to neglect mathematical subjects, which by their very utility demanded a role in his scheme. In his discussion of them he emphasised their use in daily life, while in his *Dialectic* he elucidated some links between mathematics and dialectic.[28]

T.C.D. collected not only Ramus' own works, but also those of his follower, Rodolphus Snellius, who had been responsible for a well-known Ramist text – *Snellio-Ramaeum philosophiae syntagma* (1596).[29] In 1581 Snellius had been appointed *extraordinarius professor* in mathematics at the University of Leiden, and though stigmatised for his championing of Ramism, he continued to teach his methods. Snellius agreed with the Ramist imperative of utility – a factor possibly apparent in the attention paid to the geography collections of the College library. The inclusion in the early collection of T.C.D. of five of Snellius' works, plus works by the well-known Ramist, Freigius, is indicative of the strong hold the Ramist approach held in T.C.D.

Evidence from the notebooks of Ussher, and the library catalogue itself, suggests that mathematical subjects were considered of far more importance than their physical counterparts. The branches of mathematics delineated by Ussher were primarily concerned with the study of geometry and arithmetic, and here the tried and trusted sources, such as Euclid and Archimedes held sway, though in his lists of contemporary authors Ramus was evidently thought to be the chief author of note.[30] Indeed, although very few works of a mathematical nature were borrowed from Challoner and Ussher, the most popular text was Ramus' *Geometria* while Ussher possessed a copy of Ramus' *Scholae mathematicae* (Basle 1569).[31]

28 Philippe Desan: *Naissance de la méthode. Machiavel, La Ramée, Bodin, Montaigne, Descartes,* Paris 1987, pp. 82. 85.

29 John Gascoigne: A reappraisal of the role of the universities in the scientific Revolution, in: D. C. Lindberg, R. S. Westman, eds.: *Reappraisals of the Scientific Revolution,* Cambridge 1990, p. 223.

30 T.C.D. MS 790 fol. 36r.

31 See for example, T.C.D. MS 357 fol. 8r-v.

Challoner, in the reorganisation of his library in 1608, purchased another mathematical textbook by an adherent of Ramism.[32] Christianus Urstitius (Wurstisen, 1544-1588) had been professor of mathematics and professor of theology at the University of Basle. He probably met Ramus during the latter's stay in the Swiss city and the meeting proved to be an important one for Urstitius as it continued to influence his approach to arithmetic when he devised his *Elementa arithmetica.*[33] This book proved so popular that Thomas Hood decided in 1596 to translate it into English, giving it the title *The Elements of Arithmeticke Most Methodically Delivered,* a rendition of the original title that points to its chief source.[34] For in reality Urstitius' text-book was little other than a rehash of Ramus' own deliberations on the subject. Indeed, in his dedicatory epistle, Urstitius was eager to acknowledge his debt to Ramus and Freigius.[35] Urstitius, Ramus and Freigius headed Ussher's suggestions for contemporary reading in arithmetic.[36] The College library contained a copy of Freigius' *Quaestiones geometricae* of 1583 – continuing this trend.[37] Urstitius' *Elementa* continued to be popular in England well into the 1640s as its use by Milton testifies. Milton's own commentary on Ramist was influenced by George Downame – another author whose works were collected by the fellows of T.C.D.[38]

The T.C.D. music collection, the fourth area of the quadrivium, also sheds light on the spread of Ramism in the curriculum of the early College, as one of the few works to be collected was the *musica* of Fredericus Beurhusius, actually the *Erotematum musicae* of Friedrich Beurhaus (1536-1609). This work was first published in 1573, and due to its popularity went through three further imprints.[39] What is noteworthy for us here is not his innovatory arrangement of material, but, rather, the fact that Beurhaus

32 T.C.D. MS 357 fol. 14r.

33 See Walter J. Ong: Christianus Urstitius and Ramus' New Mathematics, in: *Bibliothèque d'Humanisme et Renaissance* 36 (1974) pp. 604-610.

34 Ibid. p. 605.

35 Ibid. p. 607.

36 T.C.D. MS 790 fol. 36r.

37 T.C.D. MS 2.

38 Ong: Urstitius (as note 33) p. 606.

39 Stanley Sadie, ed.: *The New Grove Dictionary of Music and Musicians,* London 1980, vol. 2, p. 665.

was a follower of Ramus and used Ramist precepts in his work. It is therefore extremely illuminating to find Ussher recommending "Beurhusius" as the chief authority in musical education.[40] Thus, Ramist influence is not only readily apparent in the subjects of the trivium in early T.C.D., but went much further than this, in informing the attitude of the early members of T.C.D. in many areas of scientific interest, and particularly in the quadrivial subjects of arithmetic, geometry and music.

Other areas were also affected. In ethics, for example, Melanchthon's choice of Aristotelian over Platonic interpretations inevitably ensured a measure of acceptance of Aristotelian ideas on this subject in Protestant university circles, but again, it could be ameliorated by Ramist perspectives. This trend is evident in Zwinger's interpretation of Aristotelian ethics. The inclusion of this work in the T.C.D. collection, along with that of Rodolphus Snellius on ethics, is indicative of the Ramist framework in which Aristotle's *Politics* was examined. In the areas of ethics and politics Aristotle reigned supreme – but it is viewed through the Ramist prisms of Zwinger and Freigius.[41] Indeed, Freigius appears to have been a favourite author, as his name crops up on the lists of authors relating to jurisprudence.[42] Even in theology Ramism could also be identified as a utilitarian handmaiden. Certainly this was the theory elucidated by the fourth provost, Temple, who had already outlined its usefulness in the study of theology in his *Commentary on twenty psalms,* published in 1605.

Perhaps yet another indication of a strong adherence to a Ramist model is the absence of any real metaphysical section in the library collections. Metaphysics did not loom large in the early seventeenth century curriculum of T.C.D., no doubt because of the well-known antagonism of Ramus to metaphysical studies.[43] Metaphysics, as a subject category, was not distinguished as an identifiable body, and works in this sub-division were scattered in both the humanities and theology sections. This trend concurs with tendencies on the continent, especially in Protestant universities, where

40 T.C.D. MS 782 fol. 5r.

41 T.C.D. MS 790 fol. 37r.

42 Ibid.

43 Hugh Kearney: *Scholars and Gentlemen. Universities and Society in pre-industrial Britain, 1500-1700,* London 1970, p. 49.

metaphysics was not considered as a part of philosophy.[44] It is illuminating to recognise that Ussher did not consider metaphysics important enough to register as a subject grouping, in his various divisions on these lines in his notebooks, nor do any of his notebooks concentrate on metaphysical issues.

The evidence, then, for a Ramist supremacy in the curriculum of T.C.D. in its early years is overwhelming, but it is also intriguing when compared with developments elsewhere in Europe. In mainland Europe in the period 1580 to 1630 one continues to find Ramism in a fairly strong position – though usually one finds an accommodation with the Melanchthonian interpretation of the Aristotelian system.[45] In the British Isles, however, the picture is somewhat different. Though Ramism continued to arouse debate within academic circles until 1618 most commentators agree that by 1600 the English Ramist experiment was in decline. Exactly why this was the case is a much-disputed topic. Kearney points to a play of 1598 which pokes fun at the Ramist scholar *Stupido* and states that "the revival of scholasticism may be dated from the 1590s and its progress traced within several colleges".[46] This revival is, in his eyes, an essentially negative phenomenon as his pejorative description of Whitgift's role demonstrates: "John Whitgift had become the Torquemada of the church in Cecil's eyes and Elizabeth seemed content with the role of Isabella."[47] Todd agrees with this analysis but suggests that though "after 1600 Oxbridge students were increasingly being exposed to the writings of Jesuit and other neo-scholastic theologians and dialecticians" their readings of these authors were seen as a "necessary evil".[48]

44 Joseph S. Freedman: Philosophy Instruction within the Institutional framework of central European schools and universities during the Reformation Era, in: *History of Universities* 6 (1985) p. 117.

45 For a discussion of these trends see Joseph Freedman: The Diffusion of the Writings of Petrus Ramus in Central Europe, c. 1570 - c. 1630, in: *Renaissance Quarterly* 46 (1993) pp. 98-152; Jarmila Pešková: Ordines Lectionum jako pramen poznáni výuky na artistické fakulte Prazské Univerzity v Letech 1570-1619, in: *Acta Universitatis Carolinae* 30 (1990) pp. 9-30. I am grateful to Drazen Nozinic for translating this article. More material on this subject is present in this volume of essays.

46 Kearney: *Scholars* (as note 43) pp. 50. 83.

47 Ibid. p. 79.

48 Margo Todd: *Christian Humanism and the Puritan Social Order,* Cambridge 1987, p. 72.

Both interpretations are challenged by Mordechai Feingold who reinterprets developments as a rise of neo-Aristotelianism (rather than a simple rehash of medieval scholastic authors).[49] Far from being a negative development, this process is seen as the coming of age of the academic community. Now scholars eschew the trivial simplicities of the Ramist system and are more discriminating: "the clampdown on Ramus can be interpreted as a sign of intellectual integrity and broad-mindedness."[50] The latter analysis has much to recommend it but it forces one to investigate why it is that T.C.D., in the first two decades of its existence resembles more the Cambridge of the 1580s than the early seventeenth century, which was moving beyond Ramism.

There are a number of ways of explaining this phenomenon but surely the most important factor must be the influence of Luke Challoner, vice-provost and one of the chief book buyers. Challoner himself had gone to Trinity College, Cambridge in the 1580s – just at the height of Ramist influence in Cambridge.[51] Ramus' own dialectical textbooks had been popular in Cambridge during Challoner's stay, judging by inventories of booksellers and stationers.[52] Indeed, from the spread of Ramist works throughout his catalogue, one may surmise that Challoner was deeply influenced by this author. He was there during the famous debates of Temple and Digby and his university connections would have been with well-known Ramists.

Friendship networks play an important role in explaining the Ramist dominance in T.C.D. Temple, for instance, would have been well known to Challoner not only from his days in Cambridge but also because of his position as secretary to Philip Sidney and subsequently the Earl of Essex. It has been suggested that latter connection may have brought him to Ireland on one occasion. The connection with Sir Philip Sidney is of particular in-

49 See Feingold's chapter on the humanities at Oxford in Nicholas Tyacke: *History of the University of Oxford,* vol. 4, Oxford 1997, pp. 211-358.

50 Ibid. p. 292.

51 John Venn, J. A. Venn, eds.: *Alumni Cantabrigiensis,* Cambridge 1922, vol. 1, p. 315.

52 See Elisabeth S. Leedham-Green, ed.: *Books in Cambridge Inventories. Book lists from Vice Chancellors' court probate inventories in the Tudor and Stuart periods,* 2 vols., Cambridge 1986.

terest, as his link to the Wechel press – the main printers of Ramus' works, was the key which opened the lock of English universities to Ramist influence.[53] The fact that Temple, one of the most important supporters of Ramism in Cambridge, was later appointed as provost of T.C.D. – at a time when the College was effectively ruled by Challoner – indicates the Ramist friendship network at work between T.C.D. and Ramist adherents in Cambridge, particularly in the more Puritan Colleges of Trinity, Christ's, St. John's, Emmanuel and Sidney Sussex.

But Cambridge was not the only conduit by which Ramism seeped into T.C.D. Forming another part of the Puritan friendship nexus were the Scottish Ramists such as Andrew Melville and Robert Rollock. The role of Scottish universities in fostering Ramist studies in T.C.D. was as important as the Cambridge connection. Two well-known Scots, James Hamilton and James Fullerton, had been educated under the Ramist system in Scotland and were to prove influential in the early years of T.C.D.[54] The curriculum of T.C.D. has a number of Melvillian features, while Rollock, famous for inculcating Ramism in Edinburgh, was a very popular author in T.C.D. library collections.

Any doubts that might have been felt about the linkage of Ramism and Puritanism must surely be dispelled by an examination of the early history of T.C.D. There were numerous reasons why Puritans in particular would be attracted to Ramism. Ramus himself had stated that his third judgement was a link between dialectic and religion and had even gone so far as to say that a logical education led to salvation. As Philippe Desan has pointed out, "Pierre de La Ramée passa sa vie à promouvoir, défendre et théoriser les bases d'une méthode unique à partir de laquelle il serait possible d'analyser tous les arts en relation à Dieu."[55] William Perkins, one of the most popular authors in the moderate Puritan T.C.D. library, had applied Ramist methodologies to the study of theology in his ever popular treatise on preaching, printed in Latin in the year of the College's foundation, 1592, and in English in 1606 under the following title: *The*

53 Ong: *Ramus* (as note 6) p. 302.

54 Andrew Dalzel: *History of the University of Edinburgh from its foundation,* Edinburgh 1862, p. 17.

55 Desan: *Naissance de la méthode* (as note 28) p. 11.

Arte of Prophecying, or, a Treatise concerning the sacred and onely true manner and methode of Preaching.[56]

One of the greatest exponents of logical analysis of Scripture was the fourth provost of T.C.D., William Temple, who published his *Logicall Analysis of Twentie Select Psalms* in London in 1605. The arguments Temple uses in his preface point to the importance moderate Puritans placed on logic/dialectic:

> For to be able, first to contrive and perfit the frame of discourse: and then againe to discover and lay foorth in the pure naturals not onely the serverall members but the particular structure and conformation of the same: what estimation doth it not deserve. And this is it which Logike doth undertake and performe. Out of the former imployment this worthie art furnisheth unto us varietie and choice of arguments for cleering the cause in question: and directeth us to hold that, which is true and iustifiable: to disclaime that, which is inconsequent: to marshall each thing in his proper ranke and place.[57]

Above all it was a tool which must be used:

> It succeedeth with the Art of Logike, as it doth with a drug of rare vertues: which resting confined within the limits of the Apothecaries shop, winneth not anie opinion or credit with men: but being taken out and skillfully applied, discloseth her value & merit. So this art resting imprisoned within the lists of the schoole and the cabinets of the learned, findeth no grace or regard: but being drawn foorth as it were into the field of imployment and service, it doth then manifest to the world the proportion of worth and respect, that is due unto her.[58]

When utilised in this manner the Ramist dialectic could and did, in Temple's view, demonstrate the practical use of the Psalms. Hence it could provide an invaluable instrument in the inculcation of morality to the laity. By using Ramist techniques acquired in the university, a preacher could enhance his own understanding and application of practical divinity.

56 Howell: *Logic* (as note 18) p. 206. Howell says that Perkins "was not a thoroughgoing Ramist" and that his Ramist tendencies were limited to the form of presentation of arguments. Francis Yates disagrees with this, and states that Perkins was actually more Ramist than this implies – see Francis A. Yates: *The Art of Memory,* London 1966, pp. 260-278.

57 William Temple: *A Logicall Analysis of Twentie select Psalms,* London 1605, preface, n.p.

58 Ibid.

A dominant theme in Temple's analysis was 'utility' – in this case the utility of the Ramist system for theological studies and their eventual transmission to the laity. The Ramist emphasis on 'utility' has received much attention and has led to some startling interpretations. Kearney's linkage of it to social radicalism is highly questionable.[59] He appears to think that *only* a social-economic argument could have generated such vociferous disagreement as the Cambridge Ramist debate:

> Ramism was thus something more than an educational movement concerned with reforming the curriculum and teaching methods. Its sponsors, its implication, its general backround had radical overtones, and the appearance of Ramism in the curriculum or in a students notes is tantamount to finding revolutionary doctrines being taught in a religious seminary.[60]

His subsequent argument that separatists – surely the most radical religious group of all – had little interest in Ramism seems to undermine this theory, and certainly in an Irish context the belief that Ramism is necessarily linked to social radicalism finds very little support.[61] One can hardly call the two principal supporters of the foundation of T.C.D., the Archbishop of Armagh, Adam Loftus and his successor Henry Ussher, social revolutionaries!

Anthony Grafton and Lisa Jardine are far more circumspect in their linking of Ramist utility and social order, preferring to concentrate on the reception of Ramist ideas within the merchant classes. However, though their declaration that "Ramus deliberately discarded the difficulty and rigour of high scholastic schooling and thereby attracted those who regarded education as a means to social position rather than as a preparation for a life of scholarship" is undoubtedly true, their subsequent argument presents some difficulties when viewed from the perspective of Irish Ramism.[62] They state that Ramus "explicitly (though not necessarily deliberately) achieved the final *secularization* of humanist teaching – the transition from 'humanism' to 'the humanities'."[63] Ramism, we are told, "opened the prospect that the

59 Kearney: *Scholars* (as note 43) p. 52.

60 Ibid. p. 53.

61 Ibid. p. 73.

62 Anthony Grafton, Lisa Jardine, eds.: *From Humanism to the Humanities. Education in the Liberal Arts in Fifteenth and Sixteenth Century Europe,* London 1986, p. 168.

63 Ibid. (their italics).

purpose of education was to purvey information and skills, not to be morally improving".[64] These deductions may well be the logical result of Ramus' deliberations but if so, they were not ones which the members of T.C.D. envisaged.[65] For their interest in Ramism necessarily encompassed the view that it aided rather than hindered moral improvement. Essentially, they saw Ramism as an invaluable tool which enabled them to incorporate humanist studies into the service of theology.

Similarly, though Feingold is undoubtedly correct in suggesting that Ramism was viewed by the establishment as a threat to order, one must ask oneself whether this was the perception of the moderate Puritans actually using it as a methodology.[66] One might plausibly suggest that in Ireland the need to project an image of 'civility' and order outweighed other considerations. The Puritan members of T.C.D. were well aware that one of the perceived functions of the new College was the "civilizing of the natives" – in conjunction with the advancement of reformed religion.[67] The adoption of a Ramist curriculum not only acted as a touchstone of Puritan identity at a time of persecution by the establishment, but it could also be marketed as a framework which imposed order on the curriculum. By streamlining approaches to the queen of faculties, theology, Ramism could be viewed as a system which made the dual goals of colonization – confessionalization and civilizing of the natives – more attainable. In this regard it is interesting to see that Ramism played an important role in another Puritan colonial university – Harvard – an institution with connections to T.C.D. which sought similar ends.[68] Indeed the application of Ramism in the Harvard curriculum

64 Ibid. p. 170.

65 Elizabethanne Boran: Perceptions of the Role of Trinity College, Dublin, from 1592 to 1641, in: Andrea Romano, ed.: *Università in Europa. Le istituzioni universitarie dal Medio Evo ai nostri giorni: strutture, organizzazione, funzionamento,* Messina 1995, pp. 257-266.

66 Feingold: Humanities (as note 2) p. 290.

67 The creation of an extremely large library could also be seen as an attempt to assert superiority in the field of 'civility' – on this point see Elizabethanne Boran: The Function of the Library in the Early Seventeenth Century, in: Vincent Kinane, Anne Walsh, eds.: *Essays on the history of Trinity College Library Dublin,* Dublin 2000, pp. 39-52.

68 On the connections between the Puritan enterprise in Ireland and at Harvard see Elizabethanne Boran, Helga Robinson-Hammerstein: The Promotion of 'Civility'

of the mid seventeenth century, at a time when Ramism was well past its peak in the motherland should surely force us to investigate the role of Ramism in the colonizing process.[69]

The relative lack of secondary schools in early sevententh century Ireland meant that in T.C.D. students were also slightly younger than their counterparts in England. Hence, the simplicity of the Ramist approach was much appreciated by Puritans whose attitude to these studies was that they were means to an end (the study of divinity), rather than an end in themselves. As Morgan suggests "the Puritan search for a purer logic may well have been at least in part a response to radical insistence on simplicity in the analysis and diffusion of Scripture."[70] Time and again we are told that Ramism appealed to the younger members of universities.[71] One has only to remember Temple's age when he embarked on his now famous quarrel with Digby. The Hermogean dichotomization not only looked good in the printed text but also gave the impression that all knowledge was definable and ordered – much more manageable. By applying a single method to all areas of knowledge, the Ramist method offered a relatively quick and easy approach to the seven liberal arts and their related disciplines. Even one of Ramus' most vociferous opponents, Adrien Turnèbe, had to concede that:

> Method – no word is more popular in our lectures these days, none more often heard, none gives off a more delightful ring than that term. Everything else, if you use it often enough, will end by nauseating your readers. This is the only thing that never makes them sick.[72]

and the Quest for the Creation of a 'City of Peace': The Beginning of Trinity College, Dublin and of Harvard College, Cambridge, Mass., in: *Paedagogica Historica* 34 (1998) pp. 479-506.

69 On Ramism in Harvard see Perry Miller: *The New England Mind,* Harvard 1939. For a slightly more cautious appraisal of the influence of Ramism on the Harvard curriculum see Samuel E. Morison: *Harvard College in the Seventeenth Century,* Harvard 1935, pp. 188-189.

70 John Morgan: *Godly Learning. Puritan attitudes towards Reason, Learning and Education, 1560-1640,* Cambridge 1986, p. 110.

71 See for example, Peter Sharratt: Peter Ramus and the Reform of the University: the Divorce of Philosophy and Eloquence? in: Peter Sharratt, ed.: *French Renaissance Studies 1540-70,* Edinburgh 1976, pp. 4-20.

72 Ong: *Ramus* (as note 6) p. 228.

Added to that, the tendency of Ramus and his followers of laying emphasis on the use of the vernacular would surely have been appreciated by students – not to mention the fact that the Ramist methodology corresponded very well with the ever popular creation of commonplace-books.[73]

The admittedly crude uses that Ramism was put to by its devotees has sometimes caused commentators to deny Ramism or its practitioners any connection with 'true' humanist studies. However, Ramus' own debt to humanist critique of scholastic logic is sufficiently clear. Rather than been seen as a diminution of rhetoric, the core of any humanist study, one might argue that in effect Ramus rhetoricises logic/dialectic. Clearly this was Gabriel Harvey's understanding of what he was doing:

> Unite dialectic and knowledge with rhetoric. Keep your tongue in step with your mind. Learn from Erasmus to combine an abundance (copia) of words with an abundance of matter; learn from Ramus to embrace a philosophy which has been allied to eloquence; learn from Homer's Phoenix to be doers of deeds as well as writers of words.[74]

Ramus' detractors echoed this interpretation of what Ramus was trying to do:

> He holds that philosophy is far better taught from poets than from philosophers. In this obscure conjunction of philosophy with eloquence, he teaches no Aristotle, no philosophy, but pretty much poetry alone.[75]

What this tells us about the relationship of Puritans and humanism is much more difficult to determine. Irish Ramism, as evidenced by the denizens of T.C.D., was a strange beast. Kearney's assumption that Ramism could be considered anti-humanist is perhaps too dogmatic for the Irish case. Literature might not have loomed large but it and the relevant works were by means neglected. This probably has more to do with geography than a redefinition of Ramism though, since the creators of the College library – the principal indication of the ethos of the College – were well aware that their library had to be as comprehensive as possible. However, though Puritans

73 Ann Moss: *Printed Commonplace-Books and the Structuring of Renaissance Thought,* Oxford 1996, pp. 157. 159. 163.

74 Peter Mack: Humanist rhetoric and dialectic, in: Jill Kraye, ed.: *The Cambridge Companion to Renaissance Humanism,* Cambridge 1996, p. 82.

75 Grafton, Jardine: *Humanism* (as note 62) p. 167.

and humanists might disagree on the overall aim of the *studia humanitatis* they held similar positions on a number of issues – not least the centrality of the text and the importance of rhetoric in its transmission. The Protestant emphasis on Scripture automatically meant that greater attention was given to the means of the dissemination of the Word. The application of Ramist ideas in preaching seemed to offer a solution to some of the pressing problems of communication of the Word. MacKilwein had suggested the possible utility of Ramism in the context of the sermon, and Fenner's substitution of Biblical examples brought this stage a step further, by making the use of Ramism more acceptable to Biblically minded Puritans.

Todd's characterization of Puritans as embodiments of 'Christian humanism' may overstate the case but neither were they antagonistic to trends in humanist thought – they simply redirected them to the service of theology.[76] This guarded approach is quite clear in a contemporary pronouncement on the issue by Archbishop Loftus, in his advice to the second provost, his successor Walter Travers. Loftus was careful to emphasise that one should keep in mind the true aim of all learning – the study of theology:

> you must take heed lest they should study learned extravagancies which rather taint their owne affections with contagius profanesse then improve their understandings eyther to their owne good or the salvation of others, for prevention of which evill you are to see that the younger sort be well catechysed and that you prescribe to the rest a catalogue of approved bookes to be read of them as foundationall learning both human and devine, especially Divine.[77]

He continues this line of thought in his hope that:

> it will be your endeavour soe to train up those who are or shall be designed to the ministry of God's word in your College, that they may be disposed rather to study solid and profitable doctrines than the curiosityes of fancie smooth elocution.[78]

Much of my investigation into the intellectual interests of members of the early College supports Morgan's conclusions concerning Puritan attitudes to learning.[79] Clearly Puritans were not antagonistic to learning, recognis-

76 Todd: *Christian Humanism* (as note 48) pp. 91-95.

77 J. W. Stubbs: *The History of the University of Dublin from its foundation to the end of the eighteenth century,* London 1889, p. 17.

78 Ibid. p. 19.

79 Morgan: *Godly Learning* (as note 70) p. 98.

ing its importance in the training of a learned ministry, and they were willing to cast their intellectual net wide – not eschewing the works of ancient authors. Indeed, an example of a moderate Puritan attitude to pagan authors is Richard Sibbes' conclusion that "Truth comes from God, wheresoever we finde it [...] we may take it from heathen authors as a just possession."[80] Puritan efforts to reform the undergraduate curriculum did not seek to overthrow the established order, but to reinterpret it from a godly perspective. Thus, emphasis was laid on logic/dialectic, so that the Word would be seen to make sense to a minister's congregation; rhetoric was developed to enable the spread of the Word; grammar, especially Greek and Hebrew grammar, served as a stepping stone to Biblical study; natural philosophy as an investigation into the Book of Nature; history as an aid to polemical discussions.

One last reason why Puritans might have felt that Ramism was a congenial methodology was not only Ramus' well-publicised death during the St. Bartholomew's Day Massacre, but also a feeling that Ramus was part of a whole movement away from degenerate older sources. Helisaeus Roeslin echoes contemporary analyses by Richard Bostocke and Etienne Pasquier when he comments that

> in the fullness of time it was found that certain of the tenets and hypotheses [of Galen, Ptolemy and Aristotle] not only were inadequate, but entailed many absurdities. This moved some in this latter age to plan a defection and to rise up against them: Nicholas Copernicus against Ptolemy, Theophrastus Paracelsus against Galen and Aristotle, and Petrus Ramus against the latter in the philosophical arts. They did so at the very same time that many contentions arose in Europe in religion and sacred matters.[81]

It is interesting to note that in the library collections of T.C.D. all these authors were collected, with particular emphasis in medical matters being placed on Paracelsus.

Dating the decline of Ramism in T.C.D. is a fairly tricky business. In general, later acquisitions to the library of T.C.D. up to the 1650s demonstrate that theology was still regarded as the most important subject to develop, with an interest also being displayed in natural philosophy, history and law, though these works are far out-numbered by the additions to the

80 Ibid. p. 102.

81 Nicholas Jardine: *The Birth of History and Philosophy of Science,* Cambridge 1984, p. 282.

theological collections.[82] Interestingly, these collections demonstrate some interest in the 'neo-Aristotelianism', which had been so popular in the first half of the seventeenth century. Unfortunately the absence of price lists from the period 1620s to the 1650s makes it difficult to date exactly when this trend started in T.C.D. but there are a number of pointers which suggest that this movement away from a strictly Ramist interpretation was already happening even prior to the Laudian statutes of 1637, which are the embodiment of an official break with the system of education prevalent in T.C.D. in the early decades of the seventeenth century. The fact that MS 2 is an accessions catalogue, with accessions going up until the 1630s, gives us some indication of a change of direction.

Some additions made to the library catalogue in the years 1609 to 1616 indicate a softening of the Ramist hardline. In the 1609 book-buying expedition, for example, we find three works of Keckermann's being added to the collection: his *Systema rhetoricae* of 1608, the *Systema disciplinae politicae* of 1607 and finally the *Disputationes philosophicae, physicae praesertim* of 1606. Evidently this halfway house approach to Ramism was popular as his works continued to be added to the collection as additions made in 1612 demonstrate – now we have his *Systema geographicum,* his *Systema astronomiae* and most interestingly his *Scientiae metaphysicae compendiosum systema.* Later additions, post 1616, of metaphysical treatises by other authors seem to continue this trend, though the presence of an addition in 1612 of Ramus' *Septem liberales artes* should prevent us from being too categorical in our conclusions.

If we look at the additions made to the library from 1609 onwards and particularly from 1615 into the 1620s one can see a distinct trend. More works on metaphysics are being bought – for example Gaspar Finks' work of 1608, which was purchased by T.C.D. in 1612. Even as early as 1609 we find two of Suárez' treatises being purchased. The inclusion of other works by this author in the next decade and especially the early 1620s surely points to a sea change. It is, of course, entirely possible to argue that these purchases were not indicative of a new departure but were bought so that members of the College would be conversant with the arguments of their detractors, but I'm not so sure. A library catalogue, dated 1633, of Kildare Hall, an undergraduate teaching facility, reveals that by the early 1630s teaching had

82 T.C.D. MS 1 is a mid seventeenth century catalogue.

subtly changed from the dominant Ramist tone of the very early period. No works by Ramus himself are mentioned – in contrast to the number of copies of Aristotle's works freely available – while authors such as Paul of Venice, Hyperius and Libavius and Alsted are now the order of the day.[83]

This development may well pre-date the 1630s and there is some evidence that a date somewhere in the middle of the second decade of the seventeenth century may mark the beginning of the change. Obviously the trips made over to London, Oxford and Cambridge were beginning as early as 1609 to have an effect on the book-buying policies of Ussher and Challoner but the main impetus would have come after Challoner's death in 1613 and possibly around 1615 when, I suggest, James Ussher started re-inventing himself in order to position himself for a post in the ecclesiastical hierarchy. The evidence for this repositioning is substantial. I have argued elsewhere that Ussher's early correspondence demonstrates a markedly Puritan dimension, but this appears to be undergoing a change in 1616 when Ussher deliberatly distances himself from his more outspoken Puritan friends.[84] Ussher's growing friendship network – as evidenced by his correspondence – indicates that by the second decade of the seventeenth century, Ussher had formed strong links with intellectuals at Oxford and Cambridge and no doubt was influenced by the new developments there.[85] It is not inconceivable that he may also, at this stage, have started redirecting the T.C.D. curriculum along less Ramist lines since Ramism and Puritanism were seen to be so interlinked – a connection which he no longer wanted to emphasise. His appointment in the 1620s of men such as the philosopher Nathanael Carpenter surely indicates a divergence from his earlier views. In the absence of concrete evidence from student notebooks, however, this must remain conjecture. Clearly what was happening was a redirection rather than a revolution.

To conclude, an investigation of the early intellectual interests of members of T.C.D. reveals, above all, the dominance of the Ramist curriculum

83 T.C.D. MS 2160 no. 2a.

84 See Elizabethanne Boran: An early friendship network of James Ussher, archbishop of Armagh, 1625-1656, in: Helga Robinson-Hammerstein, ed.: *European Universities in the Age of Reformation and Counter Reformation,* Dublin 1998, p. 124.

85 For Ussher's correspondence see C. R. Elrington, ed.: *Ussher's Works,* vols. 15 and 16, London 1864. A new edition of Ussher's letters is currently in preparation.

in the early seventeenth century College. This interest in Ramism was an all-pervasive one, and was not confined to the sections of the trivium most commonly associated with the Ramist challenge. Ramist authors were also collected in fields other than the trivium, as can be seen from the strong showing of Ramist authors in sections of the quadrivium such as mathematics and music. Ramus' *Geometria* proved the most popular choice in geometry, and his works on arithmetic were also studied. It seems likely that the Ramist emphasis on utility of instruction may also have informed the choice of a large collection of works on geography, navigation and other practical subjects. It is clear that Ramist influence went beyond even these areas, which are more usually linked to Ramism, and it is particularly intriguing to recognise that Ramist authors were collected in such areas as music, ethics, politics and law. In other words, Ramism was not merely a style influencing the subjects of grammar, rhetoric and dialectic, but was developed in T.C.D., as an all-embracing methodology with which to investigate the arts subjects, a methodology moreover, which could be applied to theological study.

In those areas where Aristotelianism continued to serve as the basis of intellectual investigation – politics and ethics – it was an Aristotelianism interpreted through Ramist eyes. Those sections that did not conveniently fit into this schema, such as physics, received very little attention. One reason for this may have been that they were of less obvious use to a budding theologian. That theology was the perceived goal may be deduced from the fact that in each subject the areas that were developed were ones which could be of use to a theologian. Hence in the early curriculum of T.C.D., we see the essentially symbiotic nature of the relationship of Ramism and Puritanism. Ramism influenced the way in which the moderate Puritans of T.C.D. approached their sources, allowing them to utilise a humanist methodology, which they in turn channeled for their own ends. Walter Ong tells us that in 1572 Ramus presented a copy of his *Dialecticae libri duo* to William Fulke, who was in Paris at the time, and that this copy now resides in T.C.D.[86] Here, in this simple account, we see a microcosm of the modes of transmission of Ramism from France to Dublin, via the intellectual milieu of moderate Puritanism.

86 Ong: Urstitius (as note 33) p. 603.

"Beauty and the Beast": Nature, Logic and Literature in Ramus

KEES MEERHOFF

I have chosen to call my contribution "Beauty and the Beast". It could also have been called "*Defenders of the Text*", but that title has already been taken by Anthony Grafton. It is with reference to Grafton's work, and particularly the essay "Renaissance Readers and Ancient Texts" that I should like to offer a few reflections on the contributions made by humanism to the history of philosophy. Besides Grafton's book it would be appropriate to refer to other recent work, such as Peter Mack's *Renaissance Argument* (1993) – elements of this work can also be found in the chapter "Humanist Rhetoric and Dialectic" of *The Cambridge Companion to Renaissance Humanism* (1996). Other relevant works include Ann Moss' *Printed Commonplace-Books* (1996) and James Hankins' *Plato in the Italian Renaissance* (1990). Hankins, in his introduction, suggests a "typology of reading" which is of great interest to us.[1]

I should uncover here what my title signifies: the "Beauty" is the text, text which to the Renaissance, at least to start with, is "ancient". The "Beast" is something which has been disparaged among the greater part of Western historians of philosophy, namely humanist logic or dialectic. I shall

[1] Anthony Grafton: *Defenders of the Text. The Traditions of Scholarship in an Age of Science, 1450-1800,* Cambridge, Mass. 1991; Peter Mack: *Renaissance Argument. Valla and Agricola in the Traditions of Rhetoric and Dialectic,* Leiden: E. J. Brill 1993; Peter Mack: Humanist Rhetoric and Dialectic, in: *The Cambridge Companion to Renaissance Humanism,* ed. Jill Kraye, Cambridge: Cambridge University Press 1996, pp. 82-99; Ann Moss: *Printed Commonplace-Books and the Structuring of Renaissance Thought,* Oxford: Clarendon Press 1996; James Hankins: *Plato in the Italian Renaissance,* 2 vols., Leiden: E. J. Brill 1990.

plead the case of the Beast which for too long has been considered an unfortunate, indeed abusive, degeneration of scholastic logic. The thesis I am seeking to defend is that in tirelessly reclaiming the union between philosophy and eloquence, humanism made a considerable, coherent and worthwhile effort to transform traditional logic so that it could account for the functioning of the human mind such as this is manifested in the best texts both ancient and modern. This manifestation of the human mind in the text was the fundamental hypothesis of the whole humanist enterprise. Thus humanist logic presupposed the creation of means of methodical reading of ancient texts as a necessary preparation for the writing of new texts. Logic became essentially a discipline anchored in the text, drawing its *raison d'être* from the analysis and production of discourse.

A *bête noire* will also be found in this essay, namely the idea of *rhétorique restreinte.* This was suggested over twenty years ago by Gérard Genette and is perhaps an idea which has been upheld with a tenacity more worthy of a better cause.[2]

It will be remembered that Genette associated the name of Ramus (or Pierre de La Ramée) with what he portrayed as a dark moment in the history of Western education. It was with the publication in Paris of Ramus' 1555 *Dialectique* that the divorce of reason and discourse, *ratio* and *oratio* occurred. From that dark moment on rhetoric would be stripped of three of

2 Gérard Genette: La rhétorique restreinte, in: *Figures* 3, Paris: Seuil 1972, pp. 21-39; François Rigolot: Prolégomènes à une étude du statut de l'appareil liminaire des textes littéraires, in: *L'Esprit Créateur* 27 (1987) pp. 7-18 (p. 15, n. 8); Francis Goyet, ed.: *Traités de poétique et de rhétorique de la Renaissance,* Paris: Librairie Générale Française 1990, p. 454; Gilles Declercq: *L'art d'argumenter. Structures rhétoriques et littéraires,* Paris: Editions Universitaires 1993, pp. 7. 163ff. (annexe II: "Rhétorique et argumentation: histoire d'une atrophie progressive"); Georges Forestier: *Introduction à l'analyse des textes classiques. Eléments de rhétorique et de poétique du XVIIe siècle,* Paris: Nathan 1993, pp. 30. 91-62; Luis Merino Jerez: Aproximación al *De auctoribus interpretandis* y las *In artem poeticam Horatii annotationes* del Brocense, in: *Humanismo y pervivencia del mundo clásico,* ed. J. M. Maestre, J. P. Barea, I 2, Cadiz: P. U. de Cadiz 1994 [Actas del I Simposio sobre humanismo y pervivencia del mundo clásico, Alcatriz, 1990] pp. 621-631 (p. 630); Alfonso Martín Jiménez: La literatura en los tratados españoles de retórica del siglo XVI, in: *Rhetorica* 15 (1997) pp. 1-39 (pp. 8. 34: "las corrientes reduccionistas de la retórica") *et passim.*

the five tasks which it had traditionally encompassed. Dialectic would seize invention, disposition and memory, leaving its poor sister with only elocution and delivery. Henceforth rhetoric was reduced to stylistics, to the tropes and figures, opening the door to a narrow form of *belles-lettrism* and to degenerations of the same type.

While it is undeniable that Ramus very conscientiously separated logic, rhetoric and grammar in the textbooks containing the theory of each of these disciplines of the *trivium,* nonetheless the same man was nominated as professor of eloquence and philosophy at the *Collège des Lecteurs Royaux,* the future *Collège de France.* In other words the man who for reasons of methodological purity separated the arts concerning *logos,* never ceased throughout his life to underline the necessity of their conjunction in the practical usage of discourse, both in analysis and in the production of texts. Therefore it is in his *Dialectique* that one can best observe not the divorce between reason and discourse but, on the contrary, the enormous effort to construct a didactic theory which would most faithfully codify the discursive procedures utilised in the great "literary" texts of Antiquity, which provided essential models for any future orator, poet or historian. In short, if the historical validity of the concept of *rhétorique restreinte* has already been contested in studying the pedagogical practices of the Enlightenment, it is all the more necessary to set aside the concept in the period under examination here, a period marked more than any other by *the pursuit of eloquence,* to cite a celebrated essay by Hanna H. Gray.[3]

This, then, is the *exordium* with its argument *a minori;* it contains, as is appropriate, the seeds of my contribution. Indeed, my proposition is the following: far from being the start of an unfortunate evolution, the work of Ramus is, rather, the conclusion of a long humanist tradition during the course of which the reading of texts (a reading integrated more and more closely with a logic in its turn solidly attached to rhetoric) constituted the foundation both of creative invention and of the edifice of the human sciences.

3 Françoise Douay-Soublin: Non, la rhétorique française, au XVIIIe siècle, n'est pas 'restreinte' aux tropes, in: *Histoire Epistémologie Langage* 12 (1990) pp. 123-132; Hanna H. Gray: Renaissance Humanism: the Pursuit of Eloquence, in: *Renaissance Essays from the Journal of the History of Ideas,* ed. P. O. Kristeller, P. P. Wiener, New York: Harper and Row 1968.

Therefore I subscribe entirely to my colleague Ann Moss' recent conclusion that

> [Peter Ramus] was committed to the humanists' conjunction of analytical reading and production, and he believed that the whole universe of knowledge was encapsulated within the humanists' canon of texts.[4]

However, what for so long impeded the vision of this obvious fact was on the one hand the methodological orthodoxy of certain leaders of later humanism (including Ramus) and on the other the monodisciplinary education of the majority of traditional historians. The history of logic has been written by specialists of the history of philosophy such as C. Prantl, W. Risse and G. Nuchelmans who have condemned the dialectic of Melanchthon and Ramus without knowing the whole of their works, and without worrying about the connection between the diverse aspects of these works, without even drawing out the didactic and philosophical concepts which give sense to the elaboration of both logic and grammar. The same remarks go for the theologians who have studied the first Protestant dogmatic theology drawn up by Melanchthon without knowing the logical and rhetorical foundation of the *Loci Communes* (1521), and for those historians of literature who have distorted the perspective of rhetoric in history. It is, however, right to add that such optical illusions have been, if not created then at least aided by the imperatives of a methodology which, in the history of Western education has marked the passage *from humanism to the humanities,* to cite another well known title. It is certainly striking to see that the great leaders of pragmatic humanism displayed a didactic rigour which their less celebrated followers never observed. The leaders have separately published on the one hand textbooks of grammar, rhetoric, logic and on the other hand analyses of classical texts, both sacred and pagan. They added to these letters, prefaces and militant orations, which preach the integration of theory and text but which have hardly ever been read. Thanks to the work of Genette and Rigolot we are today much more attentive to the preliminary apparatus of texts and to the general manner in which rhetorical pragmatic – production – is intended to valorise the ideological positions of the texts which it accompanies.[5]

4 Moss: *Commonplace-Books* (as note 1) pp. 155-156.

5 Gérard Genette: *Palimpsestes. La littérature au second degré,* Paris: Seuil 1982; Gérard Genette: *Seuils,* Paris: Seuil 1987; Rigolot: Prolégomènes (as note 2).

The pupils, for their part, less strict yet still perfectly faithful to the humanist spirit as well as to the practice of their heroes, published textbooks in which logic, rhetoric and textual analysis are completely integrated. In this they constituted a unique propaedeutic for future inventive and critical work during the religious, political and moral debates in real society.[6] Among these more obscure pedagogues let us cite Barthélemy Latomus, Johannes Sturm and Francisco Sánchez or Sanctius, the Ramist from Salamanca. The diligent study of their formerly neglected works allows us to establish very precisely the existence of a tradition, a golden chain of humanism. Thus the great Northern humanist Rudolph Agricola studied in Italy where the methodical analysis of classical texts started to grow, probably under the influence of scholars coming from the Byzantine world. Nourished by the Italian source, *De inventione dialectica* is a work of logic in which rhetoric and textual analysis find themselves incorporated. From North to South we return to the North with Philipp Melanchthon, a pupil of Reuchlin, Agricola and Erasmus, and a friend of Martin Luther in Wittenberg. Agricola, Erasmus and Melanchthon were the intellectual masters of Latomus and Sturm who from Germany would go to teach in Paris, there to be listened to avidly not only by Johannes Calvin and Ignatius de Loyola, but also by Ramus. Ramism was particularly well rooted in the Northern countries such as Germany, Switzerland and the Low Countries, and it would also gain a hold in England, Scotland and Ireland before reaching the early American colleges.

From South to North and vice versa the major landmarks were Agricola and Melanchthon. On the map of Europe the centres of humanism were Wittenberg – where Hamlet went to study – Louvain in the Low Countries, Strasbourg, an Imperial Free-City and certainly Paris which was a veritable "powerhouse" of methodical humanism to use the apt formulation of Ann Moss.[7] The movement traversed religious boundaries; with his work spread everywhere, the Lutheran Melanchthon was well received in Paris, capital of the eldest daughter of the Church of Rome. Barthélemy Latomus, a Melan-

6 This attempt at reorientating rhetoric in the Renaissance has not always been appreciated and accorded its true significance; here I depart – once is not for ever – from the observations made by Marc Fumaroli in his essay "A l'origine d'un art français: la correspondance familière", inserted in: *La diplomatie de l'esprit. De Montaigne à La Fontaine,* Paris: Hermann 1994, p. 164, n. 3.

7 Moss: *Commonplace-Books* (as note 1) p. 166.

chthonian in Paris, friend of Johannes Sturm and Calvin's host in Strasbourg ended up combatting the Protestants as an important Catholic controversialist. And the debt of the Counter Reformation to the pedagogical methods refined by the Reformers as these were adopted by the Jesuits, is well known ...

What, then, were the dynamic conceptual forces which controlled this immense network, as complex as it was contradictory? How did humanism proceed to transform reading into a source of invention, and what were the hidden links between literature and logic?

If it is true that Ramus' work constitutes the conclusion of a long-maturing tradition, then it is legitimate to call it as a witness with retroactive value. So, I invite the reader to take one step forward and two steps backwards, while I shall limit myself to the Ramist phenomenon whose impact is particularly well known. There is insufficient space here to detail what I am asserting; the proof and the analyses which one is entitled to can be found in a collection which has recently been published in Quebec and is called *Autour de Ramus*.[8]

What is striking first of all when studying Ramus is that he never ceased to organise his own publications and those of the group of which he was the uncontested leader. Thus, the first version of his elementary logic, the *Dialecticae institutiones* presents exactly the same framework as that in the first version of the rhetoric, that is to say the *Institutiones oratoriae* published a little later by his friend Omer Talon in Paris by the same publisher.[9] Indeed the two works present an exposition structured according to the classical triad of *natura-ars-exercitatio*, nature-theory-practice, with explicit reference to Plato's *Phaedrus* (269 C-D).[10] In its turn the third element, exercise or practice, is subdivided according to another triad, namely *interpretatio*, *scriptio* and *dictio,* that is textual analysis, composition and declamation. In

8 Kees Meerhoff, Jean-Claude Moisan, eds.: *Autour de Ramus. Texte, théorie, commentaire,* Quebec: Nuit Blanche 1997.

9 Kees Meerhoff: *Rhétorique et poétique au XVIe siècle en France. Du Bellay, Ramus et les autres,* Leiden: E. J. Brill 1986, pp. 191ff.

10 See in addition to Meerhoff, Moisan: *Autour de Ramus* (as note 8) the interesting work by Luis Merino Jerez: *La pedagogía en la retórica del Brocense. Los principios pedagógicos del Humanismo renascentista* (natura, ars *y* exercitatio) *en la Retórica del Brocense* (memoria, methodus *y* analysis), Cáceres: Universidad de Extremadura 1992.

dialectic, as in rhetoric, the preponderant interest in the practical in its different forms is underlined. In the first textbook, Ramus claims that the great power of syllogistic theory, *vis praecipua,* resides in the analysis of texts. Under the rubric "interpretation" he presented an analysis of the logical framework of the first *Heriod* of Ovid (the letter of Penelope to her spouse), a model to be used in the other rubrics, composition and declamation.

This analytical technique of reducing a text to an underlying syllogism, a technique used before by Agricola and particularly by Melanchthon, remains a constant through Ramus' work, as demonstrated by the entire evolution of Ramist doctrine.

At first sight this would appear to contradict the thesis of the permanence of their central concepts, concepts fundamentally dependent on the tradition established notably in Germany. Indeed the cardinal moment in this evolution, or the most celebrated at least, is marked by the publication in Paris by the same house of the French edition of Ramus' *Dialectique,* prepared by himself and of the *Rhetorique* of his "brother" Omer Talon, prepared by their common student Antoine Fouquelin. For, if it is true that one can see in the simultaneous launch, as in the structure of the two works, the same concern for "orchestration" as before, then the *pièce de résistance* of my demonstration is deficient. The triad nature-theory-practice has disappeared; the whole of the third part concerning exercise has disappeared and henceforth the famous Ramist dichotomy reigns: "Dialectic has two parts: invention and judgment", and, "Rhetoric has two parts: elocution and pronunciation". The birth of a *rhétorique restreinte*? We should avoid being duped by this optical illusion.

One of the most interesting aspects of Ramism is something which has become banal in our own day when everyone composes texts on a computer and where there is a habit of shifting, suppressing or conserving the passages of text being produced. Some reuse paragraphs in a new text or insert passages which they have suppressed in an earlier text.

The same methods of production can be seen in the Ramists. In this sense the thesis, or rather the charge of Walter Ong linking Ramism with the invention and development of printing retains its relevance. Ramus and his group rarely threw texts away; these would reappear elsewhere or even resurfaced after years. This is the case of the missing pages concerning exercise which are found again in the explanatory commentaries added before long to the manuals by the group, finally to be reintegrated in the posthumous French edition of the *Dialectic*. This was published four years after the murder of the author during the Saint Bartholomew's Day Massacre, an event which was still to furnish Voltaire's nightmares in the future.

Furthermore, as I have already alluded to, a whole series of parallel publications does not allow us to conclude too hastily that this fundamental part of the doctrine disappears. The original triad, composed of the three complementary notions of nature, theory and practice never ceased, in reality, to inform on a deeper level Ramist method from beginning to end.[11]

For Ramus, all of the arts are first of all codifications of a competence. They have their common source in the human being, endowed with a natural aptitude to speak and reason well. This aptitude must be directed by theories faithful to the "nature" of the human spirit and formulated in such a way that they ceaselessly urge a return to this inexhaustible source. Contrary to what one has been able to affirm, theory is not the dominant element in the triad. Created by fallible beings, it must constantly be adjusted so that it draws as closely as possible to nature, to which it is, by definition, subject. The best way of drawing close to the source is to examine the work of superior beings, themselves never totally perfect. The assiduous reading of the great classical texts therefore presides simultaneously over both the refinement of the theory and the perfecting of a real competence in the practice of the art one is seeking to master. The innate gifts (*dotes ingenitae*) present in everyone from birth must be developed by frequent contact with the authors who knew how to exploit them with an uncommon mastery. For example, theoretically, every human being has at his disposal the resources of the syllogism, and makes use of them correctly everytime he produces a coherent discourse. Nevertheless the proper usage of the syllogism in concrete social reality remains a delicate matter. The only means of acquiring this competence is by observing the usages made by the great exemplars and to imitate these with constant individual practice.

11 See also Nelly Bruyère: *Méthode et dialectique dans l'œuvre de La Ramée. Renaissance et âge classique,* Paris: J. Vrin 1984, p. 314, confirmed by André Robinet: *Aux sources de l'esprit cartésien. L'axe La Ramée-Descartes. De la Dialectique de 1555 aux Regulae,* Paris: J. Vrin 1996, pp. 59-60.

Indeed God himself, in His goodness and mercy grants us that natural light, the *lumen naturale.* He has bequeathed to each the gift of the syllogism while leaving it to the diligence of each individual to discover its riches. The theory of logic must take account of the modes of the syllogism. That is easy, since the true Aristotle, when stripped of the false commentaries and medieval glosses, had already done a great deal. Besides, theory, if it wishes to conform with nature must necessarily reflect the innate capacity to reason properly and assume as a result the most simple and powerful form of the syllogism, that which passes directly from the general to the particular. In so far as the *lumen naturale* is called upon to clarify all of the disciplines and to inform the aptitudes in all of the areas within the human scope, these will take the same form of the regular syllogism. In that way Ramist grammar, rhetoric and dialectic were cast in the same "natural mould".

Why are these manuals always thin and elementary? Because the theory is simple and relatively easy. By definition the true interest of a theory is found outside of itself, that is in its practice and use. True life is elsewhere: such could be the formula, rather paradoxically, which characterises Ramist theory. Paradoxical because at the limit it has hardly anything to say on the essential: it can only designate it and ceaselessly return to it. Practice includes and surpasses theory. It is exactly this which explains the removal of the notion of exercise from the later versions of Ramist theory. Practice, the source and aim of theory, is found outside the field – in the manual it is located at best at the edges, in the margins, in the commentaries added.

This essential matter, irreducible to theory, is found in the Ramist *Dialectique* at the end, in the pages devoted to the method which is unique yet also dual: "of nature" and "of prudence". By all evidence the exposition of the method "of nature" or "of doctrine" maintains a mirroring connection with the manual, in which the necessity of holding to the structuring of a regular syllogism in the ordered exposition of a subject or discipline is insisted upon. There are other aspects which I will not touch upon: the "unique method" is the best known element of Ramism. It is important as a vehicle for the well-ordered communication between men who have received the gift of reason. Obviously it must be prevalent in didactic transmission.

Nevertheless, anchored as it is in "nature" and in "practice", method cannot but be confronted with the text. Ramus detailed in the same chapter how many literary texts (in the widest sense), both prose and poetry, conform to this model of simple exposition. But it is also this same confrontation with literature which brings him to admit that in the reality of discourse the regularity of normal logical development is frequently disturbed.

It is from these singular artifices that the method of prudence arose. Of course, in this area it is impossible to devise rules: it is the inventive freedom of the creator, of the orator, which it is necessary to designate in the manual. This creative freedom, it must be said, exists and is essential; it is essential to show its functioning and force elsewhere, in the course of logical analysis which escapes regular didactic exposition.[12]

With the frequent citation of literary specimens as examples of dialectic proceedings, the admitting of a method of prudence next to the method of doctrine constitutes the most visible mark of a properly humanist engagement in the Ramist *Dialectique.* I recall in passing that as examples the Ramist group made an appeal to the young poets of the *Pléiade,* translated from Latin in the *Dialectique* and borrowed directly from recently published texts in the *Rhetorique Francoyse* of Fouquelin. These examples, like the concept of nature which underlies the two manuals and which, as we shall soon see, would determine the structure of the curriculum, reveals the "trans-linguistic" character of the enterprise which curiously is not incompatible with nationalist motivations. The "systems" remain identical in Latin and in French; it is, nevertheless urgent, and historically correct to translate them into the national idiom.

The observation of practice, such as this is undertaken in the pedagogical "workshops" of textual analysis reveals, as we have seen, certain discursive strategies which at first sight seem to compromise the principle of the unique method. It notably reveals some *manœuvres* intended to manipulate the passions of the unlearned audience and some oratorical tricks which translate into a disposition which departs from the norm for reasons of

12 Note the explicit and illuminating distinction made by the Spanish Ramist Fadrique Furió Ceriol in his *Institutionum rhetoricarum libri tres,* Louvain 1554, pp. 106-107: "Rhetoricae dispositionis genera duo sunt, unum artis, alterum prudentiae, quae sic a me distincta sunt, non quod utraque non innitantur prudentia – est enim utrumque genus a prudentia profectum – sed quod unum artis praeceptionibus traditur, alterum nullis praeceptis edoceri potest sed sola iudicii acrimonia prudentiaque ponderatur." Cited by Merino Jerez: *La pedagogía* (as note 10) p. 123.

"prudence". It is obvious that Ramus is thinking, for example, of the tricks of Socrates with his interlocutors as conveyed in Plato's dialogues, but also it stands to reason that he is thinking of Virgil's poems and of Cicero's orations. All these texts demonstrate that in practice deviation is more the rule than the exception. Is this the end of the principle of the unique method? Not at all for anyone who has grasped the force of the Ramist system.

Logical analysis, attached above all to the argumentative disposition, frequently states that the regularity of habitual logical development is disturbed. The propositions which form the basic outline, the discursive syntax of the studied oration, is not found in the prescribed order; the conclusion precedes the major premise, the minor precedes the major etc. Now this statement is only the first step of analysis. The second step will be the reduction of the creative deviation, which is there for reasons of oratorical prudence, to the logical norm. This is called *resoluere* or *retexere* in Latin and *resoudre en* [*son*] *rang artificiel* in the French *Dialectique.* In other words the deviation, far from ruining the principle of the unique method, finally, on the contrary, confirms it. In the last analysis no poet or orator has the power to destroy the immutable order of logic, and no one can, nor wants to, evade the empire of the syllogism which commands the human mind: Ramist logic is indeed a *logica naturalis.* However, the confrontation with the oratorical or poetic text has necessitated the creation of a transformational model of analysis which, rather like innateness and the notion of competence, *virtus disserendi,* provokes obvious associations with Chomsky's grammar. Logical analysis undertaken in accordance with Ramist procedures implies the distinction of three discursive layers. "Beneath" the often capricious textual surface, adorned with all the charms of ornate discourse, it is necessary to acknowledge the existence of two other layers each found at a different level. At the deepest level a regularly arranged logical structure, a chain of argument reducible to one sole syllogism; at the intermediate level, between the surface and the deeper logical structure, is a layer which frequently, but not necessarily, presents an arrangement deviating in its disposition of arguments.

In concrete terms the Ramist expert penetrates the verbal "skin" in order to discover a sometimes regular, often subverted arrangement of the propositions which constitute the basic outline of argument of the page being studied, or of the whole text. Having stated the simultaneous presence of regular elements and of elements which show in particular the "prudence" of the orator in the care which he takes to hide his game by an irregular arrangement – in technical terms *crypsis* (κρύψις) – he proceeds bravely to the reduction of this to a regular order which alone commands on a deeper level the discursive sequence and which as a result is the *only* order susceptible of being codified in a manual of logic. And so the truth of the principle of the unique method is established. What is more it is confirmed by each subsequent analysis. In short, God has created with a marvellous economy of means. The elegance of Ramist logic proves this in a repetitive manner: all the riches of the literary text, all of the devices of oratorical invention celebrate the miracle of human creation – before being absorbed by the analytical machine, compared by a mischievous Anthony Grafton to a *sausage grinder.*[13]

Three texts in particular marked the nomination of Ramus as Regius Professor of eloquence and philosophy: two orations, including the inaugural lecture and an analysis of Cicero destined to prove, in the face of a reticent, even hostile world, the benefits of Ramist pedagogy. In the thunderous preface to the analysis of the Ciceronian oration *Pro C. Rabirio* Ramus presents his detractors as people who "would rather see winged whales flying in the sky than see logic in oratorical texts".[14] At the end of his inaugural lecture he had announced this same analysis in the following words:

> We have chosen an oration of Cicero of reduced proportions but remarkable in its use of rhetoric and dialectic in which we are going to savor the combined fruit of both disciplines.[15]

13 Anthony Grafton: Teacher, Text and Pupil in the Renaissance Classroom: A Case Study from a Parisian College, in: *History of Universities* 1 (1981) pp. 37-70 (p. 45); cf. Anthony Grafton, Lisa Jardine: *From Humanism to the Humanities. Education and the Liberal Arts in Fifteenth- and Sixteenth-Century Europe,* London: Duckworth 1986, p. 178. See also Peter Mack: Ramus Reading: The Commentaries on Cicero's *Consular orations* and Virgil's *Eclogues* and *Georgics,* in: *Journal of the Warburg and Courtauld Institutes* 61 (1998) pp. 111-141.

14 This 1551 preface is reproduced in Meerhoff, Moisan: *Autour de Ramus* (as note 8) pp. 351-352.

15 Ramus: *Oratio initio suae professionis habita, anno 1551*, towards the end: "[...] Orationem Ciceronis brevem, sed Rhetorica et Dialectica praestantem selegimus, in qua [primo] coniunctos utriusque artis fructus degustemus".

Previously he had already referred to the affair, famous in the history of French literature thanks to Rabelais and Du Bellay, which had formed the object of the other oration. These three texts, which Ramus took care to show the interconnections between, are evidence of a massive publicity offensive in which Ramus appears at one and the same time as an orator borne up by his own formula for success, as the theorist who demonstrates the efficacy of his method in practice and as the master of a new pragmatic school ready to reveal the secret of its success to the less fortunate. His teaching method would make their lives as easy as they would be pleasant; after a few years they would be capable, as he was, of "uniting eloquence to philosophy".

This union remains the final goal of Ramist pedagogy. We have been able to appreciate its conceptual rigor. To end, I should like to outline the practical development. Ramus presented the panorama in his vehement oration which preceded his assuming the chair at the *Collège des Lecteurs Royaux.* It is entitled *Pro philosophica Parisiensis Academiae disciplina oratio*. It is 1550, five years before the launch of his French versions of the *Dialectique* and of the *Rhetorique.* Du Bellay has recently published the manifesto of the *Pléiade*, *La Deffence et Illustration de la langue francoyse* (1549).

How does the curriculum appear to those young men entrusted to the Ramist group? Evidently the use of student's time is, from the outset, subject to the imperatives of the new didactic method which accorded the minimum of time to theoretical instruction and the maximum amount to practical exercises.

The first three years are devoted to the acquisition of Latin and Greek, with the accent on Latin which was the official language of Church and State. In the morning one hour of instruction is followed by five hours of memorisation exercises, of pronunciation and of spoken interaction. The same for the afternoon. Ramus sought to form practitioners (*artium exercitatores*) rather than narrow scholars (*artium doctores*).[16] Rhetorical competence is acquired during the fourth year, following a model identical to

[16] Ramus: *Pro philosophica Parisiensis Academiae disciplina oratio,* dated Paris, in the *Collège de Presles,* 17 February 1550 before Easter (1551 n.s.). Reproduced in *Scholae in liberales artes,* Basle 1569, repr., with an introduction by W. J. Ong, Hildesheim, New York: Olms 1970, cols. 997-1062, passage cited: cols. 1009-1010.

that used for the acquisition of language: one hour of instruction with five hours of practice. The *trivium* is crowned in the fifth year with the study of dialectic when anew the exercises of analysis, composition and declamation are by far more important than the lecture course.

Two things merit attention here. First of all, the application of the concept of nature to the didactic model. Why devote three years to a language, and only one to rhetoric and dialectic? Because the grammar of the ancient languages no longer has a natural base, for the simple reason that these are no longer anybody's mother tongue. Competence in logic and eloquence, on the other hand, are anchored in nature, which reduces considerably the effort demanded of pupils and consequently the place accorded to dialectic and rhetoric. One can see clearly here the transfer of a philosophical, even ideological, concept into the practice of education.

The second point of interest concerns the problematical relationship between the disciplines of the *trivium,* that famous but often badly interpreted "conjunction". Where does each end and the next begin? The separation is total at the level of theoretical instruction. By virtue of the methodological principles (the three laws) which he relates back to Aristotle, Ramus designated for each discipline its own exclusive and autonomous domain: not a word of grammar in the manual of rhetoric. Thus the fourth year is exclusively devoted to the art of oratory through instruction in the rules of rhetoric (drawn from the *Rhetorica* of Omer Talon) and especially through their practical use (in analytical exercise, composition and declamation). Nevertheless the didactic process is cumulative: in the fourth year the students have not lost their grammatical competence, but they add to it a rhetorical competence which allows them to speak and to write not only with accuracy but also with elegance. In other words the conjunction is made exclusively at the level of practical use, in the *usus* which is the prelude to the full use of all skills in social life. Thus it is in these "workshops" that, while combining the skills already acquired with skills yet to be perfected, the student moves little by little towards the intended goal, that of the union of eloquence with philosophy. For this is the final objective of those infinite exercises carried out at every stage of the Ramist curriculum.[17]

17 Ibid. cols. 1016. 1020. Cf. Meerhoff, Moisan: *Autour de Ramus* (as note 8) pp. 320-325.

Ramus, as it is well known, innovated in both the teaching of mathematics and of the natural sciences. But it is time to draw to a conclusion. The reader will be convinced by now of the absolutely central place which the analysis and production of texts takes in the conceptual and pedagogical system of Ramus. The reader will be sensitive to the ability of Ramus to transform abstract notions such as the triad nature-theory-practice into operational terms within a cleverly graduated curriculum, the keystone of which is logic, as this is made manifest through the text. Analysis shows, indeed, that the great authors are distinguished by the accuracy and elegance of their language, of course, but in the first instance by their subtle use of the resources of logic. It is in their capacity of superior users of that natural light instilled in everyone that they are presented as models to be imitated and, if possible, to be surpassed by an ever more masterful practice of the spoken and written word in a modern communication society.

I think equally that I have dissipated once and for all the tenacious myth of the *rhétorique restreinte* which is said to have found its origin in the pedagogical work of Ramus, with his separation of the disciplines of reason and discourse, *ratio* and *oratio.* No, Ramus is rather the man of the *coniunctio,* of the integration of discursive competences. As such he is the incontestable heir of the great humanist tradition which, proposing the methodical reading of texts as a prerequisite to creative invention, never ceased to proclaim the marriage of logic with literature, of Beauty with the Beast. It is appropriate to end on this clever *chiasmus,* a figure highly valued by Renaissance humanists.

The Influence of Ramus' Method on Historiography and Jurisprudence

GUIDO OLDRINI

The subject of this paper has been generally neglected in professional literature. Ramism has not been studied widely, except among a small group of specialized historians of ideas, with limited interest in philosophy, and who failed to transcend the boundaries of their field of study. For most experts, Ramism appears to consist of logical and rhetorical techniques used in the declining phases of the Renaissance, while Ramus' broader influence has not been recognized. Johann Jakob Brucker was the only major historian of philosophy that has given Ramus a position of importance. Ramus has been principally considered in a negative way, as a controversial figure (like Nizolius) who broke away from Aristotelian ways of thinking. Attention must to be given to the positive aspects of Ramus' work, and in particular to his method. This method was further developed by his disciples after his death and was applied to a wide range of disciplines, including historiography, jurisprudence, politics, poetics, mathematics, and natural science.

In this paper I will focus on the disciplines of historiography and jurisprudence. Both became increasingly specialized in the course of the sixteenth century and, as a result, both disciplines were in need of a new methodology. An extensive discussion of this topic cannot be attempted here.[1] Instead, in the six sections of this article I shall limit myself to discussing the current state of our knowledge of this topic as well as what is still to be done.

1 For a more extensive discussion see Guido Oldrini: *La disputa del metodo del Rinascimento. Indagini su Ramo e sul ramismo,* Florence: Le Lettere 1997.

1.

I wish first to draw attention to the problem of method. The central role of method for Ramus and his followers is well known. A distinctive trait of Ramism – as opposed to the Aristotelian, Melanchthonian, and humanist philosophical schools – is the fact that method applies to, and is required for, all disciplines without exception. This method not only serves as the general logical criterion for research in individual fields of knowledge, but it also functions as the instrument for measuring and checking their respective scientific consistency. Ramus distinguishes between "formal" and "material" laws proper to logic. The former is his "unique method" (*methodus unica*), i.e., the methodological principle according to which the arrangement of precepts (*dispositio*) must perforce follow a descending order of progression, from the general to the particular (*De forma lex unica est, ut absolute notius et clarius antecedat*).[2] There are three material laws of logic. These laws require – as a necessary condition for the validity of any assertion having a scientific nature – that logic must encompass the three qualities of universality, homogeneity and reciprocity; these three material laws (*praeceptorum singulorum leges*) are referred to as the "law of truth" (*lex veritatis* or *de omni*), the "law of justice" (*lex justitiae* or *per se*), and the "law of wisdom" (*lex sapientiae* or *universale primum*), respectively. The originality of Ramism – as opposed to Renaissance Aristotelianism – lies in the assigning of new meaning and value to these three material laws. Ramus extrapolates these three laws from the context which Aristotle gave them and extends them – in spite of unanimous opposition from orthodox Aristotelians such as Schegk and Cornelius Martini – beyond the realm of logic, thus rendering them indispensable methodological principles for any art or discipline.

The second of these laws – i.e., the "law of justice" – is particularly important for Ramism. Ramus believed that it is particularly important for the discipline of mathematics – a discipline which Aristotle did not discuss thoroughly or well. This law pertains not only to universality, but also to homogeneous statements. Within the context of any given discipline, such homogeneous statements are structured in the form of a whole that reverts back, is contained,

2 This is the formula that Ramus uses in the *Praefatio* to his *Scholarum physicarum libri octo,* in: *Rami-Talaei Collectaneae praefationes, epistolae, orationes,* Marpurgi 1599, ed. W. J. Ong, Hildesheim 1969, p. 50.

and is complete within itself, having no connection to the axioms or propositions of other disciplines; this is because each discipline consists of parts that are "essential (only) to one another" (*essentiales inter se*).[3]

One can draw two conclusions on the basis of this emphasis by the entire Ramist school on the "law of justice." First, each discipline can and must focus on subject-matter which is essential to it.[4] Second, both in theory and in teaching, the separation of individual disciplines – the domain of each discipline circumscribed in accordance with its place in the hierarchy of knowledge – must be enforced and respected; not even the *studia humanitatis* nor higher teaching in philosophy and theology can be exceptions to this. As Ramus stated in his famous 1562 speech advocating the reform of the University of Paris,

> Recte namque philosophi praecipiunt, si quid arithmeticum in Geometria, vel geometricum in Arithmetica tractetur, sophisticum esse [...]. Ideoque philosophia philosophice a Philosophis in Philosophica schola, Theologia theologice a Theologis in Theologica schola docenda est. Id verum, iustum, sanctum, catholicum esto.[5]

It is here that one discovers the true novelty of Ramism. In detaching the "three laws" from their Aristotelian context and applying them to all types of judgments, Ramist methodology progressed beyond the methodology of Renaissance Aristotelianism. Only by creating homogeneous statements is it possible to use Ramus' "unique method" to confirm the autonomy of each discipline as well as to recognize every field of research within its own appropriate context.

2.

The principal reason for the contemporary influence of Ramism lies in the realm of methodology. Ramus' methodological innovations not only pertained to the *artes sermocinales* (and to dialectics in particular) as well as to *artes reales* (especially mathematics), but also to the entire university curriculum. And these changes pertaining to methodology within the arts

3 Cf. *P. Rami Dialecticae libri duo, Audomari Talaei praelectionibus illustrati,* Parisiis 1566, p. 224; *Professio regia, hoc est, Septem artes liberales, in regia cathedra, per ipsum Parisiis apodictico docendi genere propositae,* ed. Joh. Th. Freigius, Basileae 1576, p. 105.

4 Aristotle: *Anal. post.* I 4, 73a34-35. 73b18-19. Cf. *P. Rami Scholae in liberales artes,* Basileae 1569, ed. W. J. Ong, Hildesheim, New York 1970, part 2, col. 916.

5 *Rami-Talaei Collectaneae praefationes* (as note 2) p. 385.

faculty curricula were not without ramifications for the disciplines of historiography and jurisprudence.

The Renaissance was a period of profound renewal for these disciplines. The humanist impulse contributed to this renewal, though much continuity with medieval tradition still remained, notably through the continued production of historical annals by chroniclers, and of glosses and commentaries by jurists. A consequence of humanism in the domain of historiography was the fact that ancient historical texts – most importantly those by Livy, but also those by Thucydides, Polybius, Caesar, Tacitus, and other authors – were now being read again. In the realm of jurisprudence, Alciato, Cujas, and other faculty members at the University of Bourges contributed to the development of philology. New logical and rhetorical categories (e.g., kinds of topics, the employment of *loci,* increased awareness of problems pertaining to correct ordering of words) were introduced and began to be utilized in both disciplines. These developments made possible the construction of a new theoretical foundation for law and history which would conform to the methodological requirement that every discipline adhere to precise rules.

It was during the 1560s that methodology introduced major changes to the domains of historiography and jurisprudence. Up until then, humanist historiography used the traditional, classical model of history, which had its basis in rhetoric (*opus maxime oratorium*); Pontano, Vives, Sturm, were among those authors that strictly adhered to it. But post-1560 historiography – usually referred to as "Polybian" – clearly departed from the older model; the principal focus of historical inquiry changed. The new historiography emphasized analysis of the *causes* of historical events, as well as the reciprocity of theory and practice in the writing of history. It was now argued that historians should not only be men of letters, but should have considerable experience in the management of public affairs.

Representatives of this new historiography – including Patrizi, Baudouin, Bodin, Loys Le Roy, and Montaigne – followed the ancients, and in particular Aristotle, Polybius, and Tacitus. In Aristotle's ethics, experience is considered an indispensable quality of φρόνησις, the "prudence" which Lipsius considered so important within Tacitus. But these post-1560 authors also followed the *usuaria* imperative – emphasizing the importance of utility, which became by that time an integral part of Renaissance culture. Ramus' thought can rightly be considered to be the clearest philosophical expression of that belief in the importance of utility.

The discipline of jurisprudence underwent analogous changes. Juridical humanism was criticized as too academic, as too absorbed in philological details, and as too removed from practice. Attention to philological orthodoxy was replaced by a juridical viewpoint most favorable to "national" problems – analogous to the "national" tendencies that manifested themselves in historiography. In the case of Bodin and Montaigne in France as well as Bacon and Ralegh in England, the principles of comparative studies and historiographical relativism opened up new avenues in both disciplines.

These were the consequences of the 1560s. The humanist advances, which continued to serve as a basis for scientific research, were supplemented by new forms of analysis and criticism. Historians and jurists began to pay attention to political and social (including "national") imperatives which were taking hold among the rising social classes. These scholars sometimes became involved in civil and religious controversies. In the realms of history and law it became necessary to participate actively in public affairs and to face reality directly. Henceforth, political realism replaced the rhetorical model in historiography and the philological model in law.

3.

While Ramism certainly influenced changes within the disciplines of historiography and jurisprudence, it is important to remain cautious when assessing this influence for the following two reasons. First, some of the formal attributes commonly associated with Ramism – for example, the extensive use of synoptic tables and dichotomies – were not exclusively Ramist, and scholars have often been rash in attributing their influence to Ramism. One must also look to the specific theories and the precise period involved in each individual instance in order to determine what role Ramism actually had. Second, the most noticeable contributions of Ramism to historiography and jurisprudence should not be credited to Ramus himself. He made some fragmentary references to the "order" (*distributio*) of history in his *Liber de moribus veterum Gallorum* (1559) and presented some brief comments pertaining to legal methodology in his commentaries on Cicero's *Somnium Scipionis* and *De legibus*. Yet if one excludes these and other fragmentary references found in Ramus' writings (e.g., in his *Prooemium mathematicum*), there is little evidence that he had any real interest in either historiography or jurisprudence.

It is therefore not without reason that scholars from the time of Charles Waddington[6] onwards have searched to account for the influence of Ramism, which was real yet indirect. History was not considered by Ramists to have any usefulness until dialectics provided it with topical and analytical arrangement.[7] The development of a connection between history and method was indirectly influenced by Ramism and the same indirect influence can be seen in the growth of "synthetic and systematic" tendencies in jurisprudence.

What can then be identified as the specific contributions of Ramism? As stated above, Ramism played a role in those innovations that took place in historiography and jurisprudence from 1560 onwards. Ramist requirements for establishing categories and precepts in accordance with the "three laws" and the "unique method" were quite influential, as was the Ramist view that knowledge cannot be gained in any discipline unless theory is coupled with practice. From 1560 onwards historians and jurists increasingly stressed the need for "experience," while jurists in particular emphasized the role of "custom" within law. And Ramist ideas concerning veracity and objectivity, which could be used to differentiate between truth in history on the one hand and fiction in literature on other,[8] were mirrored in the new historiography. Even though Ramism cannot be directly tied to these innovations, they are nonetheless indicative of a new cultural climate to which Ramism made an important contribution.

4.

In order to examine the methodological contribution of Ramism, one must look at the content of writings by individual authors. At the risk of oversimplification, those authors can be classified into the following three groups: 1) authors who were taught and/or influenced by Ramist logic and then moved on to the professions of historiography or jurisprudence while

6 Cf. Ch. Waddington: *Ramus. Sa vie, ses écrits et ses opinions,* Paris 1855, pp. 355-356.

7 Cf. W. Schmidt-Biggemann: *Topica universalis. Eine Modellgeschichte humanistischer und barocker Wissenschaft,* Hamburg 1983, pp. 59. 62.

8 Cf. P. Sharratt: *The Ideas of Pierre de La Ramée with Particular Reference to Poetic Theory in the Sixteenth Century in France,* Ph. D. thesis, University of Newcastle upon Tyne 1970, pp. 154-155.

retaining at least some of their penchant for Ramism; 2) authors who can be considered to have been militant Ramists and who placed Ramism at the service of historiography or jurisprudence; 3) authors who, while they may not have been disciples of Ramus – or even may have opposed Ramism on many points – nonetheless adopted aspects of Ramist methodology which they applied to historiography or jurisprudence.

Didacticism can be seen in many of these individuals, for the need to popularize Ramism was more important to them than was originality. Johannes Thomas Freigius is a good example of such a popularizer. He studied law at the University of Freiburg in the 1560s and was influenced by Zasius, Lagus, Oldendorp, Derrer, and Alciato. His *Methodica actionum iuris repetitio* (1569), his *Partitiones iuris utriusque* (1571), his *Neüwe practica iuris* (1574), his *Logica iureconsultorum* (1582), and his commentaries to the *Pandectes,* all clearly made Ramist logic an indispensable foundation for jurisprudence; his *Historiae synopsis* (1580), his *Paedagogus* (1582), and his speeches as rector of the *Gymnasium* in Altdorf all gave Ramist logic this same basic importance within the domain of historiography. Beginning in the 1570s, Freigius became recognized as a specialist in the art of creating outlines, summaries, and synoptic tables using Ramist methodology for didactic purposes. When doing so within the realms of jurisprudence and historiography he adhered to the Ramist view on logical order; he started with the general and then became increasingly specific. Freigius rarely mentioned Ramus' name and often mentioned Aristotle's name in its stead; yet there is no doubt that the form, the content, and the method inherent in his writings had their origin in Ramism. True, Ramus is hardly ever mentioned in these works, and when he is, his name is more often than not disguised behind Aristotle. Nevertheless, there is no mistaking the indebtedness of the ideas, structure and form of Freigius' manuals to Ramism.

While Freigius was an important German Ramist, he was by no means the only one. Ramism and those encyclopedic syntheses usually referred to as "Philippo-Ramist", found many adherents in Calvinist Germany. While professors of law in the University of Wittenberg such as Wesenbecius, Brederodius, and Forster tended to privilege the Aristotelian-humanist juridical tradition, others leaned toward Ramism. Included among them were Wilhelm Roding (Rodingus), the first German editor of and commentator on Ramus' *Dialectica,* and Caspar Pfaffrad. The latter, a Philippo-Ramist who was among those theologians who also discussed legal questions, was the author of *De studiis Rameis et optimae institutionis legibus commentatio* (1597).

Like Pfaffrad, Hermann Vultejus, a professor of jurisprudence at the University of Marburg, was also a Philippo-Ramist. Vultejus' views concerning jurisprudence echoed the views of Ramists; he opposed redundancy, excessive commentary, and organization which was not methodical. In his *Idea juris logica* (1586) methodological shortcomings of jurisprudence were criticized. Yet in some respects Vultejus was distant from Ramism. By strictly separating art from science as well as action from contemplation, and by rigidly confining law to the realm of simple "prudence" (*juris/prudentia*) he accepted a more traditional humanist approach taken by theorists such as Alciato, Budé, Connan, and Le Caron.

Ramism appears to have had a greater influence on jurisprudence in England than in Germany. Ramist ideas circulated widely at the Universities of Cambridge and Oxford during the Elizabethan period and also at the Inns of Court in London, and these ideas continued to circulate during the Stuart period. In reaction to the framework of English law during that period – rightly described as "a farrago of detailed instances which defied any scheme of arrangement, save perhaps the alphabetical"[9] – there is evidence of interest in methodology among English jurists from the 1570s onwards. This point has been made recently by William Sherman with respect to a late sixteenth-century synopsis of English law by John Dee.[10] In subsequent decades, William Fulbecke (who remained distant from Ramism) and John Doddridge (who alluded to Ramus' three laws while also drawing heavily on Aristotle) both nonetheless insisted on the importance of logic as well as method for law, without which "neither can the understanding be well taught, nor the memorie well directed", nor could one avoid "the great confusions, tedious and superfluous iterations, with which the Reports are infested".[11]

9 T. F. T. Plucknett: *Early English Legal Literature,* Cambridge 1958, p. 19.

10 W. H. Sherman: John Dee's *Brytannicae Reipublicae Synopsis:* A Reader's Guide to the Elizabethan Commonwealth, in: *The Journal of Medieval and Renaissance Studies* 20 (1990) pp. 304ff., reprinted in his book *John Dee: The Politics of Reading and Writing in the English Renaissance,* Amherst 1995, pp. 136ff.

11 I have just cited, respectively, W. Fulbecke: *A Direction or Preparative to the Study of the Lawe,* London 1600, fol. 83v; J. Doddridge: *The Lawiers Light: Or, a Due Direction to the Study of the Law,* London 1629, pp. 105-106, reprinted, with the

And Francis Bacon also held to this same view during the first quarter of the seventeenth century.

The Englishman who most clearly utilized Ramism within jurisprudence, however, was Abraham Fraunce. After studying logic at St. John's College in Cambridge, he entered Gray's Inn and devoted himself to jurisprudence. Fraunce believed that that practice of law consisted in using logic and that logical method both clarifies and explains law. In his *The Lawiers Logike* (1588), he presented a faithful rendering of the two books which made up Ramus' own textbook on dialectics. Fraunce illustrated those Ramist precepts by quoting important jurists, and above all theorists of English law, including Glanville, Bracton, Littleton, Fortescue, and Fitzherbert; he also quoted from some annals as well as from contemporary juridical reports. According to Fraunce, the logical organisation of law was closely related to conformity with Ramist method. As he himself stated, it is important to attain methodological coherence in law despite the difficulty of doing so in the case of England: "It is not my purpose, at this time to dispute," Fraunce concluded, "whether the law of England, being one onely proper, and singular law, appropriate to our nation, consisting of so many particularities, being subiect to such continuall change and alteration, can bee made an art: But that the confusion of writers may bee removed, the maner reformed, the matter better ordered, I doe surely beleeve, because I know assuredly."[12]

Ramism was known to French jurists, though its influence on law in France was much less pronounced that it was in England. Nicolas Bergeron and Antonie Loisel, who were the executors of Ramus' will, both were renowned for their writings on common law. Claude Mignault, who published a commentary on the *Rhetorica* of Ramus' collaborator Omer Talon, later utilized Ramism within the domain of jurisprudence. The case of the French jurist Hugues Doneau is more complex. Considered by modern scholars as a "systematic" jurist, his synthetical and rational exposition of law – aimed at

title *Methodus studendi,* in the last section of his book *The English Lawyer, describing a Method for the Managing of the Lawes of this Land,* London 1631, pp. 257-258, in which chapter 8, pp. 55-148, is also dedicated to the problem of the importance of logic for law.

12 A. Fraunce: *The Lawiers Logike, exemplifying the praecepts of Logike by the practise of the common Lawe,* London 1588, book 2, chap. 17, fols. 119v-120r.

expositing *omnia ratione et via, eademque proprie et dilucide*[13] – has been considered as evidence of his agreement with the "rationalism" of Ramus. Ramism played only a very minor role within Doneau's legal commentaries; yet his jurisprudence did reflect the methodological concerns current within Renaissance thought to which Ramism made an important contribution.

5.

It was only with the advent of Jean Bodin and Johannes Althusius that law and politics came together. The emphasis which both men placed on Ramist methodology became especially important, leading to remarkable results from a scientific point of view. In spite of the disparity between these two authors (Bodin was brought up in the heart of Paris while Althusius taught at the Calvinist *Hohe Schule* in Herborn), the similarities between them are striking. Both men gave primacy to raising law to the level of a science and wished to do so by utilizing a proper system. This is evident from Bodin's *Juris universi distributio* (1578-1580) as well as from Althusius' *Juris Romani libri duo: ad Leges Methodi Rameae conformati: et tabula illustrati* (1586). Bodin also developed a new "method" for history in his *Methodus ad facilem historiarum cognitionem* (1586). He regarded that method as valid with respect to all fields of research, including civil law, comparative law, politics (as utilized in his *République*), or the world of nature in general (as employed in his *Universae naturae theatrum*).

The *ars* which Bodin and Althusius referred to (*ars historiae/juris/politicae*) was clearly also a science which presupposed the need for systematic order and the use of dialectic, i.e., the articulation and utilization of logical concepts. This required one to begin with that which was fully comprehensive and universal and which would be defined before being broken down into sub-categories (*certis quibusdam generis*) descending from the broadest to the increasingly specific, accompanied by an extensive rendering of practical examples. Both Bodin and Althusius regarded order (*ordo conveniens*) to be the most suited disposition (*apta rerum dispositio*) by means of which the whole should be connected to its parts as well as these parts to one another. As Bodin stated in his *Juris universi distributio,* this order consisted of unbroken links on a chain (*perpetua sit et continua quaedam*

13 H. Donelli: *Opera omnia,* Lucae 1762-1770, vol. 1, p. xxix.

series) and was in accordance with the true method of all sciences (*haec illa est non modo artium tradendarum, sed etiam omnium scientiarum ratio communis*).[14]

But in Bodin's *Methodus ad facilem historiarum cognitionem* it was stated that *analysis*[15] is the headmistress of all arts (*praestans illa docendarum artium magistra*). Analysis, which teaches us how to divide the whole into parts and those parts into yet more specific parts, is identical to the analysis which is part of Ramist theory. Although Bodin nowhere mentioned Ramus in this connection, the similarities between Bodin and Ramus are so striking that some scholars – Kenneth McRae among them – even claimed that this above-cited passage pertaining to *analysis* might have actually been written by Ramus himself.[16] Scholars have also addressed the applicability of Ramist dialectics to Bodin's discussions of history, jurisprudence, and politics.

The explicit notion of Ramus or Ramism which is absent from Bodin's writings is to be found, however, in Althusius' first published work, the *Juris Romani libri duo,* which has already been mentioned above. In the preface of that work, Ramist logic is praised as the rule to which all law must conform. Ramus was gradually relegated to a secondary role within his later works – the *Politica* and *Dicaeologicae libri tres* – as well as within later editions of his earlier works. The revival of metaphysics and the increasing intensity of religious conflict at the end of the sixteenth century and the beginning of the seventeenth may have contributed to that development. Yet even in his later years, Althusius' method still remained basically Ramist. Among the signs of this persisting Ramist tendency were his continued use of the three laws, his continuing extensive use of dialectic, and the importance which order was given within all of his writings. As Carl Friedrich has pointed out, it was the manner in which Althusius

14 J. Bodin: *Œuvres philosophiques,* ed. P. Mesnard, Paris 1951, p. 71.

15 Ibid. p. 116.

16 K. McRae: Ramist Tendencies in the Thought of Jean Bodin, in: *Journal of the History of Ideas* 16 (1955) p. 315. The best study of the subject, which I could not yet take into account in my book *La disputa del metodo nel Rinascimento* (see note 1), is at present that of M.-D. Couzinet: *Histoire et méthode à la Renaissance. Une lecture de la* Methodus ad facilem historiarum cognitionem *de Jean Bodin,* Paris 1996.

organized his legal writings – combined with a high regard for systematic organization in general – that made him a leading practitioner of Ramist methodology.[17]

To summarize, in the writings of Bodin and Althusius we can see the culmination of the Ramist tendencies to link method and science. According to both authors, the subject-matters of historiography, jurisprudence, and politics could only be examined properly using Ramist method. The proper employment of that method, based on the three laws, guaranteed the universality, homogeneity, and reciprocity of the particular precepts that were to be examined. In the preface to the first edition of his own *Politica* (1603) Althusius accused Bodin himself of failing to do this.

6.

In conclusion, I would like to briefly discuss the place of methodology within the disciplines of historiography and jurisprudence at the end of the sixteenth century. In both disciplines there was a need to find balance and reciprocity between the discussion of universals (which was the principal task of the sciences considered generally) and the discussion of particulars (e.g., through applied science, the use of examples, and emphasis placed upon utility). To put this another way, there was a need to join *ratio* and *experientia.*

It was believed that neither *ratio* nor *experientia* could do without the other. In the domains of history and law, there could be no theory without *ratio,* but nor could *ratio* ignore the historical and juridical factors that were peculiar to individual nations. The universality of theory needed to include the critical examination of particulars. One could no longer rely exclusively on the juridical pseudo-universalism of Roman Law or the cosmological-genealogical providentialism of the ancient school of chroniclers. One could no longer limit oneself to examining these disciplines only from the top downwards using deduction; instead, it was now considered necessary to critically examine the particulars pertaining to both fields, which included the specific customs and traditions of individual nations, the relevance of new areas of inquiry, and the impact of recent geographical discoveries.

17 C. H. Friedrich: *Johannes Althusius und sein Werk in Rahmen der Entwicklung der Theorie von der Politik,* Berlin 1975, p. 59.

One was now required to use Bodinian comparativism in order to examine particulars at the lowest level.

Methodology also needed to fit into this new model. It could no longer – as was the case during the time of Le Douaren, Coras, or Baudouin – be understood simply as expressing the need for any kind of order (*ordo*) whatsoever. There was now a need for an *ordo* that was grounded in *methodus;* the use of method would guarantee that theoretical discussions in historiography and jurisprudence would at the same time be scientific. Method was important to Bacon and to La Popelinière, two theorists who also are considered to have stressed the scientific nature of historiography and jurisprudence. History and law could not and cannot be reduced to that which is practical alone; both required – and still require – theory as well.

Ramist logic provided the methodological tools that could be used in order to bring theoretical writings to historiography, jurisprudence, and other disciplines on the level of scientific inquiry. The three (material) laws of Ramism could be utilized in defense of universal reason (through the *lex veritatis*) while at the same time taking into account that which is particular (in accordance with the *lex justitiae*) and observing the requirement that all precepts be hierarchically arranged (observing the *lex sapientiae*) in conformity with the Ramist "unique method" (*unica methodus*). During the period that preceded the discarding of Ramism by Descartes and the scientific revolution, Ramism had a great deal to contribute to the development of method and science.

Logical Division and Visual Dichotomies: Ramus in the Context of Legal and Medical Writing

IAN MACLEAN

Division (*diairesis*) and definition belong to class logic, which is a notoriously difficult area; it has rarely been set a more difficult task than that of the following classification of animals which Jorge Luis Borges alleges he borrowed from "a certain Chinese encyclopaedia":

> (a) belonging to the Emperor (b) embalmed (c) tame (d) suckling pigs (e) sirens (f) mythical (g) stray dogs (h) included in the present classification (i) quaking like madmen (j) innumerable (k) drawn with a fine camel-hair brush (l) etc. (m) having just broken the pot (n) resembling flies when seen from a distance.[1]

What, we may ask, would Petrus Ramus have made of this, with his *via una et simplex,*[2] and how would he have reduced them to dichotomous presentation according to the rules of his method? Many historians of philosophy, in the light of much more sober challenges to his system than this, have been scornful of Ramus's dialectic, accusing it of oversimplification, superficiality and woeful ignorance of the true import of logic;[3] the embarrassing fact which they have to confront, even after their demonstration of these egregious flaws, is that Ramus was a compelling and controversial figure in his own day, and after his death, for about sixty years, an unparalleled publishing triumph, at least in the area of teaching manuals.[4] In this we may usefully compare him with that other

1 El idioma analytico de John Wilkins, in: *Obras completas,* vol. 2, Barcelona 1989, p. 86.

2 *Dialectica,* Paris 1548, p. 113, quoted by Nelly Bruyère: *Méthode et dialectique dans l'œuvre de La Ramée,* Paris 1984, p. 97.

3 See Brian P. Copenhaver, Charles B. Schmitt: *A history of western philosophy 3: Renaissance philosophy,* Oxford 1992, pp. 230-240.

4 W. J. Ong: *Ramus and Talon inventory,* Cambridge, Mass. 1958.

Renaissance original, Paracelsus; he also aroused violent passions in his lifetime, and was the subject of a formidable popularising campaign after his death, which, like Ramus, helped to define the sectarian lines in Northern Europe before the Thirty Years War. But whereas Paracelsus's legacy to medicine was an enduring one, in the attenuated forms of iatrochemistry and homoeopathic remedies, Ramus's legacy to pedagogy and to thought seems to have been far less substantial: some might say that it consists in no more that a method of visual presentation of complex material, the tree diagram, which, as it happens, he did not even create. As Copenhaver and Schmitt put it:

> It was above all the dichotomous tables frequently seem in the earlier age of manuscripts, and favoured by medical authors who published before Ramus, that he made his hallmark and raised to new levels of popularity [...]. The point of this apparatus was to reduce the student's confusion and to ease his labour. As a beginner in natural philosophy, for example, the student could divide *nature* into *organic* beings with vital souls and *inorganic* entities without souls, and *organic* beings could be further separated into *irrational* plants and animals and *rational* humans, and so on. The resulting synoptic scheme [...] was [...] in its own time [...] a pedagogic marvel that many found more useful than traditional syllogistic reasoning.[5]

What I shall set out to do in this paper is to look into these claims, by taking the topic of class logic (division, definition and taxonomy) and asking the following questions: what was the doctrine of class logic in an ideal-typical sixteenth-century arts course (insofar as this can be determined from what known syllabuses such as those of Paris and Padua have in common)? What did Ramus do to it? How did the higher disciplines of law and medicine in universities deal with it and with the related issue of its visual representation, and what relation did their solutions have to Ramus's own?

Class logic in the arts course is connected to a number of potent and difficult terms: *divisio (distinctio, diairesis), distributio, differentiae, definitio.* It is also involved in complex issues of difference, identity and similarity, and linked to epistemological and ontological problems. Plato's *Phaedrus* 265D-277C, and *Philebus* 16C-17A are often cited as an authoritative discussion of the virtues of division as a logical and heuristic process, which has its place at the very beginning of the philosophical enterprise;[6] Aristotle accords it an important (but not

5 Copenhaver, Schmitt: *Renaissance philosophy* (as note 3) p. 238.

6 See ibid. p. 234, for the relationship of *diairesis* and *synagoge,* and for the origins of method in Hippocrates's notion of art.

as important) a place, linking it in both the *De anima* and the *Metaphysics* with one of the two functions of the intellect (*dividendo et componendo*) by which all knowledge comes to us.[7] Later writers, including Cicero, Quintilian and Boethius, all write about it at some length; and Renaissance discussions of class logic are strongly influenced by their *dicta* on the subject.[8] This corpus of inherited knowledge is subject to the distinction made between science and arts, between scientific knowledge on the one hand, which is obtained by syllogistic demonstration in the first figure, and which is causal in nature (*scire est rem per causas cognoscere*); and endoxical knowledge on the other, which can call upon a wider range of demonstrative techniques, but cannot aspire to the objective certainty of scientific demonstration.[9] Arts belong to the sphere of endoxical knowledge; they are described, according to Porphyry's definition, as "the finite doctrine of infinite things".[10] Division has a place in both sciences and in arts, but its nature is correspondingly different. Scientific division is illustrated by Porphyry's tree, which is the source in turn of scientific definition, and employs the *differentiae* on which this is based. The standard example of this tree is given by the division of the term "substance":

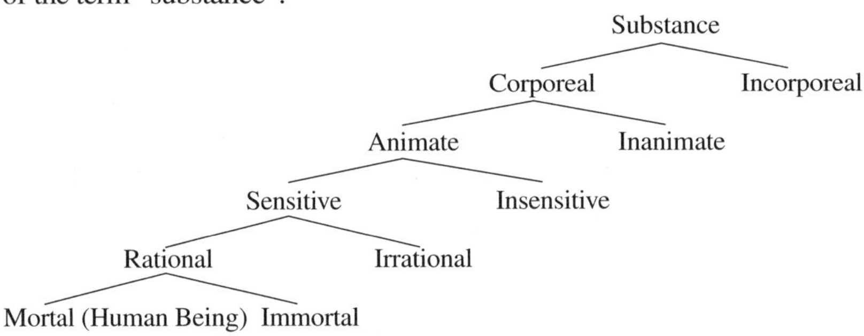

7 *De anima* III 6, 430a26ff.; *Metaphysics* X 1, 1051b1ff.; also *Physics* I 1, 184b10. Aristotle's critique of *divisio* is to be found in *Posterior analytics* II 6-7, 92a28ff.; also II 13-14, 96b13ff. The differing attitudes of the two philosophers are recorded in Giambattista da Monte: *Universa medicina,* ed. Martinus Weindrichius, Frankfort on the Main 1587, p. 95. See also Artur von Fragstein: *Die Diairesis bei Aristoteles,* Amsterdam 1967.

8 See Ian Maclean: *Interpretation and meaning in the Renaissance: the case of law,* Cambridge 1992, pp. 72-74. 106ff.; Cicero: *Topica* 5,28; Quintilian: *Institutio* 7,11; Francis Goyet: *Le sublime du 'lieu commun': L'invention rhétorique dans l'Antiquité et à la Renaissance,* Geneva, Paris 1996, pp. 553-554.

9 Maclean: *Interpretation and meaning* (as note 8) pp. 72-74.

10 [...] de infinitis finita doctrina [...], ibid. pp. 113-114.

We may note here that for the division to be scientific, the *differentiae* must be thematically appropriate and in the correct order (one could not, for example, insert "high/low" or "soft/hard" before "corporeal/incorporeal"), both petition of principles and over-division must be avoided, and there can be no remainder; the division of the term must be perfect and necessary.[11] For this to happen, three often-invoked conditions are set down *Posterior analytics* I 4 which relating to the types of proposition that can appear in a demonstration, namely the conditions κατὰ παντός, καθ' αὑτό and καθόλου πρῶτον (73a28ff.). Walter Ong, Guido Oldrini and Nelly Bruyère have written in an illuminating way about these conditions in relation to Ramus's somewhat eccentric designation of them as rules;[12] I refer to them here in a more orthodox Aristotelian context, because they help define the difference between the scientific and endoxical orders of division. The first and third conditions are standard Aristotelian ways of expressing the universally quantified proposition that every A is B; the second specifies what can be said of B in itself. The third condition combines the conditions of the first two rules, and provides the link with definition (of universals) by *genus* and *differentiae* illustrated in Porphyry's tree. This is able to supply the scientific definitions of stone (corporeal, inanimate substance), plant (corporeal, animate, insensitive substance), animal (corporeal, animate, sensitive, irrational substance) and man (corporeal, animate, sensitive, rational, mortal substance), which in each case satisfy the three conditions. Another necessary condition of adequate definition is the limit imposed on predicability by the ten predicaments or categories: substance, quantity, quality, relation, time, place, position, possession, acting, being acted upon. This does not mean that all that is knowable is confined to the ten predicaments as though these had the force of Kantian categories; only that the intellect's ability to discourse about objects is so confined.

The standard discussion of endoxical division is found in a number of Boethius's topical works, notably the possibly spurious *De divisione* and

11 Boethius: *In Categorias* 1,169D-170C.

12 Walter J. Ong: *Ramus: method and the decay of dialogue,* New York 1979, pp. 258-259; Guido Oldrini: *La disputa del metodo nel Rinascimento: indagini su Ramo e sul ramismo,* Florence 1997, pp. 85-101; Bruyère: *Méthode et dialectique* (as note 2) pp. 267-275; and below, p. 235. See also Maclean: *Interpretation and meaning* (as note 8) p. 73 (quoting Goeddeus).

the *De topicis differentiis*. He specifically stresses the usefulness of *divisio bimembris* or simple disjunction,[13] but concedes also that there are other forms of classification. The *De divisione* breaks division into four classes: *genus* into *species* (e.g. animals into rational and irrational animals); whole into parts (also known as enumeration: e.g. house into roof, walls and foundations); polysemic words or propositions into single senses (which in the scholastic system becomes the *sic distinguitur* by which contradictions are resolved by appeal to *ratio, ordo, modus,* etc.); and finally the *species* of *divisio secundum accidens,* which is itself divided into three classes (subjects divided by accidents: e.g. some men are black, some white); accidents divided by subject (things desired are either desired by the body or the soul) and accidents divided by accidents (some white things are solid [pearls] others liquid [milk]).

These divisions begin, like Porphyry's tree, with the *genus generalissimum* and end with the *species infima,* but unlike it, they may progress not by necessary and sequential steps but by intuitive dissimilarity/similarity decisions; that is to say, there may be no causal knowledge of the objects divided, and hence, no adequate account (in Aristotelian terms) of their identity and difference. Endoxical division may be heuristic in nature; it is an art, not a science, which accommodates judgements of plausibility and marginal cases. For all that, it is subject to error and falsification, as Galen points out.[14] Like Ramist dichotomies, it is pedagogically useful;[15] it can be, but does not have to be, exhaustive; but although less rigorous than Porphyry's tree, it cannot accommodate exceptions in the same synoptic table. It calls to mind Aristotle's comments on provisional taxonomy in his various treatises on animals.[16] The term *divisio* (of *genus* into *species*) is sometimes distinguished from *partitio* (of wholes into parts), as it is by Cicero (*Topica* 5,28). *Partitio* is usually taken to be potentially exhaustive, but Cicero allows for instances

13 See *In Isagogen Porphyrii commentarius* 1,9, and Ong: *Ramus* (as note 12) p. 199.

14 See Galen: *Ad Glauconem de methodo medendi* I, quoted by Giovanni Argenterio: *Opera,* Venice 1592, II 24: omnes errores qui in medicina nascuntur, ex malis divisionibus provenire.

15 See Neal Ward Gilbert: *Renaissance concepts of method,* New York 1960, pp.104-105 (quoting Giovanni Manardi).

16 See *Parts of animals* I 3, 643b9ff; also *Generation of animals* II 1, 732a1ff.; *History of animals* I 6, 491a4ff.

which are imperfect or incomplete.[17] *Partitio* is sometimes distinguished in turn from *enumeratio,* which can also be an open class. Quintilian's term for this is *anacephaleosis,* which he defines both as "a summing up", and as a "division into headings"[18] The term is used in both senses in the Renaissance, and indeed many of the tree diagrams provided in books are no more than a synoptic table of chapters grouped by subject matter.[19]

The question whether such divisions are real or nominal may be discussed in the context either of *divisio* or of *definitio:* in Cornelius Valerius's (1512-1578) *Tabula totius dialectices* composed in 1545, a much reprinted rival to Ramus's *Dialectique,* both *divisio* and *definitio* are subjected to distinction by *res et nomen;* and Boethius's categories of *divisio* are distributed between *divisio nominis ambigui* and *divisio rei* (that is, *divisio generis in species, totius in partes, subjecti in accidentia, accidentis in subjecta, accidentis in accidentia*).[20]

Before turning to the uses made of these logical instruments in law and medicine, it will be helpful to summarize what Ramus himself made of them. I am more concerned here with what was made of his thought by contemporaries, rather than any evolution it may undergo, and so I have accepted the argument made by Bruyère and others that his concept of

17 See also *Topica* 5,28f. 7,31. 8,33 (non est vitiosum in re infinita praetermittere aliquid); and Dieter Nörr: *Divisio und Partitio: Bemerkungen zur römischen Rechtsquellenlehre und zur antiken Wissenschaftstheorie,* Berlin 1972, pp. 20ff.

18 See Goyet: *Le sublime* (as note 8) pp. 553-556 (quoting Melanchthon); Quintilian: *Institutio oratoria* VI 1,1.

19 Charles B. Schmitt: *Aristotle and the Renaissance,* Cambridge, Mass. 1983, p. 58, has an illustration of *anacephaleosis* as a table of chapters (of the *Physics*).

20 Valerius's text exists in two forms: one, a broadsheet *Anacephaleosis seu enumeratio capitum omnium totius artis dialecticae summam complectens digesta,* Antwerp n.d., reproduced in Leon Voet: *The Plantin Press (1555-1589),* Amsterdam 1982, V (illustr. 50); the other, a textbook, in which *definitio nominis* is undivided, and subsequent to the *quaestiones simplices* (the predicables and predicaments) through which division and definition are possible; and *definitio rei* is subdivided into *1. essentia, 2. caussarum, 3. effectuum, partium et accidentium;* the last is clearly not scientific in nature: *Tabulae totius dialectices,* Antwerp 1573, pp. 63-66. Boethius also uses *divisio* to categorise *res* in his commentary on the Categories (1,169D-170C); and Cicero (*Topica* 5,26) distinguishes in a similar way between things that are (*quae sunt*) and things such as we perceive them (*quae intelleguntur*).

method does not change radically after 1546.[21] Ramus attacks Aristotle, Galen, Cicero, Quintilian and Porphyry for their accounts of logic and dialectic, accusing the first not only of confusing the separate activities of invention and judgement and of misattributing division and definition to the latter, but also of wrongly distinguishing between demonstrative and endoxical knowledge and hence of failing to see what is a necessary and what a non-necessary division. The latter three are accused of over-division, clumsy use of *diairesis,* confusion and (in the case of Porphyry) of omitting the *individuum* from the list of predicables.[22] It is not surprising that even well-disposed Renaissance logicians such as Rudolph Goclenius (1547-1628) and Bartholomäus Keckermann (1572/73-1609) had such difficulty in formulating a moderate critique of such iconoclastic measures directed against the whole edifice of logic as it was then understood.[23] What Ramus is left with is an art of dialectic which falls into the two parts of invention and judgement; it begins with the *genus generalissimum* and passes not to the *species infima* but the individual, which, as will be remembered, is un-

21 Bruyère: *Méthode et dialectique* (as note 2) pp. 64ff.; Ong: *Ramus* (as note 12) p. 203, claims that the main lines of Ramus's method were in place even earlier.

22 Ramus: *Rhetoricae distinctiones in Quintilianum* (1549), trans. C. Newlands, ed. J. J. Murphy, Dekalb 1986, pp. 79-80. 104. 108. 131-133; *Scholae in tres liberales artes,* Basle 1569; *Schola dialecticae* I 5, col. 17 (on Stoic division). III 3, col. 82 (against Porphyry's predicables). IV 5, cols. 120-121 (against Aristotle's categories). VII 14, cols. 258-261 (against Aristotle's division). See also Bruyère: *Méthode et dialectique* (as note 2) pp. 108. 153-155, on Ramus's attacks on Galen, and Ong: *Ramus* (as note 12) p. 203, quoting *Dialecticae institutiones* (1543), on Porphyry's failure to realize that *species* and individuals are the same thing, a point made *inter alia* in a characteristic passage of the *Scholae dialecticae* III 4, col. 87: Platonis methodus ad amplificandum tantum nugas, non ad eius usum, a Porphyrio confunditur. Plato in Philebo docet omnia aeterna uno et multis constare, terminumque et infinitatem habere, communemque intermediam ideam dividendam esse, donec ad infinitatem perventum sit, quae ipsa omittenda sit. Haec illic Plato, quae Porphyrius hic suis, non Platonis verbis interpretatur, a generalissimo per subalterna ad specialissimas species descendendum, ut individua secludantur. – There is also a discussion of definition and division in Ramus's *Defensio pro Aristotele adversus Jacobum Schecium* (1571), on which see Bruyère: *Méthode et dialectique* (as note 2) p. 185.

23 Keckermann: *Opera,* Geneva 1614, cols. 1608-1611; Goclenius: *Analyses in exercitationes aliquot Julii Caesaris Scaligeris de subtilitate,* ed. J. Schroderus, Marburg 1599, pp. 11-13.

knowable (*inintelligibilis*) in an orthodox Aristotelian sense;[24] division or distribution and definition are parts of invention, and distribution in the 1576 *Professio regia* is divided in what appears to be a Ciceronian way between *partitio* and *divisio.*[25] The method of dialectics is governed by the three laws drawn from *Posterior analytics* I 4 to which I have already alluded; they are renamed *lex veritatis, lex justitiae* and *lex sapientiae,* and function as a means to define the *materia subjecta* of an art. By them, only that which is true and necessary may be included in any argument; all those things, and only those things, which belong to the art must be included; and general things must be dealt with in a general way.[26]

I shall now examine briefly law and medicine in turn, to see what place the doctrine of division has in both the content and the visual presentation of their teaching manuals, and what this owes to the arts course and to Ramus. *Divisio* (often called *distinctio*) is especially praised by twelfth-century jurists for its pedagogical usefulness. Azo's *Summa Institutionum* claims that "doctrine is transmitted more easily by divisions [...] for partition or division stimulates the mind of the reader, provides sense to the intellect and cleverly moulds the memory."[27] It is noticeable that he makes no explicit distinction between division and partition, although he does distinguish between *distinctio directa* and *distinctio utilis* (the first arising from the literal text of the law, the second only from the imputed inten-

24 See Bruyère: *Méthode et dialectique* (as note 2) p. 96; Ong: *Ramus* (as note 12) p. 203, who points out that a consequence of Ramus's logic is the claim that men and women are of different *species,* against Aristotle: *Metaphysics* X 9, 1058a31-32, on which see Ian Maclean: *The Renaissance notion of woman,* Cambridge 1980, pp. 12-13.

25 See Ramus: *La dialectique,* ed. Michel Dassonville, Geneva 1964, pp. 89-96; Peter Mack: *Renaissance argument: Valla and Agricola in the traditions of rhetoric and dialectic,* Leiden, New York, Cologne 1993, p. 337.

26 *Scholae dialecticae* II 1, cols. 31-32. Ramus is followed by others who use these rules to justify the autonomy of an art; e.g. Johannes Goeddeus and Jean Bodin, who like Ramus links the logical rule καθ' αὑτό with the definition of equity *suum cuique tribuere* by prefacing these words with *necesse est:* see Maclean: *Interpretation and meaning* (as note 8) pp. 27. 73.

27 Facilior per divisiones trad[it]ur doctrina. Partitio sive divisio animum legentis incitat, mentem intelligentiae praeparat, memoriam artificiose reformat. Quoted by Gerhard Otte: *Dialektik und Jurisprudenz: Untersuchungen zur Methode der Glossatoren,* Frankfort on the Main 1971, p. 95.

tion of the legislator).[28] Division makes visible that which otherwise would not become known; in his discussion of the concept *obligatio,* for example, Rogerius says that "its nature cannot be understood unless by means of description and division."[29] Indeed, Placentinus goes so far as to claim that the more anything is subjected to division, the clearer it becomes, although this optimistic belief in subtle distinctions is strongly criticised both by his contemporaries and by Renaissance humanists such as Montaigne.[30]

After the rediscovery of Cicero's texts of forensic theory and the complete text of Quintilian in the early fifteenth century, a number of adaptations of topics and dialectics written specifically for legal students appear from the 1490s onwards, some of them influenced also (after 1520) by Rudolph Agricola's (1443/44-1485) *De inventione dialectica;*[31] these all treat definition before division, since it is argued that simple objects can only be distributed once their nature is known. Two arts of law widely circulated in the middle years of the century, by Jean de Coras (1500-1572) and Joachim Hopper (1523-1576), are typical of the treatment of this topic: Coras, for example, treats the method common to both definition and division (from general to particular) before dealing with each in turn, and divides division itself into *divisio* (of *genus* into *species*), *partitio* (of wholes into parts) and *enumeratio,* which is "the distinction and recension of the many things that go to make anything up": the example he gives is Ulpian, who was a *prudens,* a *iurisconsultus* and a *Christianorum persecutor.*[32] This third category is particularly interesting, since it is an open class; it repre-

28 See Otte: *Dialektik* (as note 27) p. 95; see also Placentinus: *Summa Institutionum* II 1, cited by Otte, and Aristotle: *Metaphysics* I 2,15, 983a11.

29 Natura obligationis aliter non potest cognosci, nisi per descriptionem et divisionem. Rogerius: *Summa codicis* IV 12,1, quoted by Otte: *Dialektik* (as note 27) p. 95.

30 Quanto magis res omnis distinguitur, tanto melius aperitur. Placentinus: *Summa Institutionum* II 1, quoted by Otte: *Dialektik* (as note 27) p. 95; see also Ian Maclean: Montaigne et le droit civil romain, in: *Montaigne et la rhétorique,* ed. John O'Brien, Malcolm Quainton, James J. Supple, Paris 1995, pp. 163-176.

31 See Maclean: *Interpretation and meaning* (as note 8) p. 79 (referring to Gammaro, de Federicis, Hegendorff, Cantiuncula, Oldendorp and Apellus); also Mack: *Renaissance argument* (as note 25) p. 299.

32 Multorum, quae uni alicui insunt, distinctio, et recensio. Coras: *Tractatus de iuris arte* (1560), in: *Tractatus iuris universi,* Venice 1584, I 62r.

sents the ability to deal with the infinitely variable nature of mankind and human actions without falling into the Sorites or Antisthenes paradoxes.[33]

An unexpected complement to enumeration or the open class is the rule. This is derived from existing law, as we learn from *Digest* 50,17,1;[34] yet although the terms *generaliter* and *regulariter* are used to indicate that which must necessarily and in all circumstances be the case, the medieval jurist Irnerius in his gloss on *Institutes* 1,3 states that "it is not absurd to find something which does not fall under the rule."[35] He does not mean by this that the rule is not of universal application, but that it remains a rule even when exceptions have been stated, provided that they are all made explicit; the full version of the well-known maxim that the exception proves the rule (nowadays used in precisely the opposite sense to that accorded to it in the sixteenth century) reads *exceptio firmat regulam in casibus non exceptis* (the exception confirms the rule in respect of cases not excepted). The rule (which can also be a legal norm) is immutable; the exception (which here denotes a procedure invoked by a defendant to nullify the rights of the plaintiff) is expressed by the verbal formulae *nisi*, *si non*, *si nihil*, *si tantum*, *at*, *nonnunquam*, *aut si*, *vel si*, *extra quam*. It can be applied to itself recursively (and invoked alternately by defendant and plaintiff), for there are exceptions to exceptions (*replicationes*), exceptions to exceptions to exceptions (*triplicationes*), even exceptions to exceptions to exceptions to exceptions (*quadriplicationes: Digest* 44,1,2).[36] This procedure comple-

33 On the knowledge of these paradoxes in this period, see Lisa Jardine: Humanism and the teaching of logic, in: *The Cambridge history of later medieval philosophy,* ed. Norman Kretzmann, Anthony Kenny, Jan Pinborg, Cambridge 1982, pp. 798-807. There is also much written about the dangers of definition in respect of the infinity of human actions, of which the *locus classicus* is D 50,17,202 (Javolenus): omnis definitio in iure civili periculosa est: rarum est enim ut non subverti posset. On this rule, whose source could be Cicero: *De oratore* 2,25,108, see Maclean: *Interpretation and meaning* (as note 8) pp. 104ff.

34 D 50,17,1 (Paulus): Regula est quae rem breviter enarrat: non [ut] ex regula ius sumatur, sed ex iure quod est, regula fiat.

35 Non est absurdum aliquid extra regulam inveniri. *Glossa,* ad Inst. 1,3, quoted by Otte: *Dialektik* (as note 27) p. 95.

36 D 44.1.22 (Paulus): Exceptio est condicio, quae modo eximit reum damnatione, modo minuit damnationem. Replicatio est contraria exceptio, quasi exceptionis exceptio. D 44,1,2 (Ulpianus): [...] exceptio actorem excludit, replicatio reum. Sed et

ments Coras's enumeration, by which an open class of things can be invoked in law; the exception to the rule reveals the rule or distinction to be itself incomplete, yet what it does not include remains both knowable and usable as knowledge; it represents a clear case of the law exceeding any Ramist method in flexibility and inclusiveness, even though it simultaneously demonstrates the intolerance found among legal theorists of mixed categories and mediate contraries.[37]

A further point concerns the presentation of legal argument by divisions or partitions, and the relationship of this to tree diagrams. Coras and Hopper are discursive texts without diagrams; there are also dichotomous presentations of the legal art directly inspired by Ramus, of which Johannes Thomas Freigius's (1543-1583) *De logica iurisconsultorum* (1582), Hermann Vultejus's (1565-1634) *Idea iuris logica* (1586), and Johannes Althusius's (1557/63-1638) *Juris Romani libri duo: Ad leges methodi Rameae conformati: et tabulâ illustrati* (1586) are examples; and Christoph Hegendorff's (1500-1540) *Dialectica legalis* of 1536 precedes Ramus himself in using a systematized dichotomous presentation of the law.[38] I have argued elsewhere

contra replicationem solet dari triplicatio, et contra triplicationem rursus et deinceps multiplicantur nomina, dum aut reus aut actor obiicit. On exceptions and rules, see Johannes Oldendorp (1480?-1567): *Lexicon juris,* Frankfort n.d., s.v. regula: Regulae iuris sunt notitiae quaedam semper firmae et immutabiles: ut non solum lapideis olim tabulis, sed etiam cordibus hominum indelebili characteri a Deo ipso sunt insculptae, veram vivendi honestatem exprimentes. Regulae officium est, exhibere nobis generaliter definitionem Iuris, quod alioquin per singulas negotiorum circumstantias latissime patet. Qua in re plurimum commodatur humanae fragilitati; Quae in numerosa legum multitudine, quasi in labyrintho quodam periclitari solet, nisi summas earum sententias et primaria (ut sic dicam) consilia prae oculis habeat. François Hotman (1524-1590): *Commentarius de verbis juris,* Lyon 1569, p. 324: Differt [...] Lex a Regula: quod illa dicitur, quam populus certo et cognito omnibus iure constituit: Regula vero, quam Iurisconsulti ex pluribus legum capitibus notatis inter seque comparatis elicuerunt: ut non omnino incommode dicere possumus, leges esse tanquam voces, et sonos singulos: Regulam vero esse tanquam harmoniam et concentum. Ibid. p. 139: Exceptio (ait Ulpia. [D 44,2,1,2]) dicta est quaedam exclusio [...] Excludendum autem dixit, pro Eiiciendum, et amputandum, nam Excipere proprie detrahere est [...].

37 On the intolerance of law to mediate contraries, see Maclean: *Interpretation and meaning* (as note 8) p. 39.

38 Ibid. pp. 83-86.

that the practice of *distinctiones* is well-established in law, and that it could be, and was, represented by means other than tree diagrams.[39] As Ramus never declared that spatial representation of logic was intrinsic to his method, this point is directed more against Ong, who claimed for Ramus and for the Renaissance print culture in general a breakthrough in class logic associated with the visualisation of division.[40] Not only were these and other spatial diagrams used in medieval texts, but also (and more importantly) they were only one of a number of means by which divisions were communicated to readers, and continued to be until well after the Renaissance.[41]

Medicine offers a rather different pedagogical picture, as is to be expected; for the foremost of its authorities, Galen, recommends a set of methods particularly appropriate to the art which is not identical to those codified by scholastic writers from the *Organon,* bearing traces as they do of both Platonism and Stoicism. In this art division, sometimes described as "Galen's precept",[42] has a particularly important role to play; it is described as "the mother of demonstration and definition" and its practice is what defines a good rational doctor from methodists and empirics.[43] Galen's various *dicta* are scattered throughout his vast *œuvre,* and have to be brought together and made coherent by Renaissance scholars; and to further complicate the picture, the major school of medical pedagogy, Padua, produces (and exports) a number of metholodogies in the course of the sixteenth century which are closely allied to Aristotelian natural philosophy and psychology (the *Physics* and the *De anima*). I shall take for convenience Giro-

39 Ibid. pp. 40-43. 83-86.

40 Ong: *Ramus* (as note 12) pp. 199ff.; a thesis developed later in *Orality and literacy: the technologizing of the word,* London 1982.

41 See Schmitt: *Aristotle and the Renaissance* (as note 19) p. 56; M. Evans: The geometry of the mind, in: *Architectural Association Quarterly* 12/4 (1980) pp. 32-55; K. J. Höltgen: Synoptische Tabellen in der medizinischen Literatur und die Logik Agricolas und Ramus', in: *Sudhoffs Archiv* 49 (1965) pp. 371-390. For an early example of a table prepared for pedagogical purposes in a philosophical context, see Gregor Reisch: *Margarita philosophiae,* Fribourg 1504.

42 Cardano: *Opera omnia,* ed. Charles Spon, Lyon 1663, III 627ff. (*De subtilitate* XVI). Ramus attacks Galen's logic and method, although not specifically on the issue of division: see Bruyère: *Méthode et dialectique* (as note 2) pp. 108. 153-155.

43 Giambattista da Monte: *Universa medicina* (as note 7) pp. 231. 82ff.

lamo Capo di Vacca's (1523-1589) account of methodology composed in the middle years of the century as an example of modified Galenism, and Santori Santori's (1561-1636) Aristotelian critique of this, first published in 1603, as a means of situating the use of division in medical texts, before looking at examples of discursive divisions and tree diagrams in medicine and related natural science.

Capo di Vacca, following Giovanni Manardi (1462-1536), rightly accuses the medical theorists who had dealt with the question of method before him of confusing instrumental doctrine (a pedagogical method) with a heuristic process (of passing from the known to the unknown, the manifest to the hidden); the first of these equates to the three methods (resolutive, compositive and definitive) discussed at the beginning of Galen's *Ars parva;* the second is the fourfold system of Peripatetic origin (setting out the definitive, demonstrative, divisive and resolutive methods).[44] As a professor of practical medicine, Capo di Vacca wants to reintroduce endoxical knowledge (and especially *experimentum* or the fruits of practical experience) into the second category of method, and does so by admitting the knowledge of individuals to be part of the resolutive method;[45] like the lawyers, he treats definition before division, and divides definition into nominal and real modes. The divisive method has itself no (sub)division; it is an "art" by which we come into the knowledge of anything *quotuplex sit* (in as many sorts as it is found) through *differentiae* which we already know. The procedure runs two risks, according to Galen's report of Hippocrates's words on the subject: that of excess division, and that of under-division.[46] To avoid these faults, it is essential to apply Porphyry's rules, and ensure that the *differentiae* are thematic (e.g. that the *differentia* rational/irrational is applied to animal); that there is no progress beyond the *species infima* (i.e. that no division is made of the accidents of anything); and that the divisions are correctly hierarchised (e.g. apply incorporeal/corporeal before

44 Manardi has a more complex division of the meanings of "doctrine": see Gilbert: *Renaissance concepts of method* (as note 15) p. 104; Capo di Vacca (Capivaccius): *Methodus practicae medicinae,* Lyon 1596, pp. 1014-1038.

45 Ibid. pp. 1038-1040.

46 Over-division is of course related to Ockham's razor: qui methoda divisiva est brevissima via, non debemus entia sine necessitate multiplicare. Giambattista da Monte: *Medicina universa* (as note 7) p. 23.

oviparous/viviparous). Divisions can also be achieved through the four forms of opposition (affirmation/negation, privation/possession, immediate/mediate, contraries/correlation); as we shall see, the discipline of medicine makes greatest use of division based on mediate contraries.

Capo di Vacca follows his Paduan predecessor Giambattista da Monte (1498-1552), who in his own lectures on method delivered in the 1540s and 1550s adopted Boethius's doctrine of division, recognised the place of enumeration, and through his admission of an open class of division was able in the resolutive method to treat individuals (and their infinite temperamental variability in a humoral system) as opposed only to *species,* and give a place in his system to induction (not perfect induction, but the rhetorical kind or that implied by Aristotle's provisional taxonomies set out in his zoological works).[47] He was also able to produce endoxical divisions of medical categories such as fever in his works.[48] Capo di Vacca's successor at Padua, Santori Santori, writing at the end of the century, excludes them, and furthermore restricts the function of division to that of a handmaid of the demonstrative method. Nothing, according to him, can be discovered by division alone; a syllogistic demonstration *a posteriori* is needed if a petition of principles is to be avoided. Santori's impressive *Methodi vitandorum errorum* of 1603 operate through a syndrome of signs whose coming together provides the certainty of diagnosis to which the method aspires; there is a role in this for the enumeration of parts and the division into *genera subalterna* and *species,* but it is not heuristic and endoxical, where that of Capo di Vacca is.[49] Yet even Santori accepts that induction can provide a "sufficient" means of proving something or establishing a diagnosis; and his argument from syndromes of signs which uses negative as well

47 Giambattista da Monte's own rather different account of division, in which discusses "artificial division" (following Alexander of Aphrodisias), and makes distinctions between essential and accidental division and between speculative and "active" or "factitive" (*factiva*) division, is to be found in his *Universa medicina* (as note 7) pp. 4-17. 95-101. For the Aristotelian *loci,* see above, note 17; *Rhetoric* 1393a22ff.; and *Prior analytics* II 23.

48 Capo di Vacca: *Methodus* (as note 44) pp. 1003-1005.

49 Santori: *Methodi vitandorum errorum omnium, qui in arte medica contingunt* (1603), Venice 1630, I 7, cols. 25-29. V 2, col. 420 (referring to Boethius's *divisio bimembris*). XI 3, col. 734.

as positive evidence shows that medical theorists, like the lawyers, were able to conceive of exceptions which were none the less the subject of operative knowledge.[50]

Where the two higher disciplines diverge radically is on the issue of opposition. For medical science, all contraries are not maximally different (as required by Aristotle's *Metaphysics*)[51] but merely greatly different; and nearly all oppositions in medicine fall into the category of mediate contrary, which itself is treated in a particularly loose way; in writing about the condition of neutrality which is the mediate state between health and sickness in 1539, Jacques Dubois of Paris (1478-1555) lists all of the following versions of mediacy: not altogether healthy nor altogether sick; not more healthy than sick; between healthy and sick; partly healthy, partly sick; sometimes healthy, sometimes sick.[52] This demonstrates the flexibility and adaptability of the logical procedures adopted in the realist (as opposed to legally stipulative) context of medicine, which exceeds the resources available to Ramists.

Santori's views, strongly influenced by the Aristotelian natural philosophy taught at Padua, are a reminder of the close association of medicine and the more general investigation of nature at this time. But his strict Aristotelianism is not shared by all sixteenth-century natural philosophers. An example of a more flexible use made of division by a doctor-scientist is provided by the figure of Girolamo Cardano (1501-1576), who was greatly addicted to this procedure. He shows how *differentiae* and oppositive divisions can be produced in novel ways, which often reach areas of enquiry theoretically closed to the intellect by the constraint placed upon it by logic or even by religion. One example of this is provided by the opening pages of his *De rerum varietate* (1557),

50 Ibid. II 2, cols. 258-264; but cf. the hostility to induction expressed in XII *passim,* cols. 767-804. On the connection (or confusion) between induction and division, see also Ramus: *Schola dialecticae* XVIII col. 551, and Averroes: *Colliget* I, quoted by Giambattista da Monte: *Universa medicina* (as note 7) p. 14 (ars dividendi inductioni assimiletur, et [est] valde imperfecta via), with which da Monte does not agree, arguing that "usus divisionis maxime facit ad omnes artes et actiones, quae circa particularia versantur". For an example of Cardano's use of the exception in respect of conjectural knowledge, see Nancy Siraisi: *The clock and the mirror: Girolamo Cardano and Renaissance medicine,* Princeton 1997, p. 131.

51 *Metaphysics* V 10,2, 1018a27-28.

52 Jacques Dubois (Sylvius): *Methodus sex librorum Galeni in differentiis et causis morborum et symptomatum in tabellas sex ordine suo coniecta,* Paris 1539, C1r.

where he employs a division like that of Porphyry to investigate the term *universum* by asking the following questions in turn: Is there one universe or are there many? If there is only one, is it finite or infinite? If it is finite, does it have anything outside itself, or nothing? If there is something outside itself, is it mobile or immobile, and is it created (*genitum*) or uncreated?[53] There are two manifest transgressions of method (and of the papal bull of 1516)[54] here: the question whether the universe has anything outside it or beyond it is a contradiction in terms (since the universe by definition must include everything), yet it is an admissible question, as it exposes the ambiguity over the limits of the sublunary world which is inherent in the Christian compromise doctrine of the universe;[55] and the application to one *genus subalternum* (the finite, single universe) of two *differentiae* (*genitum/ingenitum, mobile/immobile*) which cannot be subordinated one to the other, exposes the weakness of Porphyry's tree and even the *divisio bimembris* of Boethius. Another example of Cardano's flexible thinking is provided by his employment of oppositional or combinational division in his earlier work the *De subtilitate* (1550), in which man is divided by the attribute deception, treated actively and passively; Cardano posits that there are three orders of man: the divine (which does not deceive and is not deceived), the human (which does deceive and is not deceived), the brutal (which does not deceive and is deceived). The fourth combination (which deceives and is deceived) is perfectly coherent, but is not given a separate name, perhaps because a mixture of human and brutal is for Cardano a conceivable mediate category.[56] This combinatory logic, better known in the writings of Lull, is here used heuristically by Cardano, rather than exhaustively, as is more usual.[57]

As has already been pointed out by other historians, tree diagrams and dichotomies are common in medical writing from the 1530s, following the

53 Cardano: *De rerum varietate* I 1, in: *Opera omnia* (as note 42) III 1: but in the *De libris propriis* (ibid. I 128), he declares a preference for *divisio perfecta.*

54 On this, see the literature cited by Copenhaver, Schmitt: *Renaissance philosophy* (as note 3) p. 108.

55 The issue of a space beyond the cosmos arises in the Paris condemnation of 1277 by Bishop Tempier of certain Aristotelian propositions in natural philosophy; see *The Cambridge history of later medieval philosophy* (as note 33) pp. 536-538.

56 Cardano: *De subtilitate* XI, in: *Opera omnia* (as note 42) III 551.

57 Ibid. VI. XIII, in: *Opera omnia* (as note 42) III 452. 571 (examples of speculative numerological analogies); see also Paolo Rossi: *Clavis universalis: arti della memoria e logica combinatoria da Lull a Leibniz,* Bologna 1983.

translation of Galen's semiological works in 1528 at Paris.[58] K. J. Höltgen's article of 1965 pointed to the tabular presentation of Galen's *De differentiis morborum* by Leonhard Fuchs (1501-1560) published at Basle in January 1536/37 as the earliest large-scale case of such publication; but Fuchs himself acknowledges that a certain Stephanus Dutemplaeus preceded him, with his *Tabulae sex in sex Galeni libros de morbis et symptomatis,* which Baudrier records as being published by Sebastian Gryphius at Lyon in 1530.[59] This publication also stimulated Johannes Agricola [Ammonius] (ca. 1490-1570) to plan to produce at Augsburg a similar *tabella* for a different work of Galen in 1534, and probably inspired Christoph Heyll's limited use of visual dichotomies in his edition of the medieval medical writer Alphonsus Betrutius at Mainz in the same year.[60] One of the most elaborate early examples of such *tabellae* is Jacques Dubois's *Methodus sex librorum Galeni in differentiis et causis morborum et symptomatum in tabellas sex ordine suo coniecta,* published at Paris in 1539 by the Wechel publishing house, whom Ramus himself was later to employ.[61] This

58 The semiological works of Galen published in Latin by Simon de Colines in 1528 were the *De differentiis symptomatum* and the *De differentiis morborum.* It may be the case that the idea of producing these in tabular form was in part inspired by the illustrated broadsheets which gave a synoptic representation of the human body, together with other information disguised behind flaps of paper, which were a common teaching aid throughout Europe at about this time.

59 For Höltgen: Synoptische Tabellen, see above note 41. The acknowledgement of Dutemplaeus is in the note to the reader on the first page of Fuchs's work. See also Henri Louis Baudrier: *Bibliographie lyonnaise,* Lyon 1895-1921, VIII 57; Michael Maittaire: *Annales typographici,* The Hague 1722, II 2, p. 574.

60 On Agricola Ammonius, see Peter Dilg: Johann Agricola Ammonius' Kommentar zu Galens *Methodus medendi,* in: *Galen's method of healing*, ed. Fridolf Kudlien, Richard J. Durling, Leiden, New York, Copenhaven, Cologne 1991, pp. 190-198 (esp. 194). See also Heyll's edition of Betrutius's *Methodi cognoscendorum tam particularium quam universalium morborum,* Mainz 1534, pp. 88-89, where the diagram is introduced by the words: quod ne obscurum sit methodicae artis studiosis (nam amethodos, et non diligentissimos prohibeo plane a librorum meorum lectione) placuit subiicere quae quibus locis inveniatur ad hunc modum.

61 On Wechel, see Maittaire: *Annales typographici* (as note 59) pp. 412ff. The differences between Fuchs's and Dubois's dichotomies of the same works by Galen show how difficult the art of endoxical division was: cf. the comment by their Pisan colleague Giovanni Argenterio, written about a decade later: quantum enim difficile sit bene dividere docet Aristoteles *initio operis de partibus animalium* tantoque profecto difficilior nobis est dividendi ratio, quod nemo hactenus ex an-

not only has a vast array of Galenic *loci* organised in bracketed dichotomies, but also has exhaustive combinatories such as the following:

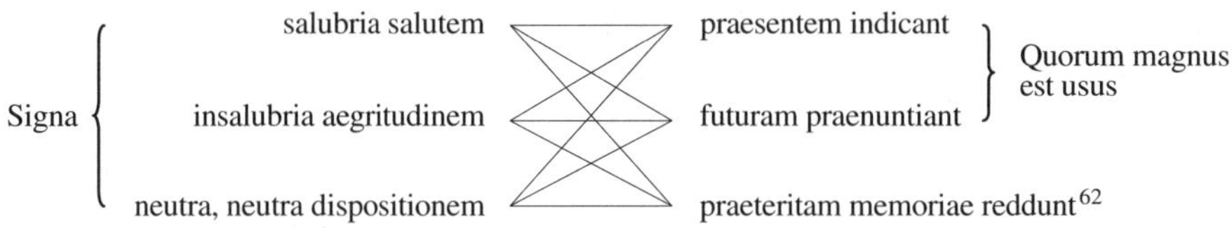

Dubois chooses not only to order in the endoxical mode miscellaneous *dicta* on signs scattered throughout the Galenic corpus, but to deal also with more restricted and even scientific (i.e. causally demonstrative) material in this work, which is only matched in its scale to my knowledge by Theodor Zwinger's (1533-1588) *In artem medicinalem Galeni tabulae et commentarii,* printed in Basle in 1561.

These and other examples had a marked effect on medical publishing. By the 1570s, all major European medical publishing centres apart from those in Spain and Italy (which for some reason seem to eschew nearly all use of the tree diagram)[63] were making free use of the dichotomous table for a variety of purposes: both for *anacephaleoses,* and for divisions of material according to both scientific and endoxical principles. Authors and publishers stressed the pedagogical usefulness of this mode of presentation, and there is even evidence that it was used in lectures.[64] A late and very ambitious use of the diagram in medical writing was that made by Ernestus Fridericus Fabricius in 1626, who sets out to reduce all of Galenic and Paracelsian medicine into

tiquis scriptoribus eam satis explicare sit conatus (*De morborum differentiis,* proemium, in: *Opera,* Hanau 1610, col. 1463).

62 Dubois: *Methodus in tabellas coniecta* (as note 52) C1r.

63 The work of Santori referred to above (see note 49) and Jacopo Zabarella's *Tabulae logicae,* Venice 1580, are rare exceptions to the rule.

64 A surviving copy of a medical disputation delivered in Wittenberg in 1601 under the presidency of Jacobus Cocus contains manuscript notes which include not only the transcription of the dichotomies found in André Du Laurens's *De crisibus* of 1596 but also a presentation by tree diagram of the disputation subject (*De signorum discretione*) by the *praeses* himself; the copy is to be found in the Universitäts- und Landesbibliothek Halle. See also note 57 above, and the claims made on the titlepage of Leonhard Fuchs's *Universae medicinae compendium,* Basle 1537.

comparative tables.[65] But just as in the case of law, discursive divisions are just as much used as visual representations,[66] and even one of the few medical texts to declare itself to be written in a Ramist mode (David Crusius's [1589-1640] *Theatrum morborum hermetico-hippocraticum seu methodica morborum et curationis eorundem dispositio* of 1615, announced in 1612 as *Idea morborum hermetico-hippocratica methodo Ramea adornata*) contains in fact only one synoptic table, and a great deal of material presented in commentary or *quaestio* form. For doctors as much as for lawyers, division is also partition and enumeration, and as such both a closed and an open class, not excluding the possibility of knowledge beyond the field delimited by the division's head term or *genus.*

From this brief survey of the evidence, it seems to me that the senior disciplines, armed as they were with Aristotelian and Galenic methodology, were more subtle and more flexible than Ramism, and had the added advantage of being reflexive (they were able to discourse about their operative concepts such as division and definition using other operative concepts, and *vice versa,* without having to create a formal metalanguage or a hierarchy of abstraction).[67] They were also able to admit of the existence of knowledge which lay beyond the confines of their *diaireses* but was not excluded from their awareness and even their use. Because law and medicine were arts, and an art was considered to be "the finite doctrine of infinite things", they could remain true to their vocations as practical systems, and at the same time aspire to theoretical formulation of their operative and logical procedures. I do not think therefore that in spite of the widespread use of the tree diagram, they owed much, if anything to Ramism; their evolution over the course of the sixteenth century suggests that a body of university teachers

[65] Fabricius: *Medicinae utriusque Galenicae et hermeticae anatome philosophica,* Hamburg 1626.

[66] Two examples of doctors eschewing the use of diagrams while using *diairesis* are Hieremias Thriverus: *Universae medicinae brevissima absolutissimaque methodus,* Leiden 1592, and Jean de Varanda: *Physiologia et pathologia,* Montpellier 1620.

[67] An example of such sophistication is provided by Capo di Vacca's *Methodus* (as note 44), in which the author makes it clear that he is both using and mentioning analytical terms. Cf. Also Giambattista da Monte's shrewd comment in his *Universa medicina* (as note 7) p. 101: Aristoteles bene dixit, absurdum es[se] simul et scientiam et modum sciendi quaerere.

of law and medicine were able to reflect on and modify the work of their predecessors without having recourse to Ramus's iconoclastic and simplistic method. And, if I am not mistaken, they could have faced up better to the challenge posed by Borges's classification of animals with which I opened this lecture by deploying Boethius's endoxical division. It is true that "innumerable", "etc." and "included in the present classification" could not appear in one of their synoptic tables; but it could appear in their account of what lay beyond it, as the residue of their endoxical division which they were able to recognise; "sucking pigs" and "stray dogs" are for them *species* of the *genus,* and "tame" a *genus subalternum;* "belonging to the emperor" is an accident defined by a subject; "embalmed", "having just broken the pot", "sirens", and "mythical" are subjects defined by an accident (if we can take non-existence, perhaps contentiously, to be an accident, in contravention of the rule *non entis nulla differentia*);[68] and "drawn by a fine camel hair brush" "quaking like madmen" and "resembling flies when seen from a distance" might be seen as accidents of accidents. Cardano, I am sure, would even have been able to accommodate this list into one of his aberrant divisions, constructed on Porphyryan rather than Boethian lines. I should of course have no excuse for this flippant *excursus,* had it not been provoked precisely by Foucault's use of Borges's list to illustrate what he took to be the unscientific taxonomy of sixteenth-century natural philosophers such as Ulisse Aldrovandi.[69] I am tempted to suggest that far from displaying the blindness of such writers, such construction of complex classes demonstrates their sophistication in that it shows them to be aware of some of the logical and taxonomic problems with which the simplified Ramist system was not equipped to deal. However great Ramus's influence on schoolroom teaching may be claimed to be, I do not myself think therefore that it contributed much, if anything, to the logical sophistication of the higher education of his day.

68 See Aristotle: *Physics* IV 8, 215a10; *Topics* IV 6, 128b9; *Metaphysics* I 9, 990b9ff.

69 Michel Foucault: *Les mots et les choses,* Paris 1966, pp. 7. 54.

A Challenge to the Reader: Ramus on *Astrologia* without Hypotheses

NICHOLAS JARDINE AND ALAIN SEGONDS

1. Introduction

In 1563 Ramus wrote to Georg Joachim Rheticus, mathematician, astronomer and astrologer, then practising medicine in Cracow. In his letter Ramus condemned all astronomical hypotheses as impious, contrary to the laws of logic, and at odds with the simple nature of the heavenly bodies.[1] The ancient Babylonians, Egyptians and Greeks before Eudoxus had, he maintained, been able to "continue the heavenly motions", that is, predict celestial phenomena, without hypotheses.[2] He urged Rheticus to devote himself to the task of creating an astronomy free of hypotheses and able to "show the light of her stars to the minds of men with the same splendour as that with which nature presents itself to our eyes for contemplation"; should Rheticus succeed in cutting this Gordian knot, he would merit not the kingdom of Asia but that of *astrologia*.[3] In his *Prœmium mathematicum* of

1 J. Th. Freigius, ed.: *P. Rami professio regia,* Basle 1576, sig. [hand] 1r-2v. Recent editions of the letter with translations: M. Delcourt: Une lettre de Ramus à Joachim Rheticus (1563), in: *Bulletin de l'Association Guillaume Budé* 44 (July 1934) pp. 3-15; K. H. Burmeister: *Georg Joachim Rhetikus 1514-1574. Eine Bio-Bibliographie,* vol. 3: *Briefwechsel,* Wiesbaden 1968, p. 176. Delcourt's version is more accurate, as noted by M.-P. Lerner: L'humanisme a-t-il sécrété des difficultés au développement de la science au XVIe siècle? Le cas de l'astronomie, in: *Revue de Synthèse,* 3e sér., nos. 93-94 (1979) pp. 48-71, n. 52. A translation with corrections to Delcourt's text is in *Georgii Joachimi Rhetici narratio prima,* ed., transl., comm. H. Hugonnard-Roche and J.-P. Verdet with M.-P. Lerner and A. Segonds, Wrocław 1982 [Studia Copernicana XX] pp. 238-245.

2 Delcourt: Une lettre de Ramus (as note 1) p. 11.

3 Ibid. p. 15.

1567, in the context of a similar but more verbose account of the history of ancient astronomy, Ramus reiterated his censures of hypotheses, his attribution to the ancients of the capacity to predict celestial phenomena without them, and his challenge to restore hypothesis-free astronomy.[4]

> So the fabrication of hypotheses is absurd. However, the fabrication is quite simple in Eudoxus, Aristotle and Calippus, who thought the hypotheses were true: indeed, they were venerated as gods of the orbs without stars. But in later times there was a far more absurd tale of demonstrating the truth through false causes. For this reason logic, in the first place, then the mathematical elements of arithmetic and geometry would provide the greatest assistance in establishing the purity and dignity of this magnificent art. And would that Copernicus had rather set his mind on the establishment of such an astronomy without hypotheses. For it would have been easier by far for him to describe an astronomy corresponding to the truth of its stars, than as if with the labour of a giant to move the earth, so that in consequence of the motion of the earth we might observe the stars at rest. Why does there not rather arise from so many noble schools of Germany someone who is at once a philosopher and a mathematician to gain amidst eternal praise the prize that is wagered. And if a benefit of transitory use can be proposed as a reward of such virtue, I will promise you a Regius professorship at Paris as the prize for an astronomy without hypotheses. I shall most happily fulfil this promise, even by the giving up of my own professorship.[5]

To this he added a specific appeal to August, Elector of Saxony, to commission some "new German Ptolemy or Copernicus" to restore the ancient astronomy, consisting simply of generalisations about what is constant in the heavens derived logically from observations.[6]

Ramus' plea for an astronomy without hypotheses achieved considerable notoriety. It is delightfully commemorated in Johannes Thomas Freigius' epitaph prefaced to his edition of Ramus' *Praelectiones in Ciceronis orationes octo consulares:*

> In terris Ramus mortalia corpora linquens,
> Venit ubi ad summi cœlica tecta Dei.
> O stulti Astronomi, quae vestra insania dixit?
> In domibus nusquam est circulus aethereus.[7]

4 *Procœmium mathematicum,* Paris 1567, pp. 210-217; a slightly emended version is in *Scholae mathematicae,* Paris 1569, pp. 49-50.

5 *Procœmium mathematicum,* pp. 216-217; *Scholae mathematicae,* p. 50. This is the passage quoted by Kepler in his *Astronomia nova* of 1609 (see below).

6 *Procœmium mathematicum,* pp. 299-300; *Scholae mathematicae,* pp. 70-71.

7 Basle 1573, unnumbered page before 45 bis.

But these are puzzling texts, posing at once literal challenges and – by reason of their enigmatic claims about the ancient astronomy, their puzzling interpretations of classical sources, and their obscure pronouncements on the goals of astronomy – hermeneutic challenges, both for us and, as we shall see, for Ramus' contemporaries.[8]

In this paper we shall start by relating Ramus' programme for the renewal of the arts to his astronomical challenges. In the light of Ramus' classical sources, we shall then speculate briefly on the content of Ramus' ancient astronomy. In the body of the paper we shall look at the extraordinarily diverse responses of Rheticus, Christoph Rothmann, Tycho Brahe, Nicolaus Ursus, and, finally, at Kepler's elaborate and playful claim to have met Ramus' challenge in full. The paper closes with some reflections on challenges, responses and open texts in the early modern period.

2. *Astronomy Without Hypotheses*

What was Ramus' astronomy without hypotheses? Why did he believe the ancients to have possessed it? And by what means did he envisage it as being restored? Partial answers are, we think, to be got by considering Ramus' pronouncements on astronomy and its history against the background of his general views on the reform of logic and the arts.[9] For Ramus the precepts or rules of an art are rules for its exercise and practice. Thus the theorems of geometry are rules for measuring, the truths of arithmetic are rules for counting, and the doctrines of theology are rules for living well.[10] Whatever in an art does not contribute to its practice is superfluous and to be rejected.[11]

8 On the astronomy without hypotheses see Lerner: Humanisme (as note 1) pp. 59-63; E. Aiton: Johannes Kepler and the astronomy without hypotheses, in: *Japanese Studies in the History of Science* 14 (1975) pp. 49-70; R. Hooykaas: *Humanisme, science et réforme. Pierre de la Ramée (1515-1572),* Leiden 1958, ch. 9; N. Jardine: *The Birth of History and Philosophy of Science. Kepler's A Defence of Tycho Against Ursus, with Essays on its Provenance and Significance,* Cambridge 1984, pp. 234-236.

9 The following remarks draw on Hooykaas: ibid. ch. 5.

10 See the passages from Ramus cited by Hooykaas: ibid. pp. 24-25.

11 The second of Ramus' famous laws of method (adapted from Aristotle's *Posterior Analytics*), the *lex kath'auto* or *lex homogenei,* required exclusion from an art of everything irrelevant. On Ramus' logical laws see W. J. Ong: *Ramus. Method, and*

Men have an innate capacity for the arts, their precepts being formed in the mind through the operations of logic or "natural reason" from observation and experience. Thus the arts are best learned not from books, but from the company of skilled artisans (if books must be studied, then the works of poets and orators are preferable to academic treatises and disputations). All the arts had been practised to perfection among the Jewish patriarchs, passed on *via* Abraham to the Egyptians, and thence acquired by the Greeks. Far from having been improved by book learning in later ages, the arts have been impoverished by it.[12] Ramus' basic claims about astronomy – that it was perfectly known to the ancients, that since hypotheses are unnecessary for the practice of astronomy they should be rejected, that the precepts of astronomy are derived by logic from experience – are therefore of a piece with his general views on the history and nature of arts.

However, the astronomy without hypotheses remains puzzling. Did Ramus suppose, like Francesco Patrizi a generation later, that the planets move of their own accord in a liquid ether precisely as they appear to move to us earthlings?[13] Did he, as Christoph Rothmann thought, merely reject epicycles and eccentrics, accepting a simple system of planetary orbs as in the pseudo-Aristotelian *De mundo*?[14] Was he, perhaps, as Reijer Hooykaas has

the Decay of Dialogue, Cambridge, Mass. 1958, pp. 258ff.; N. Bruyère: *Méthode et dialectique dans l'œuvre de La Ramée,* Paris 1984, ch. 3; G. Oldrini: *La disputa del metodo nel Rinascimento. Indagini su Ramo e sul ramismo,* Florence 1997.

12 In *Scholae mathematicae* (as note 4) pp. 99-100, Ramus argues that all the arts were instituted by logic in ancient Europe, Africa and Asia; that this natural logic has been almost lost; and that it is best learnt from "poets, historians, orators, writers of georgics, and jurisconsults, and indeed the holy prophets". On the tradition of mathematical arts from the ancients see Ramus: *Proœmium mathematicum* (as note 4) pp. 2-17; *Scholae mathematicae* (as note 4) pp. 1-5, which draws on Pliny, Diodorus Siculus and Josephus' *Jewish Antiquities;* also *Petri Rami Professoris Regii, et Audomari Talaei collectaneae praefationes, epistolae, orationes,* Paris 1577, pp. 234ff.

13 F. Patrizi: *Nova de universis philosophia,* Ferrara 1591. On Patrizi's cosmology see B. Brickman: *An Introduction to Francesco Patrizi's 'Nova de universis philosophia'*, New York 1941; E. Rosen: Francesco Patrizi and the celestial spheres, in: *Physis* 26 (1984) pp. 305-324; M.-P. Lerner: *Le monde des sphères, II: La fin du cosmos classique,* Paris 1997, pp. 104-108.

14 "[...] You appear to believe in real concentric orbs, not indeed as many as Eudoxus and the rest, but only seven or eight, just as Aristotle too did in his book *De mundo*

intimated, a crypto-Copernican?[15] Or did Ramus, as one of us has argued, reject outright the quest for hidden causes prior to the phenomena?[16] This last view has the advantage of allowing us to take at face value Ramus' earliest and least ambiguous account (in his *Oratio de professione liberalium artium* of 1563) of the astronomy without hypotheses as an *historia* and of its methods as geometrical measuring and arithmetical calculation.[17] It fits with Ramus' denial of the Aristotelian distinction between what is prior or better known to us and what is prior in nature; and it comports well with Tycho Brahe's assessment of Ramus as unaware of the extent of irregularity in the celestial phenomena.[18] However, it faces the not inconsiderable difficulty that Ramus does, on occasion, mention celestial physics, citing Ptolemy and Copernicus as sources of "solid" doctrine, which comports uneasily with the denial of all hidden causes of celestial phenomena.[19]

But there is something a bit fishy about these attempts to tie down the sense of Ramus' astronomy without hypotheses. For, as we shall see, Ramus' sixteenth-century readers attributed widely, indeed wildly various significance to Ramus' riddling pronouncements. With our speculations we are, it seems, rising too easily to Ramus' bait.

ad Alexandrum, ch. 2". Rothmann: *Observationum stellarum fixarum liber primus* (Murhardsche Bibliothek der Stadt Kassel, 2' Ms. Astr. 5 nr. 7) fol. 56r, ed. M. A. Granada: L'eliminazione delle sfere celesti e ipotesi astronomiche in un inedito di Christoph Rothmann. L'influenza di Jean Pena e la polemica con Petrus Ramus, in: *Rivista di storia della filosofia* 52 (1997) pp. 785-821. The reference is to *De mundo* 329a15-24.

15 Hooykaas: *Humanisme* (as note 8) pp. 71-72.

16 Jardine: *History and Philosophy of Science* (as note 8) pp. 234-236.

17 Ramus: *Oratio de professione liberalium artium* [Paris 1563], in: *Rami et Talaei collectaneae praefationes, epistolae, orationes* (as note 12) pp. 512-529. p. 526: [...] esset enim Astrologia, si locum sideris cujusque & motum geometricè, si motum & tempus arithmeticè, è nudis ac simplicibus experientiis metiatur, non multo difficilior mathemata tenenti, quâ quaevis historia [...] sublatis infinitarum tabularum centuriis, totoque astrologico opere ad geometricum radium & arithmeticum calculum revocato.

18 See below.

19 *Rami et Talaei collectaneae praefationes, epistolae, orationes* (as note 12) p. 96: physica materia de rebus cœlestibus solida est in Ptolemæo & Copernico [...].

3. Rheticus' Accommodation

In 1568, that is, eight years after the despatch of Ramus' challenge, Rheticus finally responded:

> In the fourth place, I come finally to that task which you too had in mind, that I should liberate [reading *liberem* with Simler, not *liberarem* with Burmeister] the astronomical art from hypotheses, being content with observations alone. And if only we possessed the observations of all ages, and the same carried out according to our method of making observations, which I judge to be entirely the same as those which were used by the first investigators of our art; then we might find out such a means of tabulation as would not be in need of perpetual emendation.
>
> To these things we shall add tables of uneven motions, from which we derive the place of the stars and all phenomena with the same facility as from ephemerides themselves. In all these things I am far more hostile to Ptolemy than you to Euclid, and I beat him more than you Euclid. For just as the little houses which small boys build of mud and sand are to the buildings of Vitruvius or the palaces of flourishing Rome, so the great constructions of Ptolemy, which you may rather call the great destructions, relate to the true and solid doctrine concerning the motions of the stars, which you indeed call the astronomy of the Egyptians, who treated these sciences with their measuring rods (which the Greeks from ignorance called obelisks) evidently inspired by the divine mind.[20]

Ramus' exchange with Rheticus has occasioned much puzzlement. Was Ramus a crypto-Copernican? Had Rheticus abandoned his Copernicanism?[21] Simpler explanations are, however, to hand. From his letter it is evident that Ramus' hopes for a reform of astronomy at the hands of the Copernican Rheticus rested in part on the computational skills displayed in Rheticus'

20 J. Simler: *Bibliotecha instituta et collata a Conrado Gesnero,* Zurich 1574, p. 228; Burmeister: *Rhetikus* (as note 1) vol. 3, p. 188. Simler does not print the entire letter, of which the original is not now extant, though there is an eighteenth-century ms. copy in Cracow (Biblioteka Jagiellońska: ms. 925, fol. 5), which we have not seen. The date of the letter is given by Simler as 1568, as conjectured in Hugonnard-Roche's and Verdet's edition of Rheticus' *Narratio prima* (as note 1) p. 246.

21 Burmeister (*Rhetikus,* as note 1, vol. 1: *Humanist und Wegbereiter der modernen Naturwissenschaften,* Wiesbaden 1967, pp. 161-162) has read Rheticus' letter as marking his move beyond Copernicus to an astronomy entirely free of hypotheses. For balanced assessments see R. Hooykaas: *G. J. Rheticus' Treatise on Holy Scripture and the Motion of the Earth,* Amsterdam 1984, ch. 7: "Rheticus, Ramus, and the Copernican hypotheses"; R. S. Westman: The Melanchthon Circle: Rheticus and the Wittenberg Interpretation of the Copernican Theory, in: *Isis* 66 (1975) pp. 165-193.

Canon.[22] Further, he may well have known that Rheticus shared his belief in the superiority of ancient Egyptian astronomy. He could have known this from Rheticus' introduction to his edition of Werner's books on spherical triangles; and he could perhaps have guessed it from the titlepage of Rheticus' *Canon* which figures an obelisk of the kind he believed the Egyptians to have used as a gnomon.[23] Finally, it is evident that Ramus mis-attributed to Rheticus the infamous anonymous preface to Copernicus' *De revolutionibus;* and Ramus certainly shared the view expressed in that preface that only a fool would accept the truth of astronomical hypotheses, though he did not endorse its treatment of them as indispensable for purposes of calculation of the celestial motions.[24]

In the light of all this, Rheticus' response becomes more comprehensible. Hooykaas has noted that Rheticus "was always eager to please his correspondents";[25] and a reading of his correspondence confirms that he was indeed a master of accommodation of his beliefs to those of others as to divine writ. And in this case, addressing a famous and senior figure, who had tried to secure for him an appointment in Paris, he had good reasons to be accommodating.[26] Accordingly, in his response he confined himself to points of agreement with Ramus, namely the need for accurate astronomical tables not in need of perpetual emendation, and the opposition to

22 "By the publication of [your] *Canon* you have excited extraordinary expectation of renown [...] and since I have entered once already into the treasures of your mind, hear what further *largesse* of this kind I require you to give with your habitual munificence." Delcourt: Une lettre de Ramus (as note 1) p. 7. Rheticus' *Canon doctrinae triangulorum* had been published in Leipzig in 1551.

23 *Ioannis Verneri De Triangvlis Sphæricis libri Qvatvor* [...] *Nunc primum Studio & Diligentia Georgii Ioachimi Rhetici in lucem editi,* Cracow 1557; facsimile: Abhandlungen zur Geschichte der mathematischen Wissenschaften XXIV, Leipzig, 1907. The Egyptian obelisk on the titlepage, destroyed in riots in the year of Rheticus' death, was the one placed by Augustus in the circus in Rome; see Hooykaas: *Rheticus* (as note 21) pp. 152-153.

24 "And unless I am mistaken your letter prefaced to Copernicus shows clearly [...]". Delcourt: Une lettre de Ramus (as note 1) p. 13. That Ramus was mistaken was pointed out by E. Rosen: The Ramus-Rheticus correspondence, in: *Journal of the History of Ideas* 1 (1940) pp. 363-368.

25 Hooykaas: *Rheticus' Treatise* (as note 21) p. 162.

26 On Ramus' invitation of Rheticus to work in Paris see Rosen: Ramus-Rheticus (as note 24) pp. 367-368.

Ptolemy's hypotheses. Indeed, practising a modicum of mental reservation by taking "hypothesis" with the connotation, widespread in the period, of false or illegitimate postulate, he was even able to endorse the project of liberating astronomy from hypotheses. On the points of disagreement – the truth of a heliocentric cosmology, the fact that he had not written the sceptical preface to Copernicus' *De revolutionibus,* the fact that he had considerable admiration for Ptolemy – he remained diplomatically silent.[27]

4. Tycho's Denigration

We know of Ramus' exhortation of Tycho, at a meeting in Augsburg in 1570, to construct an astronomy without hypotheses only through Tycho's report of the matter in a letter to Rothmann of 1587. Following a reference to his own "reasoning concerning the disposition of the celestial revolutions" differing from both the Ptolemaic and the Copernican, Tycho declares:

> What that most celebrated philosopher of our age, Petrus Ramus, supposed, namely, that astronomy could be established without hypotheses through logical reasoning, is without foundation. He did indeed propose this view to me sixteen years ago, when we were in Augsburg, and at the same time he urged me, when I had brought exact order to the courses of the stars through hypotheses, to attempt the same without them. He added as a reason that this could be done, that he had read that the Egyptians once possessed very easy astronomical knowledge, and that, since the *ratio* of hypotheses seems difficult and intricate, they must have known the courses of the stars in some quicker and simpler way, and therefore without any hypotheses. But I resisted him, showing that the celestial phenomena cannot be rendered into certain science without hypotheses, nor explained so as to be understood.[28]

Tycho goes on to downplay the astronomical knowledge of the Egyptians as limited to certain recurrences in the celestial phenomena, and to disparage Ramus as one who "seemed not to have seen far into the recesses of this art, and not to have noticed the variety in the motions of the star in no way re-

27 In his *Narratio prima,* Danzig 1549, fol. Ci,r-v, Rheticus had written of Ptolemy's "indefatigable diligence in calculating, superhuman accuracy of observations, and truly divine method of investigating all the motions and appearances".

28 J. L. E. Dreyer, ed.: *Tychonis Brahe Dani opera omnia,* Copenhagen 1913-1929, vol. 6, pp. 88-89.

curring at fixed times of the year".[29] Slight though Tycho's response may be, it is damaging. For it really does seem from Ramus' remarks on the astronomy of the ancients that he supposed that a predictively adequate astronomy was to be got through a compilation of regularities in the apparent motions.[30]

5. Rothmann's Refutation

In his *Procemium mathematicum* Ramus had expressed great admiration for the astronomical observatory and instruments of Wilhelm, Landgraf of Hessen-Kassel.[31] According to the Landgraf's mathematician, Christoph Rothmann, Wilhelm admired Ramus as a philosopher;[32] and Calvinist Marburg, the site of Hessen-Kassel's newly founded university, was indeed a centre of Ramist teaching and publication.[33] This mutual admiration failed, however, to deter Rothmann from a vigorous denunciation of Ramus' astronomy without hypotheses.

The refutation is to be found in Rothmann's unpublished *Observationum stellarum fixarum liber primus,* in the chapter entitled "That the matter of

29 Ibid. p. 89.

30 It should, however, be remembered that Tycho was in the habit of denigrating others, including such competent mathematicians as Paul Wittich, Nicolaus Ursus and Duncan Liddel: see, for example, Tycho's dedicatory preface to his *Epistularum astronomicarum liber primus* of 1596, in: *Opera omnia* (as note 28) vol. 6, p. 12; and Tycho's letter to Kepler of December 1599, in: *Johannes Kepler: Gesammelte Werke,* ed. W. von Dyck, M. Caspar, Munich 1937ff., vol. 14, pp. 89-98.

31 *Prooemium mathematicum* (as note 4) p. 287; *Scholae mathematicae* (as note 4) p. 67.

32 "So now then, Peter Ramus, philosopher, in the judgment of our prince by far the most excellent in our century [...]". Rothmann: *Observationes stellarum* (as note 14) fol. 55v.

33 Marburg was one of the three major German centres for the printing of Ramist textbooks: see Ong: *Ramus* (as note 11) p. 296. Notable teachers of Ramist doctrines at Marburg included Lazarus Schöner (professor of mathematics 1575-1578), Hieronymus Treutler (professor of rhetoric 1591-1593), and Rudolph Goclenius the Elder (professor of logic 1581-1589, professor of physics 1589-1603); see F. Gundlach: *Catalogus professorum academiae marburgensis,* Marburg 1927. Goclenius was not, as indicated by W. J. Ong: *Ramus and Talon Inventory,* Cambridge, Mass. 1958, p. 519, a follower of Ramus; but he did teach the dialectics of Ramus and Melanchthon, together with Aristotelian logic; see W. Risse: *Die Logik der Neuzeit,* vol. 1: *1500-1640,* Stuttgart - Bad Cannstatt 1964, pp. 156ff.

the celestial spheres is not a solid which carries around the firmly attached and embedded planets, but that it is subtle and liquid [matter], which easily yields to the motions of the planets: and what view one should take of astronomical hypotheses".[34]

Rothmann opens with an accurate point-by-point account of Ramus' positions, and proceeds to a point-by-point demolition. First he concedes that "hypotheses of eccentrics and epicycles are false and absurd fabrications", at least if Ramus intends real orbs – and he insists that though Eudoxus and others did postulate them, they were not introduced either by Ptolemy or Copernicus[35] (Rothmann may have been aware that Ramus himself regarded Ptolemy and Copernicus as authorities in matters of celestial physics).[36] However, Ramus' rejection of all hypotheses shows inadequate understanding of astronomy. Hypotheses in the form of geometrical and arithmetical descriptions of the motions of bodies are indispensable for astronomy as for other mathematical disciplines.[37] Rothmann then turns to Ramus' appeal to a fully adequate ancient astronomy without hypotheses:

> Nor is it an obstacle that astronomy was once without hypotheses and nevertheless eclipses were predicted. For how meagre and how crude it was appears not only from the beginning of this chapter [see fol. 51v: *fuisse enim veterum astronomia ...*] and from the motion of the moon in eighteen- and nineteen-year periods and the period of Calippus, constantly and repeatedly corrected [*semper magis ac magis sucessive correcto*]; but also indeed from book 9, ch. 2 of *Megalé syntaxis,* where Ptolemy from Hipparchus and others reports that the mathematicians before Hipparchus demonstrated the motions of the five planets as if in the case of one single anomaly and equal [literally, in Ptolemy's Greek, "direct"] motion, as if they possessed one inequality only, that which depends on the sun. So, at that time eclipses were predicted by any old means, as also was the motion of the luminaries; and it is not unknown that this can be done by those moderately versed in this art. Nor did they bother about partial eclipses or exact precision; rather, they thought it sufficiently divine that these things could be predicted in whatever manner, and if places of the remaining planets were predicted within ten or more degrees.[38]

34 Rothmann: *Observationes stellarum* (as note 14) fol. 51v.

35 Ibid. fol. 56r.

36 Cf. note 19 above.

37 Rothmann: *Observationes stellarum* (as note 14) fol. 56r-v.

38 Ibid. fol. 56v.

Next Rothmann tackles Ramus' unfavourable comparison of astronomy with grammar, rhetoric, dialectic, and other liberal arts. Hipparchus and Ptolemy, the princes of mathematicians, derived their hypotheses about the "forms of the motions" of the heavenly bodies from observations, just as the principles of other arts are derived from observations.[39] And the relatively greater difficulty in restoring astronomy after a book-burning would arise not from the presence of hypotheses in astronomy, but from the time needed to secure a sufficient body of observations from which to derive hypotheses.[40] Against the charge of Proclus, as adduced by Ramus, that astronomers proceed topsy-turvy, deriving conclusions from hypotheses rather than *vice versa,* Rothmann proceeds by pointing out other instances of Proclus' incompetence in astronomy, for example his claim that the shadows of the gnomons increase before the summer solstice but decrease before the winter solstice.[41] He also notes that he cannot find the material cited by Ramus in Proclus' *Hypotyposis,* and speculates that his own copy may be defective (rightly so, since he was evidently using the defective translation by Giorgio Valla rather than the *editio princeps* of Grynaeus).[42] Rothmann next suggests that Ramus himself is inconsistent in his attitude to hypotheses, for in defending (against Aristippus) the use of arithmetic and geometry to measure the celestial motions he admits the very use of hypotheses that he elsewhere rejects.[43] The peroration, in which Rothmann first intimates that Ramus, once better educated, would convert to his view, and then claims the prize, is a *tour de force* worthy of quotation in full:

> Indeed, because I have seen that you lack the doctrine of triangles, I have time and again heartily wished that you had it, especially when I read your commentaries on the Ciceronian *Somnium Scipionis.* For without it we are indeed children in this art, nor do we see more than the blind themselves. Nor should that most excellent mathematician of all ages, Copernicus, have been neglected by you, although he in-

39 Ibid. fols. 56v-57r.

40 Ibid. fol. 57r.

41 Ibid. fol. 57r-v. Proclus: *Hypotyposis* VII 42, ed. C. Manitius, Leipzig 1909, pp. 233-234. In citing Proclus' errors, Rothmann quotes roughly from Giorgio Valla's Latin translation (Venice 1501), appended to *Claudii Ptolemæi Pelusiensis Alexandrini omnia qua extant opera,* Basle 1551.

42 Simon Grynaeus, ed., Basle 1540. The passage to which Ramus had referred is missing from the garbled translation by Valla.

43 Rothmann: *Observationes stellarum* (as note 14) fol. 57v.

troduced the motion of the earth. He alone would have deterred you from behaving blindly like a mole. But I do not doubt, my good Ramus, that you are going to be on our side, and that now after all this time you will, because of your love of truth, change your opinion concerning the removal of hypotheses, and that you will be sufficiently persuaded that I have removed so many axes, so many real orbs, and that kind of monstrosities, which enormously obscured astronomy. And if prizes are still available for virtue, I have invented a way by which the motions of all the planets could very easily be calculated not by means of tables, but without astronomical tables, only from the sources of geometry, using the doctrine of triangles by addition and subtraction alone: so that you would not any more have to complain about those hundreds of tables. But we will deal with this matter in due course, if we find that gratitude and benevolence are still surviving in this world.[44]

In form as in substance this is a powerful counter-blast. It is dialectically expert, making cogent use of the strategies proper to a formal *refutatio*. Thus Ramus is successively convicted of ambiguity (in his use of the term *hypothesis*), of ineptitude (in the art of astronomy in general, and in the doctrine of triangles in particular), of false analogy (in his comparison of the methods and principles of astronomy with those of the other liberal arts), of appeal to incompetent authority (Proclus), and of self-contradiction (concerning hypotheses and the uses of mathematics in astronomy).[45] In fact, this is a truly Ramean refutation of Ramus, the style and tone being strikingly reminiscent of Ramus' demolitions of Aristotle.[46]

As for content, Rothmann effectively uses Ramus as a foil for a cogent and original statement of his own view that the primary goal of astronomy is not to seek out causes, but to establish hypotheses in the form of accurate mathematical descriptions of the true motions of the heavenly bodies.[47]

44 Ibid. fol. 58r.

45 Many of the strategies proper to refutations in the period derive ultimately from Aristotle's *De sophisticis elenchis.* On the history of the doctrine of fallacies see C. L. Hamblin: *Fallacies,* London 1970.

46 We are particularly reminded of the manner of Ramus' *Scholarum physicarum libri octo,* Paris 1565.

47 On Rothmann's Copernicanism and his rejection of material spheres see B. T. Moran: Christoph Rothmann, the Copernican theory and institutional and technical influences on the criticism of Aristotelian cosmology, in: *Sixteenth Century Journal* 13 (1992) pp. 85-108; B. R. Goldstein, P. Barker: The role of Rothmann in the dissolution of the celestial spheres, in: *British Journal for the History of Science* 28 (1995) pp. 385-403; M. A. Granada: *El debate cosmológico en 1588. Bruno, Brahe, Rothmann, Ursus, Röslin,* Naples 1996, ch. 3; Lerner: Humanisme (as note 1) pp. 58-66.

6. Ursus' Invective

The setting of the onslaught by Ursus (Nicolaus Reymers Baer) is a scurrilous work, *De hypothesibus astronomicis* [...] *tractatus* [...], published without a licence in Prague in 1597. In this work the Imperial Mathematician defended himself against Brahe and others who had charged him with plagiarism of the Tychonic world system.[48] In this work Ursus' stance is sceptical, and he adopts one of the positions rejected by Ramus, namely that hypotheses are fictions indispensable for the purpose of saving the celestial phenomena.

> Now it is evident that without the assistance and help of such fictitious suppositions, or hypotheses, there is no way in which the motions of the heavenly bodies can be kept track of, the appearances of the heavenly motions saved, or, finally, the calculations of the appearances carried out; even though certain men ignorantly affirm and state the contrary, dreaming and pretending about I know not what antique astronomy of the ancient Chaldaeans and Egyptians. So the madness and insanity of certain men is to be hissed off the stage and laughed out of court – men who rashly deny the need for hypotheses and affirm that astronomy can be altogether established, completed and constructed by logical means alone (why, I ask, don't they do it, if they are able to?), as if logic alone would suffice for understanding everything and there were no need for the individual arts, arts for whose handing on logic ought to be considered only as a catholic organon, that is, the general and universal tool, since a distinction ought to be maintained between an instrument (*instrumentum*) and a doctrine (*documentum*), that is, between the form and matter of philosophy, nor ought everything to be compounded and mixed together and every art stuck into one jar labelled "logic".[49]

Though Ramus is not mentioned, the content and wording of the attack – notably the reference to the establishment of astronomy by logical means alone, and the echoing of Ramus' use of the term *documentum* for the doctrines obtained by the *instrumentum* of logic – leave no doubt that Kepler was right in identifying Ramus as the target.[50] Ursus goes on to concede the

48 On Ursus' attack on Brahe and its background see Jardine: *History and Philosophy of Science* (as note 8) ch. 2; Granada: *El debate cosmológico* (as note 47) ch. 4.

49 Ursus: *De hypothesibus astronomicis* [...] *tractatus* [...], Prague 1597, sig. Ci,v.

50 Ramus: *Prœmium mathematicum* (as note 4) p. 212; *Scholae mathematicae* (as note 4) p. 49: "Grammaticae, Rhetoricae, Logicae, Arithmeticae, Geometriae, atque omninò praeter astrologiam, cujusvis disciplinae principia sunt experientiae & observationes primae in generalia & universalia documenta à logicis artificibus collectae & inductae." Kepler identified Ramus as the target in his *Apologia Tychonis contra Ursum,* ms. Pulkowo V, fol. 265r, ed. Jardine: *History and Philosophy of Science* (as note 8) p. 86.

superiority of ancient learning in astronomy as in medicine and music – a puzzling concession, since in the history of hypotheses which follows, and incidentally draws quite heavily on Ramus, he speculates that the most ancient astronomy used a simple but inadequate system of concentric planetary spheres.[51] He then promises to restore the ancient astronomy and to produce a sample of its foundations at the end of the work, a sample which turns out not to consist of astronomical hypotheses at all, but rather of techniques for the calculation of planar and spherical triangles.[52]

Returning from this not altogether coherent digression, Ursus reasserts the necessity of hypotheses in astronomy:

> But, to return to the point, I declare that no argument could lead me to believe that without hypotheses that old astronomy of the ancients was preserved and handed on, and the calculation of the heavenly motions duly carried out. For, I ask you, what sort of astronomy would there be if that Platonic pair of wings, I mean arithmetic and geometry, were taken away, cut off or rejected? Surely it would be lame, mutilated and altogether imperfect. For in the calculation of the heavenly motions one has to know the periodic time and hence the periods themselves. But these very periods are most suitably portrayed and represented by imaginary circles. And, to conclude, those very circles and their mechanical delineation comprise and constitute the hypotheses themselves. Behold, the utter necessity of hypotheses is now clear and evident.[53]

7. Kepler Claims the Prize

In a letter of October 1597 to his former teacher at Tübingen, Michael Maestlin, Kepler jokingly claimed that he was declining his appointment to Ramus' Regius professorship at Paris, offered to him on the strength of his recently published *Mysterium cosmographicum.* He explained his entitlement as follows:

> If Ramus wants exterminated hypotheses which, though believed, are postulated, not proved, and if he praises that astronomy without hypotheses which is content with the apparatus of celestial orbs of nature alone, as indeed [in passages] before and

51 Ursus: *De hypothesibus* (as note 49) sig. Ci,v-Ci,r. Ursus' *Logicae metamorphosis,* Strasbourg 1589, treats Ramist logic with respect, combining elements of it with Aristotelian precepts.

52 Ursus: *De hypothesibus,* sig. Cii,r. sig. Hiv,v-Hv,v.

53 Ibid. sig. Cii,r.

> after he seems altogether to imply, then I, or Copernicus, or both together, have won, and the Ramean professorship is owed to us. If, on the other hand, Ramus altogether rejects all hypotheses, whether true and natural or false, then he is what I said above: namely, a fool, and that, I think, in your judgment too. Truly, to preserve the honour of both of us, I prefer to call myself Regius Professor than Ramus a fool.[54]

Humourless as ever, Maestlin failed to get the joke, replying that he had as yet heard nothing about Kepler's appointment.[55] Remorselessly jocular, Kepler responded with a quibble:

> At first sight your letter greatly worried me, most excellent teacher. For when I saw that you had taken seriously what I had written as a joke about the proffered Ramean professorship, I could not but expect that you by way of congratulation would communicate it to others. Accordingly, so that you get the joke properly, I shall embody the matter in an argument.
>
> Whoever constructs an astronomy without hypotheses, he can claim a Regius professorship from Ramus. I construct an astronomy without hypotheses. So I can claim from Ramus a Regius professorship.
>
> The major I prove from Ramus' magnificent promise, which is to be found in the cited folio and book. The minor I prove by dilemma. For Ramus either wants an astronomy altogether without hypotheses, or without such hypotheses as cannot be demonstrated. If the former, Ramus is a fool. If the latter, clearly Copernicus has such hypotheses, which I have demonstrated by my discovery, and thus I have satisfied Ramus. From this lucid account I think you will understand the course of the joke. Naturally in the midst of anxious contemplations aroused in me concerning the contradictions of the entire world, it pleased me to digress in this pleasant dream about the Regius professorship when by chance I came across this passage of Ramus' *Scholae mathematicae.*[56]

A decade later, Kepler published, on the reverse of the titlepage of his *Astronomia nova,* Ramus' pledge of his chair for an astronomy without hypotheses together with his own claim for the redemption of the pledge.

> Conveniently, Ramus, you have forsaken this pledge by departing from both life and the professorship, which if you now held I would indeed claim as mine by right, as I shall prove by this work even according to the judgment of your own logic. You request help for the magnificent science only from logic and mathematics; I ask you not to exclude physics, which it can in no way do without. Indeed, unless I am mistaken you readily grant [this], for in fact you surround your constructor with philo-

54 Kepler: *Gesammelte Werke* (as note 30) vol. 13, p. 141.

55 Ibid. p. 151.

56 Ibid. pp. 164-165.

> sophy as well as mathematics. With the same readiness, now hear philosophy defending a thing most absurd indeed to the multitude, not with a gigantic effort, but with the best reasons: and in so doing she does nothing new, nothing immoderate, but performs the office for which she was invented.
>
> It is a most absurd tale, I declare, to demonstrate natural things through false causes: but the tale is not in Copernicus, for he considered his hypotheses to be true no less than those ancients of yours considered theirs: nor did he only think them, but also demonstrated them to be true: as witness I offer this work.
>
> Don't you want to know the author of this tale, at which you are so greatly annoyed? Andreas Osiander is named in my copy, in the hand of Hieronymus Schreiber of Nuremberg. So this Andreas, when he had charge of the publication of Copernicus, placed at the front of the book that preface which you declare most absurd, but he (insofar as one can gather from his letters to Copernicus) thought most prudent, Copernicus himself being either already dead or entirely unaware. So Copernicus does not mythologise but seriously paradoxologises, that is, philosophises, which you require in an astronomer.[57]

Kepler, serious as always in his joking, was punctilious in meeting the terms of Ramus' challenge. In particular, he maintained that heliocentric astronomy and the cosmographical mystery of the interposition of the planetary circuits in the nested regular solids had formed a part of the esoteric doctrine of the Pythagoreans. In the poetic greeting to the reader on the reverse of the titlepage of his *Mysterium cosmographicum* the mystery is attributed to Pythagoras,[58] and in his *Apologia Tychonis contra Ursum* (written 1600/01, but not published until the nineteenth century), Kepler argued that in *De caelo* Aristotle had unwittingly transmitted Pythagorean heliocentrism.[59] In *Paralipomena ad Vitellionem* of 1604 Kepler attributed Copernican astronomy to the "Pythagorean" Euclid on the strength of the propositions of Euclid's *Optics* relating perceived motion to the motion of the observer.[60]

57 Kepler: *Gesammelte Werke* (as note 30) vol. 3, p. 6, quoted from Ramus: *Scholae mathematicae* (as note 4) p. 50.

58 Kepler: *Gesammelte Werke* (as note 30) vol. 1, p. 4: Hic te Pythagoras docet omnia quinque figuris / Scilicet exemplo docuit, nos posse renasci.

59 Kepler: *Apologia Tychonis* (as note 50) fols. 280r-281v, ed. Jardine: *History and Philosophy of Science* (as note 8) pp. 103-104. On Kepler's interpretations and translations of passages from *De caelo* see N. Jardine, A. Segonds: Kepler as reader and translator of Aristotle, in: *Conversations with Aristotle,* ed. C. Blackwell, S. Kusukawa, Aldershot 2000, pp. 205-232.

60 Kepler: *Gesammelte Werke* (as note 30) vol. 2, pp. 284-285.

In his *Harmonice mundi* of 1619 Kepler went further. Proclus' *Commentary on the First Book of Euclid's Elements* had presented the treatment of the regular solids in Book X as the goal of the entire work and he had attributed to the Pythagoreans the view that the regular solids have a special cosmic significance;[61] and in his *Scholae mathematicae* Ramus had denounced Euclid's Book X and the "Pythagorean superstition" attached to it by Proclus as confused and unworthy of inclusion in the work.[62] In the introduction to *Harmonice mundi,* Kepler first quotes the offending passage in Ramus, then proceeds to defend Euclid and Proclus, his principal classical authority for the cosmic role of the five regular solids, against Ramus' slurs. Kepler chides Ramus for having failed to make enough effort to understand Book X.[63] He then reports proudly how the reading of his *Mysterium cosmographicum* has converted Ramus' follower Lazarus Schöner to a proper appreciation of Euclid's treatment of regular solids and of their cosmological significance.[64] He suggests that Ramus has dismissed Proclus out of contempt for the Pythagorean sect to which Proclus belonged. In relation to this, he asks "what if the Pythagoreans had taught the same as me, but had hidden the opinion beneath a covering of words?"; and by way of confirmation of this conjecture Kepler offers a yet more elaborate reading of passages from Aristotle's *De caelo,* in which he finds sufficient traces of the Pythagorean doctrines that can be decoded as hiding the cosmographical mystery *sub ænigmate.*[65]

A further respect in which Kepler at least tacitly adhered to the terms of Ramus' challenge had to do with the way in which his astronomical doctrines were formed. Ramus had urged the restoration of astronomy through "logic in the first place, then the mathematical elements of arithmetic and geometry", and he had called on "someone who is both philosopher and mathematician" to perform the task.[66] As we have seen, in *Astronomia nova* Kepler offered to prove his claim "even according to the judgment of your

61 *Procli Diadochi in primum Euclidis elementorum librorum commentarii ex recognitione Godofredi Friedlein,* Leipzig 1873, pp. 19. 70-71.

62 Ramus: *Scholae mathematicae* (as note 4) pp. 257-258.

63 Kepler: *Gesammelte Werke* (as note 30) vol. 6, pp. 16-17.

64 Ibid. p. 17.

65 Ibid. pp. 17-18.

66 See above, p. 249.

own logic", asking Ramus on the strength of his recognition of the relevance of philosophy to astronomy to allow appeal to physics. Kepler proceeded in the introduction to the work to defend his Copernican system against physical objections to the earth's motion, and to sketch the dynamic conjectures to be used in chapters 27 and 63 in support of his account of the circuit of Mars and the other planets.[67]

Of course, Kepler's responses were playful and speculative. He did not really set out to meet the challenge as Ramus had posed it, but only as he conjectured that Ramus might have been persuaded to re-pose it. Kepler's decoding of the cosmographical mystery hidden in the exoteric doctrine of the Pythagoreans as garbled by Aristotle was frankly conjectural; and so too was the dymanics that supported his hypotheses describing the planetary circuits. But for Kepler such playful conjecture emulated the divine creative play, and as such it could be hoped to yield the profoundest insights.[68]

8. Conclusion

Ramus' calls for an *astrologia* without hypotheses were, as we have seen, both real challenges and interpretative challenges. Strange though it may appear to us, this kind of challenge, at once literal and literary, was characteristic of the learned discourse of the period. Emblems were often consciously ambiguous and esoteric, posed as brain-teasers for learned and competitive discussion.[69] And the erudite – Ramus and Kepler, though not perhaps Ursus, included – delighted in philologically difficult and recondite passages in classical authors. Kepler's learned correspondence with his powerful patron, Herwart von Hohenburg, Chancellor of Bavaria, on the astrological significances of passages from Lucan's *De bello civili,* provides typical examples.[70]

67 Kepler: *Gesammelte Werke* (as note 30) vol. 3, pp. 18-35.

68 On playful conjecture and divine creativity in Kepler see D. P. Walker: Kepler's celestial music, in: *Studies in Musical Science in the Late Renaissance,* Leiden 1978, pp. 34-62; Jardine, Segonds: Kepler (as note 59).

69 Cf. M. Biagioli: Galileo the emblem-maker, in: *Isis* 81 (1990) pp. 230-258.

70 On Kepler's philological skills see A. Grafton: *Defenders of the Text. The Traditions of Scholarship in an Age of Science, 1450-1800,* Cambridge, Mass. 1991, ch. 7: Humanism and science in Rudolphine Prague: Kepler in context.

Ramus' text was, and remains to us, provocatively obscure. Whether through accident, design, incompetence, or some happy combination, it is an open text with a vengeance. Of course, no text is entirely open; and the responses we have just examined rest on a modicum of common ground. All, save perhaps Tycho, took on the challenge at least at the literary level, and hence attempted, as was proper in duels of the period, to satisfy the challenger on his own terms – though as we have seen, Rothmann and Kepler were opportunists in their interpretations of the terms of the challenge.[71] Further all, save perhaps Rothmann, shared or at least purported to share with Ramus a vision of the goal of astronomy as a restoration of ancient wisdom. But within this minimal framework Ramus' text allowed an extraordinary range of readings, ranging from Rheticus' accommodation, through Tycho's dismissal, Rothmann's rigorous refutation and Ursus' invective, to Kepler's playful appropriation. In the eclectic culture of the period, such diversity and copiousness of response was a mark not of the failure, but of the splendid success of Ramus' challenge.

71 On the proprieties of early modern duelling see F. Billacois: *The Duel: Its Rise and Fall in Early Modern France,* New Haven 1990.

Index of Names

Aarsleff, Hans 162 n. 71
Abraham (Old Testament patriarch) 251
Acontius, Jacobus (died 1566) 50 153
Actius, Simon 152
Adam, Charles 50 n. 43
Adam, Günther 100
Adama, Lollius (1544-1609) 41
Adelung, Johann Christoph (1732/34-1806) 91
76 n. 37 80 n. 49 87 n. 79
Aelhuysen, J. van 47 n. 31
Aemylius, Paulus (died 1529) 17 n. 37
Agricola, Nicolaus 30
Agricola, Rudolph (1443/44-1485) 42 52 105 132 133 143 181 183 204 206 236
10 n. 7 43 n. 15 105 n. 49
Agricola Ammonius, Johannes (ca. 1490-1570) 244
244 n. 60
Aidius, Andrew (died 1630) 65 66
Ainsworth, Henry (1571-1623) 144
145 n. 36
Aiton, E. 250 n. 8
Akkerman, Fokke 54 n. 3 56 n. 9
Alberius → Aubery
Alciato, Andrea (1491-1550) 218 221 222
Aldrich, Henry (1647-1710) 136
Aldrovandi, Ulisse (1522-1605) 247
Alexander of Aphrodisias (2nd/3rd century AD) 241 n. 47
Alsted, Johann Heinrich (1588-1638) 88 135 198
Althusius, Johannes (1557/63-1638) 224-226 238
Amerbach (printer) 21 n. 54
Amerbach, Basilius (1534-1591) 11
11 n. 10
Ames (Amesius), William (1576-1633) 40 53 141
40 n. 7 8 141 n. 29
Ammonius → Agricola Ammonius
Anders, Stefan 62 n. 27
Annerstedt, C. 108f. n. 1-3 116 n. 11
125 n. 14
Antisthenes (ca. 445 - ca. 365 BC) 237
Apel(lus), Johannes 236 n. 31
Aquinas → Thomas Aquinas
Archimedes (ca. 287-212 BC) 184
Aretius (Marti), Benedictus (ca. 1522-1574) 12f.
Argenterio, Giovanni 232 n. 14 244 n. 61
Aristippus (ca. 430 - ca. 355 BC) 166 258
Aristotle (384-322 BC) 13 15 16 19 20 27 29 30 35 36 37 38 41 42 43 45 47 48 51 52f. 58 60 65 71 74f. 82f. 88 89 91 92 93 94 95 96 99 103 104 115 116 126 127 131 132 133 134 135 140 141 142 143 146 147 150 151 154 155 157 159 160 161 164 166 167 168 169 170 171 175 178 183 186 194 196 198 208 213 216 218 221 222 229 232 234 241 242 249 259 263 264 265
10 n. 7 15 n. 29 19 n. 44 20 n. 52 21 n. 53 22 n. 70 25 n. 99
26f. n. 104-106 29 n. 126 53 n. 49

77 n. 40 82 n. 58 83 n. 60 61 87 n. 80 82 88 n. 85 89 n. 92 90 n. 95 92 n. 1 96 n. 18 98 n. 28 99 n. 30 104 n. 47 143 n. 33 148 n. 43 149 n. 47 161 n. 69 162f. n. 73 217 n. 4 230 n. 7 234 n. 22 235 n. 24 236 n. 28 244 n. 61 246f. n. 67 68 250 n. 11 251 n. 14 259 n. 45
Arnauld, Antoine (1612-1694) 158 n. 62
Ascham, Roger (1515-1568) 159f. 160 n. 67 68
Asclepiades (1st century BC) 16 n. 33
Aubery (Alberius), Claude 183
August, Elector of Saxony (1526-1586) 249
Augustine (Augustinus, Aurelius) (354-430) 51
Augustus (1st Roman emperor, 63 BC - 14 AD) 254 n. 23
Austin, Warren B. 139 n. 21
Averroes (1126-1198) 242 n. 50
Azo (12th century jurist) 235

Bacon, Francis (1561-1626) 153 168 170-173 219 223 227 168 n. 83 171ff. n. 91-98
Bangs, Carl 155 n. 58
Barea, J. P. 201 n. 2
Barebone, John 146f.
Barker, Ambrose (died 1583) 132
Barker, P. 259 n. 47
Barlaeus, Jakob 58 n. 18
Barlow, Thomas (1607-1691) 140 n. 26
Baron, Frank 76 n. 36
Barrow, Henry (died 1593) 139
Barrow, Isaac (1630-1677) 174 175 175 n. 101
Báthory, Stephan 63
Baudouin, François 218 227
Baudrier, Henri Louis 244 244 n. 59
Bauer, Barbara 68 n. 1
Bauhin, Friedrich 30
Baxter, Nathaniel 141 141 n. 27
Baxter, Richard (1615-1691) 139 140 n. 24
Beale, John (1603?-1683) 176
Bechel (printer) 21 n. 54
Beckmann, Christian (1569-1606) 100 87 n. 82
Bedell, William (1571-1642) 178 178 n. 4
Bedwell, Thomas (died 1595) 175
Bedwell, William (1561-1632) 175 n. 103
Beeckman, Abraham (1607-1663) 39 n. 4
Beeckman, Isaac (1588-1637) 39 39f. n. 4 5 48 n. 33
Bellay → Du Bellay
Bendall, Sarah 148 n. 43
Bergeron, Nicolas 223
Bergius, Conradus 88
Berkel, Klaas van 39 n. 4
Bernhardt, Wilhelm 73 n. 21
Bernus, Auguste 9 n. 2 11 n. 12 12 n. 16
Bersmannus, Gregorius 90 n. 95
Betrutius, Alphonsus 244 244 n. 60
Beurhaus (Beurhusius), Friedrich (1536-1609) 80 88 89 100 101 103 111 119 144 182 183 185f. 80 n. 51 88 n. 86 101 n. 33 34 103 n. 43
Beyer, Michael 73 n. 22
Beyssade, J.-M. 48 n. 32
Beza, Theodor (1519-1605) 13 150
Biagioli, M. 265 n. 69
Billacois, F. 266 n. 71
Bilstenius, Johannes 111 113 111 n. 5
Bindseil, Heinrich Ernst 76 n. 35-38 77 n. 40
Birch, Thomas (1705-1766) 176 n. 104

Blackwell, C. 263 n. 60
Blair, Ann 162 n. 71
Blom, H. W. 42 n. 14
Bodin, Jean (1530-1596) 218 219
224-226
225 n. 14 15 235 n. 26
Boethius, Anicius Manlius Severinus
(ca. 480-524) 39 88 132 230 231
233 241 243 247
231 n. 11 233 n. 20 241 n. 49
Böhme, Jakob (1575-1624) 80 n. 49
Bolduanus, Paulus 86 n. 78
Bolsack, Johannes (1600-1674) 66
Bolton, Joseph S. G. 168 n. 82
Bolton, Robert (1572-1631) 178 n. 4 7
Booth, Cornelis 49 n. 40
Bootius (Boate), Arnold (1600-1653) 44
44 n. 19
Bootius (Boate), Gerard (1604-1650) 44
44 n. 19
Boran, Elizabethanne 177 n. 1
192 n. 65 67 68 198 n. 84
Borges, Jorge Luis 228 247
Bos, E. P. 47 n. 27
Bostocke, Richard (ca. 1530-1605) 196
Bouman, H. 46 n. 25
Boyle, Robert (1627-1691) 35 176
176 n. 104
Bracton, Henry de 223
Brahe, Tycho (1546-1601) 173 250 252
255f. 260 263 266
255f. n. 28-30 260 n. 48 50 263 n. 59
Brandmüller, Johannes (1533-1596) 12
12 n. 14
Brederodius (van Brederode), Petrus
Cornelius 221
Brerewood, Edward (1565?-1613)
163 n. 73
Bretschneider, Karl Gottlieb 75
75 n. 29 76 n. 33 77 n. 39 92 n. 1
94 n. 9 98 n. 28
Brickman, B. 251 n. 13
Briggs, Julia 166 n. 79
Brooke, Christopher (died 1628)
148 n. 43
Brooke, Samuel (died 1632) 167f. n. 82
Browne, Robert (1550-1633) 139
Brucker, Johann Jakob (1696-1770) 215
146 n. 39
Brundell, Barry 131 n. 6
Brunnemann, Johannes 88
88 n. 84
Brutus, Marcus Iunius (85-42 BC) 30
30 n. 137
Bruyère, Nelly 89 95 110 130 231 233
75 n. 28 89 n. 90 95 n. 16 130 n. 4
207 n. 11 228 n. 2 231 n. 12
234 n. 21 22 235 n. 24 239 n. 42
251 n. 11
Bryllinger (printer) 21 n. 54
Budé, Guillaume 222
Bullard, M. R. A. 138 n. 20
Bullinger, Heinrich (1504-1575) 12
12 n. 15
Bumann, Carl 100
Burgersdijk, Franco (1590-1635) 26 36
47 49 135
32 n. 144 47 n. 29 163 n. 73
Burghley → Cecil, William
Burmeister, K. H. 253
248 n. 1 253 n. 20 21
Burton, Robert (1577-1640) 151 159
137 n. 18 152 n. 53 159 n. 65
Buscher, Heizo (1564-1598) 100 101
101 n. 33 102 n. 35
Bush, Sargent 134 n. 10
Busing, Georg 100
Butler, Charles (died 1647) 181

Caesar, Gaius Iulius (100-44 BC) 218
17 n. 36 22 n. 75
Caesarius, Johannes (1460-1551) 183
Calippus (Kallippos of Kyzikos) 249
257
Calvin, Johannes (1509-1564) 204 205
Camden, William (1551-1623) 171f. n. 92

Camerarius, Joachim (1500-1574) 76
76 n. 36
Cantiuncula, Claudius (ca. 1490/1500 - ca. 1564) 236 n. 31
Capo di Vacca (Capivaccius), Girolamo (1523-1589) 239-241
240 n. 44 45 241 n. 48 246 n. 67
Cardano, Girolamo (1501-1576) 242f. 247
156 n. 60 239 n. 42 242 n. 50
243 n. 53 56 57
Caron → Le Caron
Carpentarius, Jacobus 141 155 183
Carpenter, Nathanael (1589-1628?) 198
Cartwright, Thomas (1535-1603) 149
Casaubon, Isaac (1559-1614) 155
Casaubon, Méric (1599-1671) 163f.
164 n. 76
Case, John (1546?-1600) 137 150 183
151 n. 49
Caselius, Johann (1533-1613) 92 102
Casmann, Otto (1562-1607) 29 100 119
Caspar, M. 256 n. 30
Cato, Marcus Porcius (Censorius) (234-149 BC) 15
16 n. 30
Cecil, William, Lord Burghley (1520-1598) 148 171
Chaderton, Laurence (ca. 1536-1640) 133
Challoner, Luke (1550-1613) 179 180f. 184 185 188 189 198
179 n. 8 181 n. 14
Chamber, John (1546-1604) 142
142 n. 32
Charles I. de Guise, Cardinal de Lorraine (1524-1574) 149
Charles IX, king of Sweden (1550-1611) 120 121 122
Charron, Pierre (1541-1603) 130
Chauncy, Charles (1654-1672) 135
Cheke, John (1514-1557) 168
Chomsky, Noam 210
Chytraeus, David (1530-1600) 97 98 100 122
98 n. 26 27
Chytraeus, Nathan (1543-1598) 59 100
Cicero, Marcus Tullius (106-43 BC) 21 28 30 31 45 52 57 74 82f. 88 91 114 115 116 133 157 161 167 181 183 210 211 230 232 234 236 249
10 n. 7 16f. n. 35-37 19 n. 40
21 n. 56 23 n. 78 87 30 n. 137
82 n. 58 87 n. 80 88 n. 85 89 n. 92
158 n. 64 160 n. 68 211 n. 15
230 n. 8 233 n. 20 237 n. 33
Cieslak, Edmund 62 n. 28
Clamp, Albert (1555-1604) 102
Clark, Samuel (1599-1682) 148 n. 42
Classen, J. 76 n. 37
Clauberg, Johannes (1622-1665) 34 49 51f. 52
52 n. 48
Clavius, Christoph (1538-1612) 39
Clichtove, Josse 132
Coccejus, Sigismundus 88
Cocus, Jacobus 245 n. 64
Coelius, Michael 87 n. 80
Colines, Simon de 244 n. 58
Collinson, Patrick 148 n. 43 152 n. 54
Comandinus (Commandino, Federico) (1509-1575) 39
Comenius, Johann Amos (1592-1670) 61
164 n. 76
Commelinus, Jakob 58f. n. 18
Connan, François 222
Cook, Zachary 145 n. 36
Cooper, Charles Henry 149 n. 45
Copenhaver, Brian P. 229
228 n. 3 229 n. 5 6 243 n. 54
Copernicus, Nicolaus (1473-1543) 35 39 196 249 252 254 255 257 258 262 263
252 n. 19 253 n. 21 254 n. 24
Copius, Bernardus (1525-1581?) 183
Coras, Jean de (1500-1572) 227 236 238
236 n. 32

Cosin, John (1594-1672) 159 n. 65
Courtine, Jean-François 106 n. 55
Couzinet, Marie-Dominique 225 n. 16
Cramer, Andreas 87
87 n. 82
Cramer, Christoph 89
Crapulli, Giovanni 49 n. 38
Cratander (printer) 21 n. 54
Crellius, Fortunatus (died 1605) 41 66
Crusius, David (1589-1640) 246
Cujas, Jacques (1522-1590) 218
Curio (printer) 21 n. 54
Curio, Caelius Augustinus (1538-1564)
10 n. 6
Curio, Caelius Secundus (1503-1569)
10 n. 6

Dalzel, Andrew 189 n. 54
Dąmbska, Izydora 64 n. 35
da Monte, Giambattista (1498-1552) 241
230 n. 7 239 n. 43 240f. n. 46 47
242 n. 50 246 n. 67
Dassonville, Michel 235 n. 25
Decimator, Henricus 79 n. 45
Declercq, Gilles 201 n. 2
de Colines → Colines
Dee, John (1527-1609) 222
de Federicis → Federicis
de Haan → Haan
Dejordy, A. 140 n. 26
de la Primaudaye → Primaudaye
de Launoy → Launoy
Delcourt, M. 248 n. 1-3 254 n. 22 24
Demosthenes (384-322 BC) 10 157
10 n. 7
Dent, C. M. 147 n. 40
De Raey (Raei), Joannes (1622-1702)
51 52 53
52 n. 48
de Roy → Regius
Derrer, Sebastian 221
Desan, Philippe 189
184 n. 28 189 n. 55
Descartes, René (1596-1650) 35 37 39
47f. 49 50 51 52 53 153 227
48 n. 34 49 n. 38 50 n. 42 43
De Veno, Henricus (died 1613) 41
41 n. 10
de Vocht → Vocht
de Waard → Waard
De Wilhem, David (1588-1658) 49 n. 40
Dibon, Paul 45
41f. n. 9-12 43 n. 15 16 44f. n. 20 21
46 n. 26 49 n. 37 40
Dickoff, James 158 n. 62
Dieterich, Konrad (1575-1639) 29
Digby, Everard (1578-1606) 140 183
188 193
149 n. 47
Dilg, Peter 244 n. 60
Diodorus Siculus 251 n. 12
Dodd(e)ridge, John (1555-1628) 222
222 n. 11
Doneau, Hughes 223f.
Donelli, H. 224 n. 13
Donne, John (1573-1631) 167 n. 82
170 n. 87
Douaren → Le Douaren
Douay-Soublin, Françoise 202 n. 3
Downame, George (died 1634) 29 181 185
Downes, Alan S. 158 n. 64
Dresserus, Matthaeus 83 n. 61
Dreyer, J. L. E. 255f. n. 28 29
Du Bellay, Joachim (1522-1560) 157 212
Dubois (Sylvius), Jacques (1478-1555)
242 244f.
242 n. 52 244f. n. 61 62
Duchhardt, Heinz 64 n. 32
Du Laurens, André 245 n. 64
Du Moulin → Molinaeus
Duns Scotus, Johannes (1265/66-1308)
118 171
Durling, Richard J. 244 n. 60
Dutemplaeus, Stephanus 244
244 n. 59
Dyck, W. von 256 n. 30

Earle, John (ca. 1601-1665) 166 170
167 n. 81 170 n. 89
Eckart, Wolfgang Uwe 44 n. 18
Ellis, Joseph J. 135 n. 14
Ellis, Robert L. 168 n. 83
Elrington, C. R. 198 n. 85
Elsmann, Thomas 54 n. 1 3 56 n. 9
57 n. 12 62 n. 27 64 n. 33
144 n. 35
Elswich, Johann Hermann von 97
97 n. 25
Embecanus (Einbecanus), Cunradus 70
70 n. 4 5 98 n. 28
Enfield, William 146 n. 39
Epicure (343/42-271/70 BC) 42 51
Episcopius (printer) 21 n. 54
Erasmus, Desiderius (1465/66-1536)
42 79 118 151 194 204
78 n. 42 79 n. 46 151 n. 52
Euclid (fl. 300 BC) 39 60 174 175 184
253 263 264
25 n. 97 175 n. 101 264 n. 61
Eudoxus of Knidos (ca. 390 - ca. 337 BC)
248 249 257
251 n. 14
Euripides (485/84 or 480-406 BC) 74
Eusden, John D. 141 n. 29
Evans, M. 239 n. 41

Faber → Le Fèvre
Faber Stapulensis (Lefèvre d'Etaples), Jacobus (ca. 1455-1536) 141
Fabian, Bernhard 54 n. 1
Fabricius, Ernestus Fridericus 245
246 n. 65
Fabricius, Jakob (1551-1629) 64 65 66 67
Fabricius Marcoduranus, Franciscus (died 1573) 183
Fabri → Le Fèvre
Fage, Robert 144 162
145 n. 36
Farrington, Benjamin 171 n. 90
Fäsch, Johannes Rudolf 31
Federicis, Stephanus de 236 n. 31
Feingold, Mordechai 188 192
134 n. 12 156 n. 61 162 n. 72
177 n. 2 188 n. 49 50 192 n. 66
Fenner, Dudley (1558-1587) 161 181 183
145 n. 36 162 n. 71
Ferguson, Stephen 136 n. 16
Fernel, Jean 42
Ferronus, Arnoldus (ca. 1515-1563)
17 n. 37
Feugueray, Guillaume 44f.
Finé, Oronce 39f.
Finks, Gaspar (fl. 1600) 197
Fisch, Harold 137
137 n. 17
Fitzherbert, Anthony (1470-1538) 223
Flacius Illyricus, Matthias, Jr. (1547-1593)
89 98 99
99 n. 30
Fleig, Hans 9ff. n. 3-10 11f. n. 13 14
15 n. 28 21 n. 54
Fletcher, H. F. 140 n. 26
Forestier, Georges 201 n. 2
Forster, Valentin Wilhelm 221
Fortescue, John (1394?-1476?) 223
Foucault, Michel 247
247 n. 29
Fouquelin, Antoine 206 209
Foxius Morzillus, Sebastianus 89 n. 92
Fraenkel, Peter 73 n. 18
Fragstein, Artur von 230 n. 7
Fraunce, Abraham (1587-1633) 144f. 223
145 n. 36 223 n. 12
Frederic, Elector of Saxony (1463-1525)
74 n. 25
Freedman, Joseph S. 55
55 n. 6 80 n. 50 82f. n. 58 59
83 n. 62 63 84f. n. 68-71
86f. n. 73-80 88 n. 83 88f. n. 85-89
187 n. 44 45
Freigius, Johannes Thomas (1543-1583)
10 14 16-21 25 36 80 117 137 182
183 185 186 221 238 249

10 n. 6 16ff. n. 34-37 20 n. 49 52
22 n. 72 74 75 23 n. 77-79 87
39 n. 3 80 n. 51 217 n. 3 248 n. 1
Frey, Janus Caecilius (ca. 1580-1631)
162 n. 71
Friedensburg, Walter 74 n. 22
Friedlein, Gottfried 264 n. 61
Friedrich, Arnd 56 n. 8
Friedrich, Carl H. 225
226 n. 17
Froben (printer) 21 n. 54
Fuchs, Leonhard (1501-1560) 244
244 n. 59 61 245 n. 64
Fulbecke, William (1560-1603) 222
222 n. 11
Fulke, William (1538-1589) 199
Fuller, Thomas (1608-1661) 149 n. 47
Fullerton, James (died 1631) 189
Fumaroli, Marc 204 n. 6
Furió Ceriol, Fadrique (1527-1592)
209 n. 12

Galama, Sybrand 41f. n. 9-12 45 n. 22
48 n. 34 49 n. 36 37
Gale, Theophilus (1628-1678) 141
142 n. 30
Galen (129-199) 16 167 196 232 234
239 240 244
10 n. 5 16 n. 32 232 n. 14
234 n. 22 239 n. 42 242 n. 52
244 n. 58 61
Gammaro, Pietro Andrea (1480-1528)
236 n. 31
Garber, Klaus 62 n. 27
Gascoigne, John 184 n. 29
Gassendi, Pierre (1592-1655) 41 130f.
131 n. 6
Gaza, Theodorus 92 n. 1
Genette, Gérard 201 203
201 n. 2 203 n. 5
George of Trebizond (1395-1484) 132 183
Gerber, Horst Peter 87 n. 79
Gerlo, A. 156 n. 60
Gesner, Conrad (1516-1565) 253 n. 20
Gibbs, Lee W. 40 n. 7
Gibson, Strickland 151 n. 50
Gilbert, Neal Ward 159
159 n. 66 232 n. 15 240 n. 44
Gill, Alexander (1565-1535) 170
Glanville, Ranulph de 223
Glenn, John R. 166 n. 79
Glucker, J. 155
155 n. 59
Goclenius, Rudolph (1547-1628) 89 234
39 n. 3 43 n. 16 84 n. 64 89 n. 89
234 n. 23 256 n. 33
Goeddeus, Johannes 231 n. 12
235 n. 26
Goldstein, B. R. 259 n. 47
Gomarus, Franciscus (1563-1641) 155
Goodman, Godfrey (1583-1656) 154
154 n. 57
Gorlaeus (van Goorle, Goirle), David
(born 1591/92) 44
44 n. 18
Gorscius (Górski), Jacobus 132
Gothofredus (Godefroy), Haloinus (died
1609) 29
Gough, William (1578-1653) 147f. 156
148 n. 43
Goulding, Robert 174 n. 100
Goyet, Francis 201 n. 2 230 n. 8
233 n. 18
Grafton, Anthony 129 153 191 200 211
129 n. 3 150 n. 48 154 n. 56
174 n. 99 191f. n. 62-64 194 n. 75
200 n. 1 211 n. 13 265 n. 70
Granada, M. A. 252 n. 14 259f. n. 47 48
Gray, Hanna H. 202
202 n. 3
Green, Lawrence D. 143
143 n. 33
Greene, Robert (1560-1592) 167
Greidanus, Johannes (1630/31-1668) 48
49 n. 37
Greschat, Martin 73 n. 18

Grindal, Edmund (1519?-1583) 149
Grosart, Alexander B. 151 n. 49
Grosse(n), Jakob (1592-1652) 81
81f. n. 56
Grynaeus, Samuel 30
Grynaeus, Simon (1493-1541) 258
258 n. 42
Gryphius, Sebastian 244
Guarin (printer) 21 n. 54
Guise → Charles I. de Guise
Guise → Henri I. de Lorraine
Gundlach, F. 256 n. 33
Günther, Owen (1532-1615) 99f.
99f. n. 31 32
Gustavus I Vasa, king of Sweden (1496/97-1560) 120
Gustavus II Adolphus, king of Sweden (1594-1632) 120 121 122
Gutberleth, Henricus (1572-1635) 42
Gutch, John 147 n. 40
Gutke, Georg (1589-1634) 81
81f. n. 56
Gwalter, Rudolph (1519-1586) 12
12 n. 16
Gymnicus, Johannes (ca. 1480-1544) 70
70 n. 4 5 98 n. 28

Haan, A. A. M. de 42 n. 14 46 n. 25 26
Hachtingius, Johannes (1594-1630) 41 43 48
41 n. 11
Hakewill, George 161 n. 69
Hales, John (1584-1656) 169
169f. n. 87
Hamblin, C. L. 259 n. 45
Hamilton, James (1559-1643) 189
Hamilton, William 136
136 n. 15
Hammer, Wilhelm 72 n. 17 74 n. 23
78 n. 43 44 80 n. 49 82 n. 57
Hammerstein, Notker 64 n. 32
Hankins, James 200
200 n. 1
Hardenberg, Albert Rizaeus (ca. 1510-1574) 56
Hartfelder, Karl 72 n. 17 73 n. 19-21
Hartlib, Samuel (1600?-1662) 153
Harvey, Gabriel (1550?-1631) 138 150 152f. 167 194
139 n. 21 151 n. 49 152 n. 54
Hase → Lagus
Hatch, Maurice 160 n. 67
Hatton, Sir Christopher (1540-1591)
151 n. 50
Hauenreuter, Sebald (1508-1589) 66
Haussleiter, Johannes 74 n. 25
Hazlitt, W. Carew 166 n. 80
Heath, Douglas D. 168 n. 83
Hecker, Hans 62 n. 29
Heereboord, Adriaan (1614-1661) 36 47 49 135
32 n. 144 47 n. 30
Hegendorff, Christoph (1500-1540) 238
236 n. 31
Heidanus, Abraham (1597-1678)
38 n. 1
Heidemann, Julius 82 n. 56
Heiland, Samuel (1533-1592) 66
Henri I. de Lorraine, duc de Guise (1550-1588) 119 165
166 n. 79
Henricpetri (printer) 17
Henry, John 180 n. 12
Henry IV of Navarre (1553-1610) 119
Herbert of Cherbury, Edward (1583-1648) 153
Hermes Trismegistus 39
Hermogenes (ca. 160 - ca. 225) 132 133
Heron of Alexandria (fl. 3rd quarter of 1st century AD) 39
Herwagen (printer) 21 n. 54
Herwart von Hohenburg, Hans Georg (1553-1622) 265
Heurnius, Johannes (1543-1601) 42
42 n. 13

Heylin, Peter (1599-1662) 137
Heyll, Christoph 244
244 n. 60
Hieronymus, Frank 21 n. 55
Hildebrand, Johannes 59 n. 18
Hill, Christopher 128 n. 2
Hill, W. Speed 161 n. 69
Hipparchus (fl. 1st quarter of 2nd century BC) 257 258
Hippocrates (460-370 BC) 16 240
10 n. 5 16 n. 33 229 n. 6
Hirsch, Theodor 63 n. 31 67 n. 39
Hoak, Dale 156 n. 61
Hobbes, Thomas (1588-1679) 137
Hoddaeus, Conradus 81 n. 55
Hoffmann, Daniel (1538-1611) 100
Hojer, Andreas (died after 1630) 65 66
Höltgen, K. J. 244
136 n. 16 239 n. 41 244 n. 59
Holwarda, Johannes Phocylides (1618-1651) 41
42 n. 12
Homer (9th-8th century BC) 75 194
10 n. 7
Hood, Thomas (1556-1598) 185
Hooker, Richard (1554-1600) 160f. 173
161 n. 69 70
Hooykaas, Reijer 110 251 254
128 n. 2 250 n. 8-10 252 n. 15 21
254 n. 23 25
Hopper, Joachim (1523-1576) 236 238
Hospinian, Johannes (1513-1575) 10
10 n. 7
Hotman, François (1524-1590)
238 n. 36
Howell, Wilbur S. 134 146
134 n. 11 146f. n. 39 40 151 n. 49
181 n. 18 190 n. 56
Huet, Pierre-Daniel (1630-1721) 48 n. 34
Hugonnard-Roche, H. 248 n. 1 253 n. 20
Huisman, Gerda C. 54 n. 3
Hume, Alexander (1560-1609) 155
Humphrey, Laurence (1527-1590) 146
Hutton, Sarah 180 n. 12
Huygens, Christiaan (1629-1695) 35
Hyperius, Andreas (1511-1564) 198

Ignatius de Loyola (1491-1556) 204
Irnerius (12th century jurist) 237
Isengrin (printer) 21 n. 54
Isocrates (436-338) 10

Jacchaeus, Gilbertus (1578-1628)
40 n. 6
Jackson, Holbrook 159 n. 65
Jacobs, Charles M. 74 n. 25
Jaeger, F. M. 44 n. 18
Jähnig, Bernart 62 n. 29
James, Patricia 158 n. 62
James I, king of England (1566-1625) 155
Janse, Willem 56 n. 11
Jardine, Lisa 129 153 173 191
129 n. 3 154 n. 56 173 n. 98
191f. n. 62-64 194 n. 75 211 n. 13
237 n. 33
Jardine, Nicholas 173
174 n. 99 196 n. 81 250 n. 8
252 n. 16 260 n. 48 50 263 n. 59
265 n. 68
Javolenus 237 n. 33
Jeffrey, R. W. 149 n. 47
Jesus Christ 141 142
Jiménez, Alfonso Martín 201 n. 2
Jöcher, Christian Gottlieb (1694-1758) 91
14 n. 24 16 n. 34 76 n. 36 37
79 n. 48 80 n. 49 80f. n. 51-53
82 n. 56 87 n. 79 89 n. 88 111 n. 5
John, Elector of Saxony (1468-1532)
74 n. 25
Johnson, Samuel (1696-1772) 135
Josephus, Flavius (ca. 37-95) 251 n. 12

Kant, Immanuel (1724-1804) 231
Kearney, Hugh 187 191 194
128 n. 2 186 n. 43 187 n. 46 47
191 n. 59-61

Keckermann, Bartholomäus (1572/73-1609) 27 28 41 43 47 65 66 67 97 135 197 234
83 n. 62 84 n. 64 97 n. 23 145 n. 36 162f. n. 73 234 n. 23
Keen, Ralph 72 n. 17 75 n. 27 75f. n. 30-33 76 n. 35 38
Kelley, Donald R. 138
138 n. 19
Kenny, Anthony 237 n. 33
Kepler, Johannes (1571-1630) 173 250 260 261-265 266
249 n. 5 256 n. 30 260 n. 50 262f. n. 54-60 264 n. 63-65 265 n. 67 68 70
Kermode, H. S. 162 n. 73
Killy, Walter 65 n. 36
Kinane, Vincent 192 n. 67
Kingdon, Robert M. 140 n. 24
142 n. 31
Kirby, J. 175 n. 101
Kisselbach, Heinrich (1600-1673) 31
Knafla, Louis A. 128 n. 2
Knape, Joachim 90
73 n. 21 76 n. 33
Knappert, L. 44 n. 20
Kneale, Martha 47 n. 28
Kneale, William 47 n. 28
Knox, R. B. 179 n. 8
Koch, Ulrich (1525-1585) 11
11 n. 13
Kolb, Robert 68 n. 1
Kraye, Jill 194 n. 75 200 n. 1
Kretzmann, Norman 237 n. 33
Kristeller, P. O. 202 n. 3
Krop, H. A. 42 n. 14 47 n. 27
Kudlien, Fridolf 244 n. 60
Kühlmann, Wilhelm 65 n. 36
Kuin, Roger 139 n. 22
Kuiper, E. J. 46 n. 24
Kuppers, Willem J. 57 n. 12
Kusukawa, S. 263 n. 60
Laan, Adrie H. van der 56 n. 9
Lagus (Hase), Conradus 221
Lambin, Denis (1520/21-1572)
15 n. 29
Lancelot du Voisin → La Popelinière
Lange, Reiner 59 n. 18
La Popelinière, Henri Lancelot du Voisin, sieur de 227
la Primaudaye → Primaudaye
La Ramée → Ramus
Latomus, Barthélemy 204
Laud, William (1573-1645) 178 179
Launoy, Jean de 97 n. 25
Laurens → Du Laurens
Lavater, Ludwig (1527-1586) 12
12 n. 16
Lawson, Thomas (1620-1695) 141f.
142 n. 30
Layton, Thomas (died 1565) 132
Le Caron, Loys (Louis) Charondas (1534-1613) 222
Le Douaren, François 227
Leedham-Green, Elisabeth S.
132f. n. 8 9 152 n. 54 188 n. 52
Leemann, Samuel (1651-1709) 33f. 36
Le Fèvre (Faber, Fabri), Jean-Rodolphe (ca. 1580-1650) 34 36
Lefèvre d'Etaples → Faber Stapulensis
Leibniz, Gottfried Wilhelm (1646-1716)
175 n. 101
Leinsle, Ulrich-Gottfried 93 95-97 99
93 n. 6-8 95 n. 17 96f. n. 19-21
Leishman, J. B. 165 n. 78
Lenin, Vladimir Ilyich (1870-1924) 118
Leonhardt, Jürgen 58 n. 15
Lerner, M.-P. 248 n. 1 250 n. 8
251 n. 13 259 n. 47
Le Roy, Loys 218
Leu, Hans Jacob (1689-1768)
14 n. 24 26 16 n. 34
Lewis, John 158 n. 64
Libavius, Andreas (1540-1616) 198

Liddel, Duncan (1561-1623) 102
256 n. 30
Lindberg, David C. 184 n. 29
Lindeboom, G. A. 42 n. 13
Lipenius, Martin 61 n. 24
Lipsius, Justus (1547-1606) 109 155 156 163 167 218
156 n. 60
Littleton, Thomas 223
Livy (Livius, Titus) (59 BC - 17 AD) 160 218
Lobstein, Paul 11 n. 11 12
Lodge, Thomas (1558-1625) 166
166 n. 80
Loftus, Adam (1533-1605) 191 195
Lohr, Charles H. 77 n. 41 78 n. 44
Loisel, Antonie 223
Lombard → Peter Lombard
Long, Gerard 180 n. 12
Lorraine → Charles I. de Guise
Lorraine → Henri I. de Lorraine
Lossius, Lucas (1508-1582) 66 67 77
77 n. 41 78 n. 42 43 90 n. 95
Loyola → Ignatius de Loyola
Lucan (Lucanus, Marcus Annaeus) (39-65) 265
Lucianus (ca. 120 - after 180) 102
Lucretius (99/94-55/49 BC) 42 45
Lullus → Raimundus Lullus
Luther, Martin (1483-1546) 11 62 73 113 204
74 n. 25 82 n. 57
Lutz, Ludwig (1577-1642) 24 25-31 36
24 n. 94 26 n. 104 28 n. 116 117
29 n. 124 125 128 30 n. 135
Lyffe, William (died 1569) 132

Macer, Caspar 84 n. 64 87 n. 79
MacIlmaine, Roland (Rollo) 140 144 161
162 n. 71
Mack, Peter 200
73 n. 21 194 n. 74 200 n. 1
211 n. 13 235 n. 25
MacKilwein, Robert (fl. 1574) 183 195
Maclean, Ian 136 n. 16 180 n. 12
230 n. 8-10 231 n. 12 235 n. 24 26
236 n. 30 31 237 n. 33 238f. n. 37-39
Madan, F. 149 n. 47
Maestlin, Michael (1550-1631) 261f.
Maestre, J. M. 201 n. 2
Magni, Jonas (1583-1651) 123
Mahaffy, J. P. 178 n. 4 5
Maittaire, Michael 244 n. 59 61
Major, Georg (1502-1574) 78f. 84
79 n. 46
Manardi, Giovanni (1462-1536) 240
232 n. 15 240 n. 44
Manilius, Marcus 173
Manitius, C. 258 n. 41
Mao Tse-tung (1893-1976) 118
Marion, J.-L. 48 n. 32
Marlowe, Christopher (1564-1593) 145 165
145 n. 38 166 n. 79
Marti, Benedikt → Aretius, Benedictus
Martin, James, of Dunkeld (fl. 1577) 183
Martindale, Adam 162 n. 73
Martini, Cornelius (1568-1621) 100 102-106 111 216
81 n. 55 101 n. 33 34 103ff. n. 39-55
Martini, Jacob (1570-1649) 65 111 123
98 n. 28
Martini, Johannes (1558-1629) 67
Martinius, Matthias (1572-1630) 60 64
Marx, Karl (1818-1883) 118
Mather, Increase (1639-1723) 135 142
Matthisius, Gerardus 132
Maurer, Wilhelm 93
74 n. 24 93 n. 3
Maurice, Prince of Nassau (1567-1625) 39
McConica, James K. 151
143 n. 33 150 n. 48 151 n. 49 51
McKerrow, Ronald B. 147 n. 41
167 n. 82
McKitterick, David 152 n. 54

McQuillen, Connie 152 n. 53
McRae, Kenneth 225
225 n. 16
Measell, J. S. 145 n. 36
Meerhoff, Kees 110
130 n. 4 205 n. 8-10 211 n. 14
213 n. 17
Melanchthon, Philipp (1497-1560) 31
58 59 60 65 66 67 68-91 92-94 95
97 98 99 100 101 102 103 105 111
112 113 132 133 183 186 203 204
206
39 n. 3 58 n. 15 68-91 *passim*
92 n. 1 94 n. 9-11 98 n. 28
101 n. 33 102 n. 36 38 105 n. 49
112 n. 6 233 n. 18 256 n. 33
Melville, Andrew (1545-1622) 189
Menk, Gerhard 55 n. 7 64 n. 32
Mercator, Gerhard (1512-1594) 57 58
58 n. 16
Merino Jerez, Luis 201 n. 2 205 n. 10
209 n. 12
Mesnard, P. 225 n. 14 15
Meyer, Lodewijk (1638-1681) 40 n. 7
Micyllus, Jacob (1503-1558) 76
76 n. 37
Mignault, Claude 223
Mildmay, Sir Walter (1520-1589) 133
134 n. 10
Miller, Perry 135 141
135 n. 13 141 n. 28 193 n. 69
Milton, John (1608-1674) 145 170 175
185
128 n. 1 145 n. 37 170 n. 88
175 n. 102
Mirandola → Pico della Mirandola
Moisan, Jean-Claude 205 n. 8
211 n. 14 213 n. 17
Molanus, Johannes (ca. 1510-1583)
54-59 61
54 n. 2-4 57 n. 14 58 n. 16
58f. n. 18 144 n. 35
Molhuysen, P. C. 43 n. 16 44 n. 20
Molinaeus (Du Moulin), Petrus (1568-1658) 46
46 n. 26 164 n. 76
Moltmann, Jürgen 58 n. 17
Montague, Richard (1577-1641) 163 169
140 n. 25 163 n. 75 169 n. 86
Montaigne, Michel de (1533-1592) 218
219 236
Monte → da Monte, Giambattista
Moran, B. T. 259 n. 47
Morgan, John 193 195
193 n. 70 195f. n. 79 80
Morison, Samuel E. 193 n. 69
Mosellanus, Petrus 79
79 n. 46
Moss, Ann 200 203 204
194 n. 73 200 n. 1 203 n. 4 204 n. 7
Motte, Richard (died 1592) 132f.
Moulin → Molinaeus
Müller, Georg (ca. 1603-1684) 35 36
Müller, Johannes → Regiomontanus
Müller, Michael G. 63 n. 30
Mullinger, J. B. 149 n. 46
Murphy, H. L. 180 n. 11
Murphy, James J. 71 n. 9 82f. n. 59
89 n. 91 234 n. 22
Mylius, Jakob 59 n. 18

Nadolski, Bronisław 64 n. 35
Nancel, Nicolas de 157f.
157 n. 62 158 n. 64
Nashe, Thomas (1567-1601) 147 150
165 167f.
147 n. 41 165 n. 78 167 n. 82
168 n. 84
Nauwelaerts, M. A. 156 n. 60
Neander, Michael (1525-1595) 79 n. 48
Nettelbladt, C. 124 n. 13
Neumeister, Sebastian 64 n. 32
Neumeyer, Heinz 62 n. 28
Neuser, Wilhelm H. 56 n. 11
Newlands, Carole 71 n. 9 83 n. 59
234 n. 22

Newman, William R. 127 n.* 135 n. 13
Nicephorus, Hermann (died 1626) 100
Nichols, William 149 n. 47
Nicolai, Heinrich (1605-1660) 66
Nicolaus, Johannes (died 1676) 33 36
Nifo (Niphus), Agostino (1473-1538/46) 132 133
Nilsson, Sven A. 119 n. 12
Nizolio, Mario (1498-1576) 215
Nörr, Dieter 233 n. 17
Nothold, Anton (1569-1650) 100 101
81 n. 55 101 n. 33
Nozinic, Drazen 187 n. 45
Nuchelmans, Gabriel 97 203
97 n. 22
Nürnberger, Richard 77 n. 39

O'Brien, John 236 n. 30
Ockham → William of Ockham
Oekolampad, Johannes (1482-1531)
12 n. 14
Oldendorp, Johannes (1480?-1567) 221
236 n. 31 238 n. 36
Oldrini, Guido 231
215 n. 1 231 n. 12 251 n. 11
Olevianus, Caspar (1536-1587) 133
58 n. 18
Ong, Walter J. 55 75 80 90 95 110 111 112 127 128 129 131 137 143 144 199 206 231 239
21 n. 55 55 n. 5 61 n. 24 68 n. 2 71 n. 7 71f. n. 9-15 75 n. 29 30 80 n. 50-52 89 n. 91 89f. n. 93 94 95 n. 12 96 n. 18 127 n. 1 137 n. 17 144 n. 34 158 n. 64 178 n. 6 185 n. 33-35 38 189 n. 53 193 n. 72 199 n. 86 212f. n. 16 17 216 n. 2 217 n. 4 228 n. 4 231f. n. 12 13 234 n. 21 22 235 n. 24 239 n. 40 250 n. 11 256 n. 33
Oporin (printer) 21 n. 54
Orme, William 140 n. 24
Ormerod, Oliver 168f.
169 n. 85
Ornsby, G. 159 n. 65
Osiander, Andreas (1498-1552) 263
73 n. 20
Otte, Gerhard 235f. n. 27-30 237 n. 35
Ovid (43 BC - ca. 17 AD) 75 206
Oxenstierna, Axel (1583-1654) 121 125

Pacius, Julius (1550-1635) 41
26 n. 104
Pappus (3rd-4th century AD) 40 174
Paracelsus (1493-1541) 167 196 229
Parkinson, Richard 163 n. 73
Pascal, Blaise (1623-1662) 35
Pasquier, Etienne (1529-1615) 196
Patrizi, Francesco (1529-1597) 130 218 251
21 n. 53 251 n. 13
Paulinus Gothus, Laurentius (1565-1646) 108f. 112 113f. 117 122 123
Paul of Venice (ca. 1369-1429) 198
Paulus de Castro (fl. 1400) 237 n. 34 36
Penry, John (1559-1593) 147
Périon, Joachim 132 159
Perkins, William (1558-1602) 181 189
190 n. 57
Perna (printer) 21 n. 54
Perne, Andrew 147
Pešková, Jarmila 187 n. 45
Peter Lombard (ca. 1100 - ca. 1164) 132
Peterson, Rebecca 74 n. 25
Petri (printer) 21 n. 54
Pezel, Christoph (1539-1604) 64
Pfaffrad, Caspar (1562-1622) 100 221 222
81 n. 54
Pico della Mirandola, Giovanni (1463-1494) 130
Pinborg, Jan 237 n. 33
Piscator, Johannes (1546-1625) 89 132 133 183
24 n. 92 32 n. 145
Placentinus (12th century jurist) 236
236 n. 28 30

Plato (428/27-349/48 BC) 69 92 94 95 99
100 102 103 133 157 161 174 210 229
70 n. 3 92 n.1 100 n. 32 234 n. 22
Platt, John E. 43 n. 17
Platter, Felix (1536-1614) 11
11 n. 9
Platter, Felix (1605-1671) 30 31
31 n. 142
Platter, Thomas (1499-1582) 10
11 n. 9
Pliny, the Elder (Plinius Secundus, Gaius) (23/24-79) 251 n. 12
Plucknett, T. F. T. 222 n. 9
Polanus von Polansdorf, Amandus (1561-1610) 28 29
87 n. 81
Pollard, Mary 180 n. 12
Polybius (ca. 200 - after 120 BC) 218
Pompeius, Sextus Festus (2nd century AD) 160
Pompeius Trogus 160
Pontano, Giovanni Gioviano (1426-1503) 218
Popelinière → La Popelinière
Porphyry (234-301/05) 132 178 230
231 232 234 240 243
26 n. 105 232 n. 13 234 n. 22
Potter, George R. 170 n. 87
Praetorius, Ephraim 67 n. 38
Praetorius, Paulus 85 n. 73
Prantl, C. 203
Prescott, Anne Lake 139 n. 22
Primaudaye, Pierre de la 144
145 n. 36
Priscian (6th century AD) 168
Proclus (ca. 410 - ca. 485) 174 175 258
259 264
258 n. 41 264 n. 61
Ptolemy (ca. 100 - ca. 170) 39 174 196
249 252 253 255 257 258
252 n. 19 255 n. 27 258 n. 42
Pythagoras (died 497/96 BC) 263
263 n. 58
Quainton, Malcolm 236 n. 30
Quick, John 142 n. 30
Quintilian (Quintilianus, Marcus Fabius) (ca. 35-100) 28 30 82f. 161 167
169 230 233 234 236
30 n. 137 89 n. 92 160 n. 68
230 n. 8 233 n. 18 234 n. 22

Rabelais, François (1492?-1553) 157 212
Radetti, G. 50 n. 41
Raey → De Raey
Raimundus Lullus (1232-1315/16) 164
171 243
84 n. 64
Rainolds, John (1549-1607) 142f. 150 161
143 n. 33
Ralegh, Walter (1554-1618) 219
Ramus (La Ramée), Petrus (1515-1572) *passim*
Randolph, Thomas (1605-1635) 166
166 n. 80
Rasmussen, Carl J. 134 n. 10
Rawley, William (1588-1667) 173
Regiomontanus (Müller, Johannes) (1436-1476) 39
Regius (de Roy), Henricus (1598-1679)
49 50
50 n. 43
Regneri ab Oosterga, Cyprianus (1614-1687) 42
Reisch, Gregor 239 n. 41
Renaker, David 137 n. 18
Reneri, Henricus (1593-1639) 49f.
50 n. 42
Rennemann, Henning (1567-1646) 112
117
112 n. 6 115 n. 10
Reuchlin, Johannes (1455-1522) 204
Revius, Jacobus (1586-1658) 43 n. 17
Reynolds, Samuel H. 163 n. 74
Rheticus, Georg Joachim (1514-1574)
248 250 253-255 266
253ff. n. 20-27

Richter, Gregor (1560-1624) 79
80 n. 49 83 n. 61 84 n. 66
Rigerus, Johannes 112
112 n. 6
Rigolot, François 203
201 n. 2 203 n. 5
Rijen, J. B. M. van 47 n. 27
Risse, Wilhelm 93 95 110 203
71 n. 8 93 n. 2 95 n. 13 99 n. 29
256 n. 33
Rivet, André (1573-1651) 49 n. 40
Rivius, Johannes (1500-1553) 133
Robert, Edmund (died 1580) 132
Roberts, Michael 120 n. 12
Robinet, André 130
48 n. 32 33 130 n. 5 207 n. 11
Robinson-Hammerstein, Helga
192 n. 68 198 n. 84
Rodenborch, Johann (ca. 1573-1617) 65
Roding, Wilhelm 221
Roeslin, Helisaeus (1544-1616) 196
Rogerius (12th century jurist) 236
236 n. 29
Rogers, Daniel (1538-1591) 139
139 n. 22
Rollock, Robert (1555-1599) 189
Romano, Andrea 192 n. 65
Rosen, E. 251 n. 13 254 n. 24 26
Rossi, Paolo 173
173 n. 98 243 n. 57
Rotermund, Heinrich Wilhelm 91
Rother, Wolfgang 12 n. 18 25 n. 96
27 n. 110 28 n. 115 118 121
29 n. 128 30 n. 132 135 80 n. 51
150 n. 48
Rothmann, Christoph (died 1608) 173
250 251 255 256-259 266
252 n. 14 256 n. 32 257 n. 34 35
257f. n. 37-41 258f. n. 43 44
259 n. 47
Roy → Regius
Rozendael, Aemilius van 38 n. 2
Rudd, Edmund (died 1577) 132
Rüdin, Jakob (1633-1689) 31 32
Rudinger, Erasmus 83 n. 61
Rudloff, Ortwin 56 n. 10
Rudrauffius, Kilianus (1627-1690)
84 n. 64
Rufus Candidus, Remigius 84 n. 64
Rütimeyer, Markus (1580-1647) 33 36
33 n. 147 61 n. 24
Ryff, Peter (1552-1629) 25

Sadie, Stanley 185 n. 39
Sagittarius, Thomas (1577-1621) 79
79 n. 48
Sánchez (Sanctius), Francisco
(before 1552-1623) 204
162 n. 71
Santori, Santori (1561-1636) 240-242
241f. n. 49 50 245 n. 63
Saraisi, Nancy 242 n. 50
Sassen, Ferdinand 41f. n. 9-12 43 n. 16
Savile, Henry (1549-1622) 174 175
Savile, Thomas (died 1593) 149 n. 47
Scaliger, Joseph Justus (1540-1609)
150 155 173
158 n. 63
Scaliger, Julius Caesar (1484-1558)
234 n. 23
Schäfer, Alfons 68 n. 2
Scharfius, Johannes (1595-1660) 88 n. 84
Schegk, Jakob (1511-1587) 22 80 89
133 216
22 n. 68 70 80 n. 52 89 n. 92 234 n. 22
Scheible, Heinz 68 n. 1 2 70 n. 4
73 n. 18 21 22 75 n. 27 82 n. 57
83 n. 60 98 n. 28
Scheibler, Christoph (1589-1653) 65
Schmidt-Biggemann, Wilhelm 95
65 n. 36 95 n. 15 220 n. 7
Schmitt, Charles B. 229
136 n. 16 137 n. 18 141 n. 27
180 n. 12 228 n. 3 229 n. 5 6
233 n. 19 239 n. 41 243 n. 54
Schneider, Carol 135 n. 14

Schneider, Herbert 135 n. 14
Schöner, Lazarus 119 264
39 n. 3 256 n. 33
Schoock, Martinus (1614-1669) 50 n. 43
Schorus, Henricus (fl. 1572) 183
Schotanus a Sterringa, Johannes (1643-1699) 48 n. 34
Schreiber, Hieronymus 263
Schroderus, J. 234 n. 23
Schroeder, Gerhard 58 n. 18
Schwarzwälder, Herbert 56 n. 10
Scribonius, Wilhelm Adolf (fl. 1583) 182
Segonds, Alain 173
248 n. 1 263 n. 59 265 n. 68
Selden, John (1584-1654) 163
163 n. 74
Sellberg, Erland 113 n. 8 115 n. 10
Seneca, Lucius Annaeus (4? BC - 65 AD) 45
Serjeantson, Richard W. 127 n.*
145 n. 36 158 n. 62 164 n. 76
166 n. 79
Seton, John (1498?-1567) 132 133 165
Sharratt, Peter 131
72 n. 16 110 n. 4 127 n.* 131 n. 7
150 n. 48 157 n. 62 158 n. 64
193 n. 71 220 n. 8
Shaxton, John (died 1601) 133
Sherman, William H. 222
222 n. 10
Shuger, Debora K. 73 n. 21
Sibbes, Richard (1577-1635) 196
Siccama, Willem 48 n. 36
Sidney, Philip (1554-1586) 188
Sigismund III Vasa, king of Poland and Sweden (1566-1632) 63 121 122
Simler, Josias (1530-1576) 12 253
253 n. 20
Simonius, Simon (1532-1602) 133
Simpson, Evelyn M. 170 n. 87
Skytte, Johan (1577-1645) 107-109
112 114 116 117 121-126
124 n. 13
Sluterus, Severinus (1571-1648) 88 112
113 117
89 n. 88 113 n. 7 8
Smith, John (1618-1652) 163 n. 73
Smith, Preserved 74 n. 25
Smith, Robert (died 1571) 132
Snellius (Snell van Royen), Rodolphus (1546-1613) 38-40 43 44 53 89 115
117 181 183 184 186
24 n. 91 38f. n. 2 3 40 n. 5
43 n. 16 48 n. 33 115 n. 9
Snellius (Snell van Royen), Willebrord (1580-1626) 38 40
38 n. 2 40 n. 6
Socrates (ca. 470-399 BC) 115 210
Sommerfeldt, Jacobus 79 n. 45
Sorell, Tom 51 n. 46
Sparn, Walter 93
93 n. 5
Spedding, James 168 n. 83
Sperlingius, Paulus 85 n. 73
Spieler, Silke 62 n. 29
Spinoza, Baruch de (1632-1677) 51
40 n. 7
Spon, Charles 239 n. 42
Sprunger, Keith L. 40 n. 8 141f. n. 29 30
Staehelin, Johannes Jakob 31 n. 142
Starkey, George (1628-1665) 135
Steane, J. B. 145 n. 38
Stellingwerff, Frederik (died 1623) 41
43 44
Stempel, Hermann-Adolf 73 n. 19
Sterky, Jérémie (fl. 1685-1700) 35 37
Stevin, Simon (1548/49-1620) 39
Stierius, Johannes (1588-1648)
84 n. 64
Stigelius, Johannes 78 n. 45
Stigzelius, Laurentius Matthiae (1598-1676) 108f. 112
Strigel, Victorinus (1524-1569) 77f. 84
77 n. 41 84 n. 66
Strype, John (1643-1737) 140
140 n. 25

Stubbings, Frank 134 n. 10
Stubbs, J. W. 195 n. 77 78
Sturm, Johannes (1507-1589) 9 132 133 159 183 204 205 218
Suárez, Francisco (1548-1617) 106 197 43 n. 17
Sulzer, Simon (1508-1585) 11 11 n. 13
Supple, James J. 236 n. 30
Sylvius → Dubois

Tacitus, Cornelius (ca. 55/56 - ca. 120) 218
Talaeus (Talon), Audomarus (Omer) (ca. 1510-1562) 17 22 27 33 57 59 60 61 62 67 72 132 133 152 181 182 183 205 206 213 223 14 n. 23 16f. n. 35 36 21 n. 57 22 n. 62 63 71 73 23 n. 79 80 84 24 n. 89 90 92 93 95 27 n. 108 32 n. 145 33 n. 150 39 n. 3 72 n. 13 160 n. 68 216f. n. 2 3 217 n. 5 251 n. 12 252 n. 17 19
Tannery, Paul 50 n. 43
Tempier, Etienne (died 1279) 243 n. 55
Temple, William (1555-1627) 133 177 183 184 186 188 189 190f. 193 141 n. 27 149 n. 47 177 n. 3 190 n. 57 58
Thijssen-Schoute, C. Louise 48f. n. 34-37
Thomas Aquinas (1225?-1274) 113 118 132 171
Thompson, Stephen (died 1599) 133
Thorndike, Herbert (1598-1672) 139 139 n. 23
Thriverus, Hieremias 246 n. 66
Thucydides (ca. 460 - ca. 400 BC) 218
Tillman, Abraham (died 1590) 132
Timpler, Clemens (1563/64-1624) 87 n. 80
Titelmans, Franciscus (1532-1596) 183
Todd, Margo 187 195 187 n. 48 195 n. 76
Toletus, Franciscus (1532-1596) 183
Trapezuntius → George of Trebizond
Travers, John (died 1620) 179 179 n. 10
Travers, Walter (1548?-1635) 195
Trescho, Sebastian 59 n. 18
Treutler, Hieronymus 256 n. 33
Trinaeus, Fridericus 81 n. 54
Turnèbe, Adrien (1512-1565) 158 193 158 n. 64
Tyacke, Nicholas 134 n. 12 177 n. 2 188 n. 49 50
Tycho → Brahe

Ulpian (2nd century jurist) 236 237f. n. 36
Ursinus, Zacharias (1534-1583) 80 150 81 n. 53 54
Urstitius → Wurstisen
Ursus (Baer), Nicolaus Reimarus (Reymers) (1551-1600) 173 250 260f. 263 265 256 n. 30 260f. n. 48-53
Ussher, Ambrose (1582-1629) 179 179 n. 10
Ussher, Henry (1550?-1613) 191
Ussher, James (1581-1656) 179 180f. 182 183 184 186 187 198 179 n. 8 181 n. 17 198 n. 85

Valerius (Wouters), Cornelius (1512-1578) 43 133 183 233 43 n. 16 48 n. 33 233 n. 20
Valla, Giorgio (1430-1499) 258 258 n. 41 42
Valla, Lorenzo (1407-1457) 132 133 183 43 n. 15
van Aelhuysen → Aelhuysen
van Berkel → Berkel
van Brederode → Brederodius
Vanderjagt, Arjo J. 54 n. 3 56 n. 9
van der Laan → Laan
van Rijen → Rijen

van Wesenbeke → Wesenbecius
Varanda, Jean de 246 n. 66
Varro, Marcus Terentius (116-27 BC) 16
16 n. 30
Vasoli, Cesare 93 95 130
93 n. 4 95 n. 14 131 n. 6
Venn, John 188 n. 51
Venn, John Archibald 188 n. 51
Veno → De Veno
Verbeek, Theo 49 n. 39 50 n. 42 43
51 n. 46
Verdet, J.-P. 248 n. 1 253 n. 20
Vervliet, H. D. L. 156 n. 60
Vettori (Victorius), Pietro (1499-1585)
15 n. 29
Virgil (Vergilius Maro, Publius)
(70-19 BC) 45 75 116 132 210
10 n. 7 18 n. 37 158 n. 62
Vitello → Witelo
Vitruvius (1st century BC) 253
Vives, Juan Luis (1492-1540) 130 143
218
21 n. 53 43 n. 15
Vocht, H. de 43 n. 16
Voet, Leon 233 n. 20
Voetius, Gysbertus (1589-1676)
43 n. 17 50 n. 43
Voetius, Paulus (1619-1667) 43 n. 17
Vollgraff, J. A. 38 n. 2
Voltaire (1694-1778) 207
Vos, Alvin 160 n. 67
Vossius, Gerardus Joannes (1577-1649)
32 n. 144
Vossius, Isaac (1618-1689) 49 n. 40
Vultejus, Hermann (1565-1634) 222 238

Waard, Cornelis de 38 n. 2 39 n. 4
Waddington, Charles 220
9 n. 1 2 12 n. 15 17 13 n. 19-22
14 n. 25 21 n. 55 220 n. 6
Walker, D. P. 265 n. 68
Walsall, Godwin (died 1608) 133
Walsh, Anne 192 n. 67
Walton, Craig 173
173 n. 98
Wansink, H. 45 n. 22 23
Wartenberg, Günther 73 n. 22
Watts, Gilbert (died 1657) 173 n. 97
Wechel (printer) 189 244
244 n. 61
Wecker, Johannes Jacobus 132
Weindrichius, Martinus 230 n. 7
Werenfels, Johannes Jakob 29
Werner, Johannes (1468-1528) 254
254 n. 23
Wesenbecius, Matthaeus 221
Westman, Robert S. 70 n. 3 73 n. 21
85 n. 72 184 n. 29 253 n. 21
Wettstein, Johannes Rudolf (1614-1684)
29
White, N. J. D. 179 n. 8
Whitgift, John (1530-1604) 187
Widmarius (Wedemejerus), Abdias
(1591-1668) 59
Widmarius (Wedemejerus), Andreas
(1552-1621) 59f.
60 n. 22
Wiedemann, Conrad 64 n. 32
Wielema, M. R. 42 n. 14
Wiener, Philip P. 175 n. 101 202 n. 3
Wilhelm IV., Landgraf of Hessen-
Kassel (1532-1592) 256
Wilhem → De Wilhem
Wilkins, John (1614-1672) 228 n. 1
Willet, Andrew 149 n. 47
William of Ockham (ca. 1290-1349/50)
240 n. 46
Williams, John 149 n. 47
Willichius, Jodocus (1501-1552) 132 183
Wilson, John 151 n. 52
Winkelmann, E. 150 n. 48
Witelo (Vitello) (1220/30 - after 1277) 39
Wittich, Christoph (1625-1687) 51
Wittich, Paul 256 n. 30
Wolf, Hieronymus (1516-1580) 10
10 n. 8

Wolfe, D. M. 128 n. 1
Wood, Anthony (1632-1695) 147 n. 40
151 n. 50
Woolf, D. R. 137 n. 18
Wotton, Anthony (1561-1626) 140 161
162 n. 71
Wotton, Henry (1568-1639) 162f.
Wouters → Valerius
Wright, William A. 160 n. 68
Wubbena, Johannes (died 1681)
48 n. 34
Wurstisen (Urstitius), Christian (1544-1588) 22 185
22 n. 67

Yates, Francis A. 190 n. 57

Zabarella, Jacopo (1533-1589) 28 29
41 89 98 99 100 102 103 104 111
101 n. 33 245 n. 63
Zacharias, Ursinus 81 n. 53 54
Zasius, Ulrich (1461-1535) 20 221
20 n. 49 50
Zedler, Johann Heinrich (1706-1763) 91
79 n. 48 80 n. 51
Zeno of Citium (333/32-262 BC) 102
Zientara, Włodzimierz 63 n. 31
Zwinger, Jakob (1569-1610) 10
15 n. 27
Zwinger, Theodor (1533-1588) 9f.
14-16 25 36 186 245
10 n. 5 14 n. 24 25 21 n. 53
Zwingli, Ulrich (1484-1531) 12

SCHWABE PHILOSOPHICA

HERAUSGEGEBEN VON
HELMUT HOLZHEY UND WOLFGANG ROTHER

As a forum for studies in the history of philosophy the series *Schwabe Philosophica* complements the new encyclopaedia *Grundriß der Geschichte der Philosophie.* The volumes of the series uphold the standards of the *Grundriß;* they are non-sectarian in their point of view, maintain a high quality of scholarship, provide thorough historical background information, and are presented in a clear, balanced, and appealing manner. With respect to method, different approaches will be embraced. The series comprises both monographs and anthologies.

Compiling studies in the history of philosophy in a series produced by the same publisher as the *Grundriß* and modelled on the same conceptions and principles aims at invigorating the kind of methodical orientation represented by the *Grundriß* within philosophical research and teaching.

The affinity to the *Grundriß* is further guaranteed by the fact that the editors have both been engaged in the project in different functions for a protracted period of time. They will contribute their philosophical competence as well as their scholarly and editorial experience.

SCHWABE PHILOSOPHICA I
The Influence of Petrus Ramus. Studies in Sixteenth and Seventeenth Century Philosophy and Sciences. Edited by Mordechai Feingold, Joseph S. Freedman and Wolfgang Rother.
2001. 285 pages. ISBN 3-7965-1560-6

SCHWABE PHILOSOPHICA II
Martin Bondeli: Kantianismus und Fichteanismus in Bern. Zur philosophischen Geistesgeschichte der Helvetik sowie zur Entstehung des nachkantischen Idealismus.
2001. ca. 400 pages. ISBN 3-7965-1724-2

SCHWABE PHILOSOPHICA III
Arend Kulenkampff: Esse est percipi. Untersuchungen zur Philosophie George Berkeleys.
2001. ca. 160 pages. ISBN 3-7965-1731-5

INSIGHT PUBLICA

Nadakkavu, Kozhikode, Kerala
Tel:0495–4020666
www.insightpublica.com
e-mail: insightpublica@gmail.com

Vidavangalkoothu

(Malayalam)

Milan Kundera

Translation: CM Rajan

First Edition :January 2017
Second Edition: December 2018

ISBN: 978-93-85899-67-6
Printed and Published by
InsightinPublica Printers & Publishers Pvt Ltd.